Adaptive Feed-Forward Control of Low Frequency Interior Noise

International Series on
INTELLIGENT SYSTEMS, CONTROL, AND AUTOMATION:
SCIENCE AND ENGINEERING

VOLUME 56

For further volumes:
www.springer.com/series/6259

Thomas Kletschkowski

Adaptive Feed-Forward Control of Low Frequency Interior Noise

Springer

Thomas Kletschkowski
Department of Mechanical Engineering,
Mechatronics
Helmut-Schmidt-University/University
of the Federal Armed Forces Hamburg
Holstenhofweg 85
Hamburg 22043
Germany
kletsch@hsu-hh.de

ISBN 978-94-007-2536-2 e-ISBN 978-94-007-2537-9
DOI 10.1007/978-94-007-2537-9
Springer Dordrecht Heidelberg London New York

Library of Congress Control Number: 2011941767

Cover design: VTeX UAB, Lithuania

Printed on acid-free paper

Springer is part of Springer Science+Business Media (www.springer.com)

Preface

This book focuses on a mechatronic approach to active control of interior noise. It strives to comprehend the results of a five year research period as chief engineer with the chair for mechatronics of the Helmut-Schmidt-University/University of the Federal Armed Forces Hamburg.

Although the book starts with fundamental concepts, the reader is expected to be familiar with engineering mechanics and/or engineering acoustics (including experimental techniques), system theory and numerical mathematics. The target audience therefore consists of post graduate students, professional engineers, and researchers working in mechatronics, and especially in the field of active interior noise control.

At the beginning of each new chapter, an abstract contains both a short summary and, as recommendations for further reading, a brief comment on literature. The important contributions to the subject matter are quoted throughout the text. However, the list of references is far from being complete. I therefore apologize to any colleagues not mentioned in spite of their important contributions to academic and/or applied research on active noise and vibration control.

Hamburg, Germany Thomas Kletschkowski

Acknowledgements

The author gratefully acknowledges the support of the Helmut-Schmidt-University/University of the Federal Armed Forces Hamburg. The top-level conditions for research and teaching provided by this institution have been essential to finish my Habilitation as well as to write this book.

Very special thanks, the author would like to express to Delf Sachau who made me familiar with Active Noise and Vibration Control in 2004 and since then has continued to give much helpful advice. The author would also like to thank Udo Zölzer and Detlef Krahè for all of their criticisms, comments and suggestions. Furthermore, the author would like to thank Uwe Schomburg and Albrecht Bertram who always encouraged me to finish this work.

Many colleagues and friends also made useful comments and suggestions that made is possible to improve this book. The author would like to thank Sten Böhme, Harald Breitbach, Mohamed Bouhaj, Christian Gerner, Julian Greßkowski, Martin Holters, Norbert Hövelmann, Kay Kochan, Rolf Lammering, Jörg Lefèvre, Marian Markiewicz, Günter Neuwirth, Oliver Pabst, Marek Pawelczyk, Bernd Samtleben, Henning Scheel, Kai Simanowski, Jochen Sommer, Fabrice Teuma, Martin Wandel and Matthias Weber.

The academic career of the author would have been impossible without the support of his family in Hamburg, Schwerin and Den Haag. Very special thanks go to Ammerentie, Roland and Karoline, to Beate and Reinhard as well as to Adriana and Izaak. The author is grateful to all of them.

Notation

Mathematical Operations and Operators

div	Divergence operator
grad	Gradient operator
max	Maximum operator
min	Minimum operator
E	Expectation
$\mathfrak{F}$	Fourier transform
$\mathfrak{F}^{-1}$	Inverse Fourier transform
$\mathfrak{F}_d$	Fourier transform of sampled signals
$\mathfrak{F}_d^{-1}$	Inverse Fourier transform of sampled signals
$\mathfrak{F}_{DFT}$	Discrete Fourier transform
$\mathfrak{F}_{DFT}^{-1}$	Inverse Discrete Fourier transform
Im	Gives the imaginary part of a complex number
Re	Gives the real part of a complex number
$\mathfrak{T}$	Transformation
○–●	Mapping from time domain to frequency domain
●–○	Mapping from frequency domain to time domain
$\Box$	Wave operator
$\vec{\Box}$	Vector wave operator
$(\,)^T$	Transposition
$(\,)^H$	Hermitian or conjugate transpose (of a matrix)
$d(\,)$	Total derivative
$\partial(\,)$	Partial derivative
$\overline{(\,)}$	Arithmetic mean
tr( )	Trace of a matrix
$\|\ \|_2$	Euclidean norm

Conventions for Signals and Systems

Conventions for Continuous-Time Signals and Systems

t	Time
f	Frequency

ω	Angular frequency, i.e. 2π times the actual frequency in hertz
$x(t)$	Continuous-time signal
$X(j\omega)$	Fourier transform of $x(t)$

Conventions for Discrete-Time Signals and Systems

n	Discrete time step
T	Sampling time, so $t = nT$ where n is an integer
$x(n)$	Discrete-time signal
$X(e^{j\omega T})$	Fourier transform of $x(n)$
$X(n)$	Fourier transform of $x(n)$ at discrete time step n

General Conventions

$\hat{x}$	Real valued amplitude of x or approximation/model of x
x'	Filtered signal
x_∞	Steady state of x
$\bar{x}$	Arithmetic mean of x
σ_x^2	Variance of x
x_{RMS}	Root mean square of x
δx	Virtual signal
$\hat{X}$	Complex amplitude of X or approximation/model of X
X'	Filtered signal
X_∞	Steady state of X
E_x	Mean signal energy
Π_x	Mean signal power
r_{xx}	Auto correlation of x
r_{xy}	Cross correlation between x and y
h	Impulse response of a system
S_{xx}	Auto spectral density of x
S_{xy}	Cross spectral density for x and y
G_{xx}	Single-sided auto spectral density of x
G_{xy}	Single-sided cross spectral density for x and y
H	Transfer function of a system

Conventions for Linear Algebra

Conventions for Scalars

x, X	Scalar variables
X_R	Real part of X, where $X_R = \mathrm{Re}(X)$
X_I	Imaginary part of X, where $X_I = \mathrm{Im}(X)$
X^*	Conjugate complex of X, where $X^* = X_R - jX_I$
$\|X\|^2$	Squared magnitude of X, where $\|X\|^2 = X^* X$

Conventions for Column Matrices

$\mathbf{x}$	Lower-case bold variables are column matrices
$\mathbf{x}^T$	The transpose of a column matrix is a row matrix
$\mathbf{x}_R$	Real part of $\mathbf{x}$, where $\mathbf{x}_R = \mathrm{Re}(\mathbf{x})$
$\mathbf{x}_I$	Imaginary part of $\mathbf{x}$, where $\mathbf{x}_I = \mathrm{Im}(\mathbf{x})$
$\mathbf{x}^H$	Hermitian of $\mathbf{x}$, where $\mathbf{x}^H = \mathbf{x}_R^T - j\mathbf{x}_I^T$

$\mathbf{x}^H\mathbf{x}$	The inner product of $\mathbf{x}$, which is a scalar
$\mathbf{x}\mathbf{x}^H$	The outer product of $\mathbf{x}$, whose trace is equal to the inner product
$\|\mathbf{x}\|_2$	Euclidean norm of $\mathbf{x}$, where $\|\mathbf{x}\|_2 = \sqrt{\mathbf{x}^H\mathbf{x}}$

Conventions for Matrices

$\mathbf{X}$	Upper-case bold variables are matrices
$\mathbf{X}^T$	The transpose of $\mathbf{X}$
$\mathbf{X}_R$	Real part of $\mathbf{X}$, where $\mathbf{X}_R = \mathrm{Re}(\mathbf{X})$
$\mathbf{X}_I$	Imaginary part of $\mathbf{X}$, where $\mathbf{X}_I = \mathrm{Im}(\mathbf{X})$
$\mathbf{X}^H$	Hermitian of $\mathbf{X}$, where $\mathbf{X}^H = \mathbf{X}_R^T - j\mathbf{X}_I^T$
$\mathbf{X}^{-1}$	The inverse of $\mathbf{X}$
$\mathbf{X}^{-H}$	The inverse of $\mathbf{X}^H$
$\mathrm{tr}(\mathbf{X})$	Trace of $\mathbf{X}$
$\lambda_i(\mathbf{X})$	The i-th eigenvalue of $\mathbf{X}$
$\|\mathbf{X}\|_2$	Euclidean norm of $\mathbf{X}$, where $\|\mathbf{X}\|_2 = \sqrt{\mathrm{tr}(\mathbf{X}^H\mathbf{X})}$
$\mathbf{I}$	The identity matrix

Conventions for Vectors

$\vec{x}$	Vector valued variable such as position vector
$\vec{x} \cdot \vec{y}$	Scalar product between vectors

Comments on Symbols

Lower-Case Latin Symbols

$\mathbf{b}$	Cost function parameter column matrix
c	Speed of sound or cost function parameter
d	Disturbance or distance between anode and cathode
e	2.718..., error signal, acoustic energy density or additive filtered error
e_{kin}	Acoustic kinetic energy density
e_{pot}	Acoustic potential energy density
f	Frequency
$f_{x(t)}(\xi)$	Probability density function of a stochastic process
f_n	n-th eigenfrequency
f_{nR}	n-th resonance frequency
$\mathbf{f}$	Load column matrix
i	Index, normal component of sound intensity or electric current
$\vec{i}$	Sound intensity vector
j	Index or imaginary number ($j = \sqrt{-1}$)
k	Index, wave number, discrete-time delay or stiffness
k'	Alternative form of complex wave number
k_{nR}	Wave number for the n-th resonance
l	Index or length
m	Index, discrete-time delay or mass
n	Index or discrete time step
$\vec{n}$	Normal vector
p	Acoustic pressure
p_{tot}	Total pressure

p_∞	Equilibrium value of total pressure
p_p	Primary noise
p_s	Anti-noise
q	Source strength, electric charge or volume velocity
r	Damping coefficient or radial distance
$\mathbf{r}$	Residuum column matrix
Δr	Change in radial distance
t	Time
t_i	Observation time point
v	Normal component of acoustic velocity
$\vec{v}$	Acoustic velocity
$\vec{v}_{tot}$	Total value of acoustic velocity
$\vec{v}_\infty$	Equilibrium value of acoustic velocity
$\mathbf{w}$	Column matrix of control filter coefficients
w_{mki}	mki-th control filter coefficient
$\hat{w}_{mki}$	mki-th auxiliary coefficient
x	Signal or x-coordinate
Δx	Separation distance
$\vec{x}$	Position vector
y	Signal or y-coordinate
z	z-coordinate

Upper-Case Latin Symbols

A	Attenuation of analogue filter
$\mathbf{A}$	Cost function parameter matrix
B	Electromagnetic induction
C	Capacity of condenser
$\mathbf{C}$	Stiffness matrix or controller matrix
C_p	Specific heat for constant pressure
C_V	Specific heat for constant volume
D	Dimensionless damping ratio of mechanical systems
$\mathbf{D}$	Damping matrix
E	Error, Energy or Bulk modulus
I	Number of control filter coefficients or instantaneous intensity
$\bar{I}$	Mean intensity
$\bar{I}_M$	Measured mean intensity
$\bar{I}_T$	True mean intensity
J	Number of filter coefficients used for plant modeling or cost function
K	Number of reference signals
L	Number of error signals, length or inductance
M	Number of controller output signals or modal overlap
$\mathbf{M}$	Mass matrix
N	Number of time steps
P_w	Probability of a stochastic process
R	Complex reflection coefficient, electric resistance or residuum
R_Z	Impedance boundary

R_P	Pressure boundary
R_v	Velocity boundary
S	Surface area or cross section
T	Sample time or periodic time or time interval
T_{XY}	Transmissibility between X and Y
T_{60}	Reverberation time
U	Electric voltage
V	Volume
ΔV	Change in volume
$\mathbf{W}_p$	Matrix used to weight the squared sound pressure
$\mathbf{W}_q$	Matrix used to weight the control signal
Z	Acoustic impedance

Lower-Case Greek Symbols

α	Absorption coefficient
α_{nR}	Absorption coefficient for the n-th resonance
β	Weighting factor
χ	Phase angle of complex reflection coefficient
$\delta(t)$	Dirac impulse
ε_{vol}	Volume compression
ε	Filtered error signal for FEFxLMS algorithm
γ	Coherence
κ	Sensitivity, e.g. of microphone
λ	Wave length
λ_i	i-th eigenvalue
μ	Step size
$\tilde{\mu}$	Power normalized step size
ω	Angular frequency
ω_n	Angular frequency corresponding to n-th eigenfrequency
ω_{nR}	Angular frequency corresponding to n-th resonance frequency
ω_M	Modal bandwidth
φ	Phase angle
φ_0	Zero phase angle
$\Delta\varphi_H$	Phase angle between transducers
$\Delta\varphi_p$	Phase angle between two sound pressures
ϕ	Velocity potential
π	3.1415...
ρ	Change in density
ρ_{tot}	Total value of density
ρ_∞	Equilibrium value of density
σ	Decay coefficient
τ	Continuous-time delay
θ	Change in temperature
θ_{tot}	Total value of temperature
θ_∞	Equilibrium value of temperature
ξ	Stochastic process or dimensionless damping ratio in acoustic systems

ξ_{nR}	Damping ratio for the n-th resonance
ζ	Dimensionless frequency ($\zeta = L/\lambda$)

Upper-Case Greek Symbols

Π	Sound power
Σ	Uncertainty

Acronyms

ABN	Airborne noise
ANC	Active noise control
ANS	Active noise system
ACM	Auto correlation matrix
ASAC	Active structural acoustic control
AVC	Active vibration control
ACF	Auto correlation function
ASD	Auto spectral density
BC	Boundary condition
BPF	Blade passage frequency
CA	Coherence analysis
CCF	Cross correlation function
CCM	Cross correlation matrix
CGLS	Conjugated gradient least square
COA	Correlation analysis
CSD	Cross spectral density
DSP	Digital signal processor
DC	Direct current
EBN	External borne noise
EOC	Engine order cancellation
FEM	Finite element method
FeLMS	Filtered error least mean square
FFT	Fast Fourier transform
FIR	Finite impulse response
FRF	Frequency response function
FEFxLMS	Fast exact filtered reference least mean square
FxLMS	Filtered reference least mean square
KHIE	Kirchhoff-Helmholtz integral equation
IBEM	Inverse boundary element method
IBN	Internal borne noise
IFEM	Inverse finite element method
IMC	Internal model control
IMSC	Independent modal space control
IPE	Initial performance estimation
IRA	Impulse response analysis
MA	Military aircraft
MFxLMS	Modified filtered reference least mean square
LMS	Least mean square

LTI	Linear time invariant
MA	Military aircraft
NCP	Normalized cumulative periodogram
NR	Noise reduction
ODE	Ordinary differential equation
PA	Public address
PDE	Partial differential equation
PVP	Principle of virtual pressure
RMS	Root mean square
RSC	Remote sensor control
SBN	Structure borne noise
SPL	Sound pressure level
SIAF	Sound intensity probe with active free field
SVD	Singular value decomposition
THF	Technologiezentrum Hamburg Finkenwerder
TPA	Transducer placement analysis
TR	Tikonov regularization
TVA	Tunable vibration absorber
VLJ	Very light jet
VVS	Volume velocity source
WA	Working area

Contents

List of Figures

List of Tables

Part I
Introduction

Chapter 1
Introduction to Interior Active Noise Control

Abstract Before we start with a detailed analysis of active noise control systems and their application to enclosed sound fields, it is necessary to motivate both the idea and the limits of active control. Furthermore, it will be necessary to describe the interior noise problem and to distinguish between different concepts that can be used to reduce interior noise by means of active control. The upcoming chapter is therefore subdivided into four sections. The first is focused on the fundamental idea of active control, whereas the second contains remarks on active control of interior noise and, to motivate the active noise control approach, an illustrative example concerning aircraft cabin noise caused by propeller rotation. The intention of the third section is to provide some information on the state of the art, while the last section describes both objective and structure of this book.

1.1 Idea, Limit and Structure of Active Control Concepts

The idea of active control is based on the *principle of superposition* that will be explained for two continuous-time harmonic signals $d(t) = \hat{d}\cos(\omega t + \varphi_d)$ and $y(t) = \hat{y}\cos(\omega t + \varphi_y)$, where ω is the angular frequency, $\hat{d}$ and $\hat{y}$ are amplitudes, and φ_d and φ_y are zero phase angles. $d(t)$ represents the *disturbance* (also known as *noise*), whereas $y(t)$ represents the *canceling signal* (also known as *anti-noise*). The latter is used to reduce the disturbance. Both signals can be interpreted as the real part of phasors such as

$$d(t) = \mathrm{Re}\big[\hat{d}e^{j(\omega t+\varphi_d)}\big] =: \mathrm{Re}\big[\hat{D}e^{j\omega t}\big], \tag{1.1}$$

and

$$y(t) = \mathrm{Re}\big[\hat{y}e^{j(\omega t+\varphi_y)}\big] =: \mathrm{Re}\big[\hat{Y}e^{j\omega t}\big], \tag{1.2}$$

where $j = \sqrt{-1}$ is the *imaginary number*, while $\hat{D}$ and $\hat{Y}$ are complex amplitudes. The superposition of $d(t)$ and $y(t)$, known as the *error signal*, is given by

$$e(t) = d(t) + y(t) = \mathrm{Re}\big[(\hat{D} + \hat{Y})e^{j\omega t}\big] =: \mathrm{Re}\big[\hat{E}e^{j\omega t}\big]. \tag{1.3}$$

The error signal is canceled, if the complex amplitude $\hat{Y}$ is adjusted to

$$\hat{Y} = -\hat{D}, \quad \text{i.e.} \quad |\hat{Y}| = |\hat{D}| \quad \text{and} \quad \varphi_y = \varphi_d - \pi. \tag{1.4}$$

T. Kletschkowski, *Adaptive Feed-Forward Control of Low Frequency Interior Noise*, Intelligent Systems, Control and Automation: Science and Engineering 56, DOI 10.1007/978-94-007-2537-9_1,

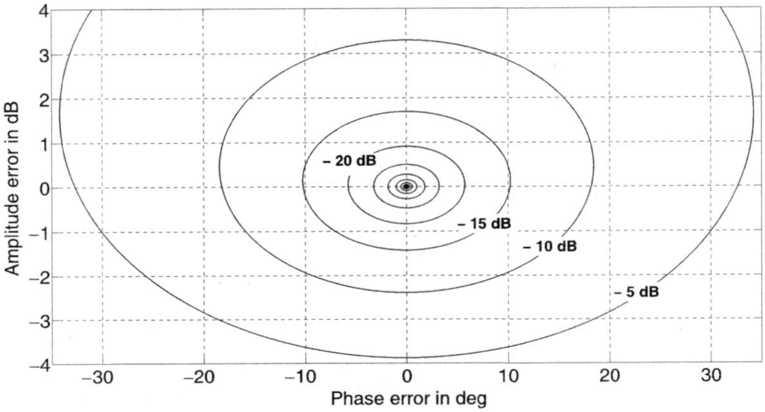

Fig. 1.1 Influence of amplitude and phase errors on the amount of active noise reduction

Equation (1.4) tells us that we are able to reject the disturbance, if (i) the magnitude of the control signal equals the magnitude of the disturbance, and (ii) the control signal is in anti-phase to the disturbance. The generation of a control signal according to (1.4) and its superposition with the disturbance is the goal of active control. However, in practical applications we will have to live with some misadjustments in both magnitude and phase that can be described by

$$\hat{Y}' = F_{\Delta} e^{j\Delta\varphi} \cdot (-\hat{D}), \tag{1.5}$$

where $\Delta\varphi$ is the phase error and $0 \leq F_{\Delta}$ is a real valued scaling factor—called amplitude error. The latter can also be expressed as $L_{F_{\Delta}} = -20\log_{10}(F_{\Delta})$. The *noise reduction* (NR) that can be achieved with misadjustments is calculated as the logarithmic ratio between the squared error and the squared disturbance such as

$$NR = 10\log_{10}\big(|\hat{D} + \hat{Y}'|^2/|\hat{D}|^2\big). \tag{1.6}$$

It is obvious that the amount of NR is infinite for a perfectly adjusted control signal, but a minor deviation from the optimal solution can cause a significant degradation of the controller performance as shown in Fig. 1.1. This figure illustrates how closely the control signal must be matched to the disturbance in its amplitude and phase to produce an appreciable reduction in the level of the error signal. For example, the phase error of the canceling signal must be within $\pm 20°$, to realize a NR of more than 10 dB, even if the amplitude error is zero.

If the reduction is expressed on the linear scale, a 10 dB attenuation of the disturbance signal (according to (1.6)) corresponds to a decay of more than 60%. This example clarifies that both measurement of the disturbance and generation of the canceling signal must be performed with high accuracy to realize a significant *control profit*.

The generation of the canceling signal can be realized with a feed-forward controller as well as with a feedback controller. The structure of both control concepts is shown in Fig. 1.2. At first we will concentrate on the information flow described by

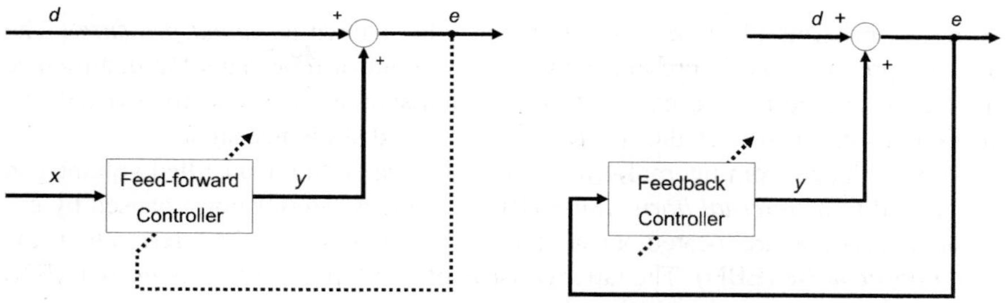

Fig. 1.2 Structure of active control concepts. *Left*: feed-forward control. *Right*: feedback control

the solid lines. It can be seen that the feedback controller, compare Fig. 1.2 (right), needs a measure of the error signal e to generate the canceling signal y. This causes a time delay, because the disturbance d arrives at the error sensor position before the first anti-noise signal is generated by active feedback control.

Control with preview is possible, if we apply a feed-forward control strategy, compare Fig. 1.2 (left). However, instead of the error signal, a reference signal x that has to be well correlated to the disturbance d is required to perform feed-forward control. It is obvious that a sufficient lead time is required for the reference signal to ensure causality. The utilization of a reference signal has two advantages:

1. It is possible to cancel only those signal components that are correlated to the reference signal that may be obtained from a noise-generating machinery to "inform" the feed-forward controller about the disturbing noise.
2. Because the reference signal is not altered by feed-forward control, the controller output has (in contrast to the feedback control concept) no effect on the controller input. Thus, instability is prevented, if the feed-forward controller acts in the steady state.

For these reasons the present book is focused on theory, design and application of *feed-forward control systems*. Please notice that the error signal can in both concepts be used to adapt the controller. This is symbolized by the dashed lines also shown in Fig. 1.2. Such an adaption enables the controller to optimize the NR by minimizing amplitude and phase errors in the control signal. This is the reason why the present book is focused on *adaptive control*. However, if we feed back the errors signal, we have to ensure that the mathematical algorithm used for the adaption process will not cause instabilities.

1.2 Remarks on Interior Noise and Active Control Approaches

1.2.1 Comments on the Interior Noise Problem

The noise exposure in an air filled volume enclosed by a structure that can be flexible (such as an aircraft fuselage) or rigid (such as the exterior wall of a house) is usually expressed in terms of the *acoustic pressure* (also known as *sound pressure*). The acoustic pressure that is present without active control represents the disturbance and can be caused by internal or external sound sources. These sources determine the time history as well as the spectral shape of the disturbance signal.

Noise radiated from internal sources that are located in the air filled volume can be referred to as *internal borne noise* (IBN), whereas interior noise caused by external sources that are located outside the air filled volume can be referred to as *external borne noise* (EBN). The latter consists of two types—*airborne noise* (ABN) and *structure-borne noise* (SBN). The associated transfer paths are the *airborne path* and the *structure-borne path*.

ABN is known as that part of the interior noise that is transmitted through the boundary structure from sources that exert pressures directly on the exterior of the air filled volume. One typical example is cabin noise in a propeller-driven aircraft caused by the pressure field that surrounds each propeller blade as a consequence of its motion and which rotates with the rotating propeller blade. Another example is cabin noise caused by the turbulent air flow over the aircraft surfaces that impinges directly on the exterior of the fuselage and transmits into the cabin (Wilby and Mixson 1991).

Examples for SBN were also described in (Wilby and Mixson 1991). The authors stated that: "Sources such as engine unbalance forces transmitted trough engine mounts and engine exhaust or propeller wakes impinging on wing or tail surfaces generate vibrational energy that is transmitted along the airframe structure and radiated into the cabin as acoustic noise." Thus, SBN can (more generalized) be defined as interior noise that is radiated from a vibrating structure into the enclosure at one point, but is excited by structural or acoustical loads at another point (and then transmitted through the structure to the point of radiation).

1.2.2 Comments on Active Control Approaches

Active reduction of interior noise is a challenging problem, because we have to consider the dynamics of the sound sources, the dynamical behavior of the noise transmission paths as well as the dynamics of the interior noise field, if we want to establish an appropriate control concept. Depending on the objective of the control task and depending on the applied actuator/sensor type, it is possible to distinguish between three different active control approaches, compare Table 1.1.

If the objective is the minimization of the interior sound pressure, measured with microphones, and, if the anti-noise is generated by loudspeakers, the approach is known as active noise control (ANC), see e.g. (Nelson and Elliott 1992).

If an active control system with structural actuators (such as electro-dynamic shakers) and acoustic sensors is used to minimize the sound radiation, active structural acoustic control (ASAC), see e.g. (Fahy and Gardonio 2007), is applied. If structural actuators as well as structural sensors (such as accelerometers) are used, and, if the objective of active control is the minimization of structural vibrations, the concept of active vibration control (AVC), see e.g. (Fuller et al. 1996), is applied.

Table 1.1 Three different approaches to active control of interior noise

Approach	Objective	Sensor	Actuator
ANC	Minimization of sound pressure	acoustical	acoustical
ASAC	Minimization of sound radiation	acoustical	structural
AVC	Minimization of structural vibration	structural	structural

The different active control approaches, listed in Table 1.1, have advantages and of course also draw backs that strongly depend on the specific application. Common to all approaches is the necessity of measuring physical values (e.g. the interior sound pressure), and the generation of canceling signals that are used to reduce disturbances by a set of actuators. The actuator driving signals are provided by a controller as a result of the information processing. The latter can be performed with an adaptive or with a non-adaptive controller as mentioned in Sect. 1.1.

The differences between the active control approaches listed in Table 1.1 are caused by the kind of sensing as well as by the kind of actuation. It can be expected that only the purely acoustic concept of ANC is capable of avoiding additional structural loads that are generated by structural actuators. Furthermore, the application of ANC provides the chance to reduce the effort of maintenance, because damaged loudspeakers and microphones can easily be replaced by new components. The disadvantage of ANC is that the noise has already entered the interior, when it is affected by this control concept.

In order to reduce the transmission of disturbing noise, ASAC and AVC must be applied to the sources and/or to the transfer paths. Typical examples are active mounts designed to reduce structural vibrations caused by engine unbalance forces, see e.g. (Billoud 2001), or distributed control based on active tuned vibration absorber as proposed in (Breitbach et al. 2006).

However, if it is impossible to block a specific noise transmission path by AVC or to suppress noise radiation from the boundary into the enclosure by ASAC, ANC is in many situations a valid approach that can directly be applied to the acoustic field by causing only minor retroaction on the surrounding structure. ANC offers the possibility to suppress acoustic resonances that can (especially in weakly damped interiors with damping ratios less than 10%) be excited with a little amount of energy. Thus, it may also be required to reduce disturbances that are not fully suppressed by passive noise treatments or active control systems based on AVC and/or ASAC. These are two reasons why this book is focused on ANC.

1.2.3 A Qualitative Comparison of Active Control Approaches

To motivate the ANC-approach, this subsection contains an introductory example that is inspired by active control of interior noise in the cabin of a propeller-driven aircraft. If we assume that the engines of such an aircraft are perfectly balanced

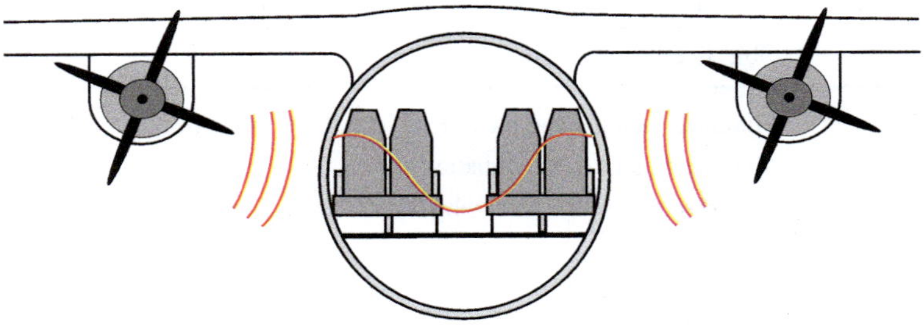

Fig. 1.3 Propeller rotation as source for interior noise (adapted from Johansson 2000). Figure also motivates the distribution of the second acoustic mode in cabin cross direction

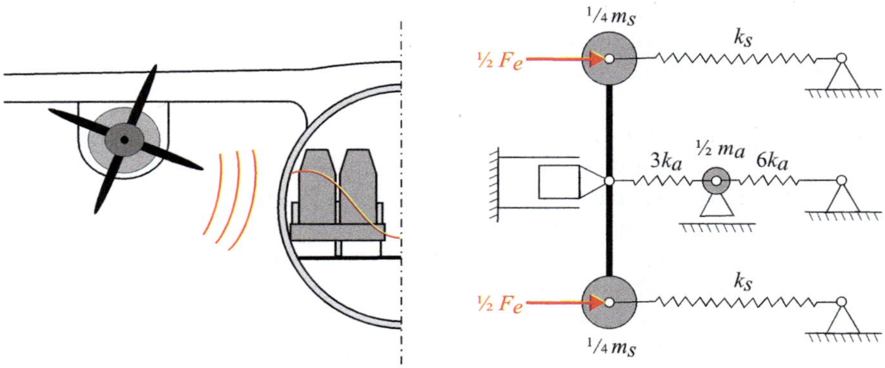

Fig. 1.4 *Left*: half of aircraft cross section with interior noise field caused by propeller rotation (adapted from Johansson 2000). *Right*: simple model of aircraft used to analyze its vibro-acoustic response for excitation close to the second acoustic mode in cross direction. Subscripts: *a*—air, *e*—excitation, *s*—structure

and that the cruise speed is low, the cabin noise will be dominated by ABN that is caused by the propeller rotation. As illustrated by Fig. 1.3 the exterior sound pressure impinges on the exterior of the fuselage and transmits into the cabin.

If we also assume that the right engine rotates clockwise and that the left engine rotates counter-clockwise (starting at the same zero phase angle as the left engine), the engines operate in a perfect reverse rotation mode. If finally the *blade passage frequency* (BPF) is close to the second acoustic mode in cross direction, as also illustrated in Fig. 1.3, we have a symmetric problem. Thus, it is (considering all these assumptions) possible to limit our investigations to the left half of the aircraft that is shown in Fig. 1.4 (left).

The goal is to suppress the acoustic resonance. The question is: What is a valid control approach—AVC, ASAC, or ANC? To answer this question we use a very simple model of the half aircraft. This model is shown in Fig. 1.4 (right). It can be seen that the mass of the half aircraft structure $m_s/2$ is represented by two mass points, each having the mass $m_s/4$. These mass points are connected by a rigid massless beam.

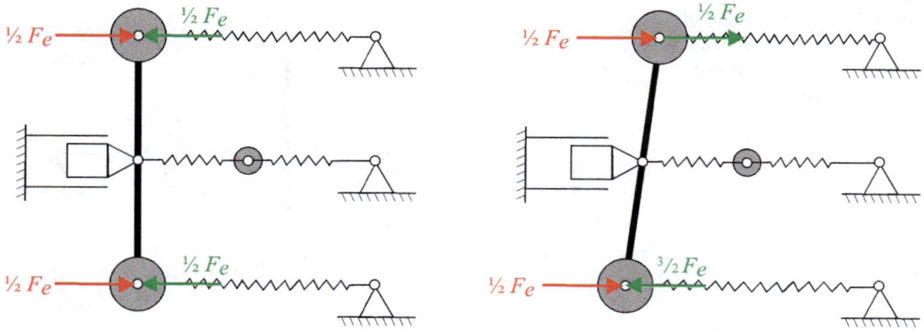

Fig. 1.5 Simple model of aircraft cross section used to illustrate two control approaches: active vibration control (*left*), and active structural acoustic control (*right*)

The connection between the beam and the mass points is also rigid. The beam is (at its central point) mounted on a bearing that allows for rotation and horizontal displacements. The rotations may be small, i.e. the torsion angle is less than 5°. The structural stiffness of the aircraft is represented by two linear elastic springs with stiffness k_s. The mass of the air in the half aircraft cabin $m_a/2$ is represented by a single point mass, whereas the compressibility of the fluid is represented by two linear elastic springs with the stiffnesses $3k_a$ and $6k_a$. All connections between springs and mass points as well as all connections between springs and bearings are rotatable and frictionless.

The external sound pressure is represented by two discrete point forces, acting on the structural mass points. Please notice that the vibro-acoustic coupling between the air filled cavity and the elastic fuselage is represented by the connection between the "air spring" with stiffness $3k_a$ and the rigid beam. The force in this spring is used to model the effect of the internal sound pressure on the structure, whereas the horizontal displacement of the connection point represents sound radiation into the fluid. To keep the calculation simple, damping effects are neglected.

Let us first analyze the effect of AVC. It is obvious that the external excitation can be compensated, if the control forces (also applied to the structural mass points) have amplitudes identical to that of the external excitation and a phase that differs by 180°, compare Fig. 1.5 (left). This control approach results in a cancellation of the structural vibration. Thus, the transfer path is blocked and both the kinetic energy of the structural model and the potential energy of the cavity model are minimized. The "costs" of this control approach are given by the sum of the control forces. This sum represents a total magnitude of $F_e = |F_e/2 + F_e/2|$. Thus, the control effort has the same order of magnitude as the excitation.

The goal of ASAC is the minimization of sound radiation that is in the simple aircraft model represented by the of the horizontal displacement of the bearing in the central point of the beam. To minimize this displacement, it is necessary to apply the control forces shown in Fig. 1.5 (right). It can be seen that one control force (acting in phase with the external load) has an amplitude identical to that of the external

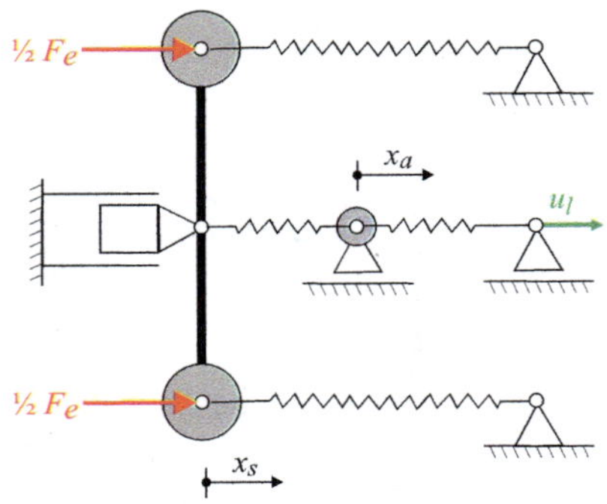

Fig. 1.6 Simple model of aircraft cross section used to illustrate the active noise control approach

excitation, whereas the other control force has an amplitude that is twice the size of the total external excitation F_e and a phase that differs by 180°.

This control approach minimizes the potential energy of the cavity model, but not the kinetic energy of the structural model. The "costs" are given by control forces that have a total magnitude of $F_e = |F_e/2 - 3 \cdot F_e/2|$. Thus, the control effort has again the same order of magnitude as the excitation.

Let us finally analyze the effect of ANC. For this purpose it is necessary to replace the fixed bearing at the right end of the "air spring" with stiffness $6k_a$ by a movable bearing, as shown in Fig. 1.6. Instead of applying a control force, we are now interested in controlling u_l, the horizontal displacement of this movable bearing. The latter can be adjusted to control the potential energy that is stored in both springs of the cavity model (this approach is known as global control) or to control only the potential energy that is stored in one of these springs (this approach is known as local control).

Anyhow, we will have to calculate both the structural displacement x_s and the displacement of the "air mass" x_a to obtain a measure for the potential energy of the cavity model that depends on u_l. For this purpose we need the equations of motion that can be obtained by applying *D'Alembert's principle* to the forces shown in the free body diagram presented in Fig. 1.7. This yields (considering a multiplication with the factor 2)

$$\begin{bmatrix} m_s & 0 \\ 0 & m_a \end{bmatrix} \begin{bmatrix} \ddot{x}_s(t) \\ \ddot{x}_a(t) \end{bmatrix} + \begin{bmatrix} 4k_s + 6k_a & -6k_a \\ -6k_a & 18k_a \end{bmatrix} \begin{bmatrix} x_s(t) \\ x_a(t) \end{bmatrix} = \begin{bmatrix} 2F_e(t) \\ 12k_a u_l(t) \end{bmatrix}. \tag{1.7}$$

If we assume harmonic fluctuations of all time-varying quantities, such as

$$\begin{aligned} F_e(t) &= \mathrm{Re}\big[\hat{F}_e e^{j\omega t}\big], & u_l(t) &= \mathrm{Re}\big[\hat{U}_l e^{j\omega t}\big], \\ x_s(t) &= \mathrm{Re}\big[\hat{X}_s e^{j\omega t}\big], & x_a(t) &= \mathrm{Re}\big[\hat{X}_a e^{j\omega t}\big], \end{aligned} \tag{1.8}$$

and, if we introduce the abbreviations

$$\begin{aligned} \omega_{0a}^2 &:= \frac{k_a}{m_a}, & \eta_a^2 &:= \frac{1}{18}\frac{\omega^2}{\omega_{0a}^2}, & m &= \frac{m_s}{m_a}, & k &= \frac{k_s}{k_a}, \\ \omega_{0s}^2 &:= \frac{k_s}{m_s}, & \eta_s^2 &:= \frac{\omega^2}{\omega_{0s}^2} = 18 \cdot \frac{m}{k} \cdot \eta_a^2, \end{aligned} \tag{1.9}$$

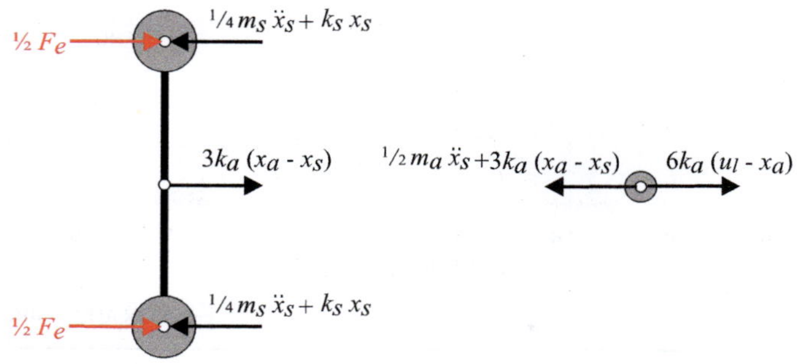

Fig. 1.7 Free body diagram of the simple aircraft model used to analyze the effect of ANC

where ω_{0a} is the natural frequency of a harmonic oscillator with mass m_a and stiffness k_a, ω_{0s} is the natural frequency of a harmonic oscillator with mass m_s and stiffness k_s, η_a and η_s are normalized frequencies, k is the stiffness ratio, and m is the mass ratio, we obtain (after some algebraic manipulations and a multiplication with $\exp[-j\omega t]$) the following set of algebraic equations

$$\begin{bmatrix} 6+4k-18m\eta_a^2 & -6 \\ -6 & 18(1-\eta_a^2) \end{bmatrix} \begin{bmatrix} \hat{X}_s \\ \hat{X}_a \end{bmatrix} = \begin{bmatrix} 2\hat{F}_e/k_a \\ 12\hat{U}_l \end{bmatrix}. \tag{1.10}$$

The solution of (1.10) yields

$$\begin{aligned} \hat{X}_s &= \frac{1}{k_a}\cdot\frac{1-\eta_a^2}{f(\eta_a)}\cdot\hat{F}_e + \frac{2}{f(\eta_a)}\cdot\hat{U}_l, \\ \hat{X}_a &= \frac{1}{3k_a}\cdot\frac{1-\eta_a^2}{f(\eta_a)}\cdot\hat{F}_e + \frac{1}{3}\cdot\frac{2(3+2k-9m\eta_a^2)}{f(\eta_a)}\cdot\hat{U}_l, \end{aligned} \tag{1.11}$$

where the frequency dependent function in the denominators is given by

$$f(\eta_a) = 2(1+k) - (3+2k+9m)\eta_a^2 + 9m\eta_a^4. \tag{1.12}$$

The first addend in the right hand sides of (1.11) describes the displacement caused by the external excitation and represents the disturbance, whereas the second addend represents the contribution caused by ANC. Based on (1.11) and (1.12) we can define a measure for the global potential energy of the cavity model

$$E_{pot,a}^{global} = 3k_a|\hat{X}_a - \hat{U}_l|^2 + \frac{3}{2}k_a|\hat{X}_s - \hat{X}_a|^2, \tag{1.13}$$

and a measure for the local potential energy stored in the "air spring" with stiffness $6k_a$

$$E_{pot,a}^{local} = 3k_a|\hat{X}_a - \hat{U}_l|^2. \tag{1.14}$$

Minimizing of (1.13) with respect to $\hat{U}_l$ yields

$$\hat{U}_l^{global,opt} = \frac{1}{k_a}\cdot\frac{2k-(3+4k+9m)\eta_a^2+18m\eta_a^4}{4k^2+a_2\eta_a^2+a_4\eta_a^4+a_6\eta_a^6+a_8\eta_a^8}\cdot\hat{F}_e, \tag{1.15}$$

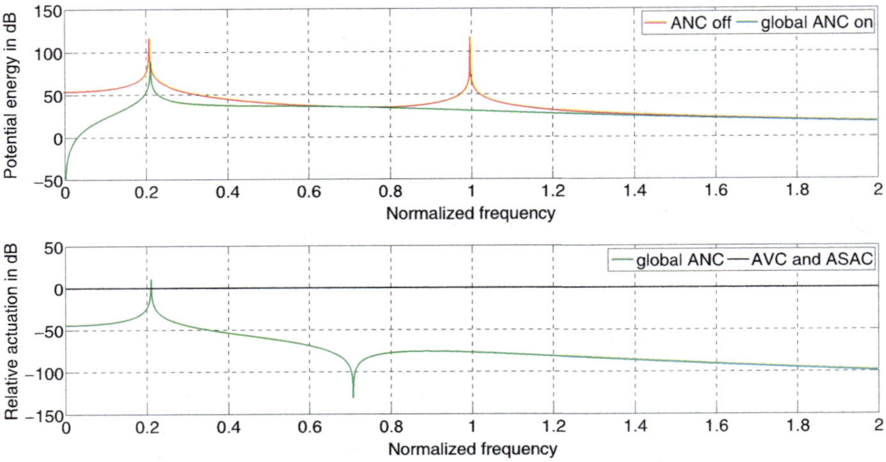

Fig. 1.8 The effect of global active noise control. *Top*: Potential energy of the simple cavity model. *Below*: Relation between actuation force and excitation force

where the coefficients a_i with $i = 2, 4, 6, 8$ are given by

$$\begin{aligned} a_2 &= -4(3+2k+9m), \\ a_4 &= 3(9+12k+4k^2+18m+24km+27m^2), \\ a_6 &= -54m(3+2k+3m), \\ a_8 &= 243m^2, \end{aligned} \tag{1.16}$$

whereas minimizing of (1.14) with respect to $\hat{U}_l$ yields

$$\hat{U}_l^{local,opt} = \frac{1}{k_a} \cdot \frac{2k-(3+4k+9m)\eta_a^2+18m\eta_a^4}{2k-3(3+2k+3m)\eta_a^2+27m\eta_a^4} \cdot \hat{F}_e. \tag{1.17}$$

The optimal solution for the global ANC-approach (specified by (1.15) and (1.16)), and the optimal solution for the local ANC-approach (specified by (1.17)) depend on the normalized frequency η_a, the mass ratio m, the stiffness ratio k and the ratio between the amplitude of the excitation force $\hat{F}_e$ and the "air stiffness" k_a. For numerical evaluation, the following parameters were used

$$k = 1000, \qquad m = 5000. \tag{1.18}$$

The effect of global ANC is shown in Fig. 1.8. The frequency dependence of the global potential energy of the simple cavity model is shown top, whereas the frequency dependence of the relative actuation is shown below. The latter was cal-

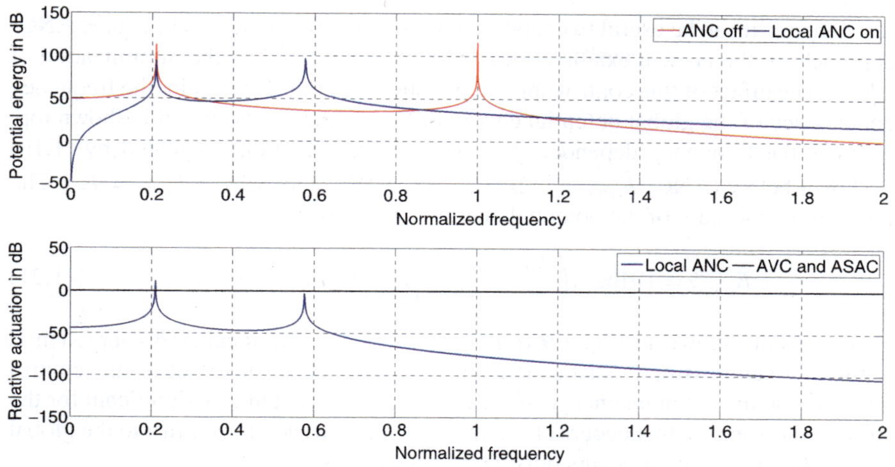

Fig. 1.9 The effect of local active noise control. *Top*: Potential energy of the simple cavity model. *Below*: Relation between actuation force and excitation force

culated as the logarithmic ratio between magnitude of the control force (ANC: $\hat{F}_{cont} := 12k_a\hat{U}_l^{global,opt}$, AVC and ASAC: $\hat{F}_{cont} = \hat{F}_e$) and $\hat{F}_e$

$$RA = 20\log_{10}\left(\frac{\hat{F}_{cont}}{\hat{F}_e}\right). \tag{1.19}$$

The logarithmic measure of the potential energy was calculated as

$$LE_{pot,a}^{global} = 10\log_{10}\left(\frac{k_a}{\hat{F}_e^2} \cdot \frac{E_{pot,a}^{global}}{E_0}\right) \quad \text{with } E_0 = 10^{-12}. \tag{1.20}$$

The results shown in Fig. 1.8 (top) prove that the global ANC-approach is capable of suppressing the acoustic resonance at $\eta_a = 1$, whereas the effect of this control approach on the potential energy of the cavity model is less significant for the structural resonance that occurs close to $\eta_a = 0.2$. Furthermore, it can be seen that amplification of the global potential energy is prevented for every analyzed frequency.

The relative actuations required for AVC, ASAC and global ANC are compared in Fig. 1.8 (below). It can be seen that the relative actuation required for global ANC is considerably smaller than the relative actuation required for the other control approaches. Above $\eta_a = 0.4$ the difference is greater than 50 dB for every analyzed frequency. For $\eta_a = 1$ the relative actuation required for global ANC is 77 dB below the relative actuation required for AVC and ASAC. Thus, the global acoustic control concept requires only 0.014% of the "costs" that are required for the other concepts to suppress the acoustic resonance.

However, global ANC requires a global measure of the acoustic potential energy. Thus, a considerable number of microphones must be installed to measure the error signal. For this reason (especially, if ANC shall be applied to suppress acoustic resonances), it can be useful to analyze the noise reduction potential of a local ANC-approach. In the present example local ANC is represented by the minimization of $E_{pot,a}^{local}$. The effect of this control approach is shown in Fig. 1.9. Again, the frequency dependence of the global potential energy of the simple cavity model is shown top, whereas the frequency dependence of the relative actuation calculated by (1.19) is shown below (ANC: $\hat{F}_{cont} := 12k_a\hat{U}_l^{local,opt}$, AVC and ASAC: $\hat{F}_{cont} = \hat{F}_e$). The logarithmic measure of the potential energy is now given by

$$LE_{pot,a}^{local} = 10\log_{10}\left(\frac{k_a}{\hat{F}_e^2}\cdot\frac{E_{pot,a}^{local}}{E_0}\right) \quad \text{with } E_0 = 10^{-12}. \tag{1.21}$$

The results shown in Fig. 1.9 (top) prove that also the local ANC-approach is capable of suppressing the acoustic resonance at $\eta_a = 1$. The effect of this control approach on the potential energy of the cavity model is again less significant for the structural resonance that occurs close to $\eta_a = 0.2$. However, in contrast to the global approach, amplification occurs outside the resonances.

As shown in Fig. 1.9 (top), the acoustic resonance is shifted from $\eta_a = 1$ down to $\eta_a \approx 0.58$. This is exactly the frequency that corresponds to the minimum of the global potential energy of the uncontrolled cavity model between the resonances specified by $\eta_a = 0.2$ and $\eta_a = 1$. Please notice that the force in the right "air spring" of the cavity model (with stiffness $6k_a$) is zero, if the displacement of the movable bearing is prescribed by (1.17).

Thus, the natural frequency of the cavity model[1] is reduced from $\omega_0^{unc} = \sqrt{18k/m}$ (uncontrolled) to $\omega_0^{con} = \sqrt{6k/m}$ (with local ANC). The ratio between these frequencies is exactly $1/\sqrt{3} \approx 0.58$. However, if local ANC is only applied to the acoustic resonance of the uncontrolled system, this approach offers the possibility to reduce the complexity control system, because the set of sensors can be reduced without lowering the control profit.

The relative actuations required for AVC, ASAC and local ANC are compared in Fig. 1.9 (below). It can be seen that the relative actuation required for local ANC is considerably smaller than the relative actuation required for the other control approaches. Above $\eta_a = 0.7$ the difference is greater than 50 dB for every analyzed frequency. For $\eta_a = 1$ the relative actuation required for local ANC is identical to the relative actuation required for the global ANC-approach. Thus, the local acoustic control concept also requires only 0.014% of the "costs" that are required for the other concepts in order to suppress the acoustic resonance. This should be a strong motivation for a scientific work on (local) ANC of enclosed sound fields.

1.3 Examples for Engineering Applications of Interior ANC

One of the earliest suggestions for active control was the "*Process of silencing sound oscillations*" invented by *Paul Lueg*, see (Lueg 1936) and (Lueg 1937). Nowadays,

[1]More precisely we have to formulate: The natural frequency of the cavity model for the case that the structural displacement is zero.

74 years after this invention, ANC is a technology that is applied to reduce interior noise in aircraft and automobiles, or to provide personal hearing protection.

Reviews on automotive applications are to be found in (Mackay and Kenchington 2004) and (Elliott 2008). Both references report on ANC-systems used for engine order cancellation (EOC) and active reduction of narrow band road noise. EOC was investigated to reduce the sound pressure levels at the engine firing frequency (second order of a 1.1 liter 4-cylinder car). The sound pressure was measured at the front seat positions using monitoring microphones separate from the error microphones used by the active control system. Reductions of up to 10 dB were achieved at the front seats for a firing frequency of about 100 Hz, whereas only minor reductions were measured for the overall SPL. For lower speeds, reductions were measured in the rear, because the first longitudinal acoustic mode (having a nodal line near the front passengers' heads) was suppressed. A commercial application was realized on Nissan's Bluebird as reported in (Hasegawa et al. 1992). The EOC-system installed in these cars used loudspeakers, amplifiers and processors separate from the audio system, and so was relatively expensive.

Honda demonstrated a mainly feedback system to control a 40 Hz boom in the front of an Accord wagon car, see (Sano et al. 2001). This system was used in combination with a fixed feed-forward system. The latter was applied to prevent the noise in the rear of the car being amplified.

This excitation was relatively narrow band and the control microphone was placed close to the loudspeakers, compared with the wavelength at 40 Hz. For this reason good performance was obtained in suppressing this resonant boom by about 10 dB. Elliott stated that: "An important aspect of reducing the cost of this system, so that it could be used in mass production, was the integration of the loudspeakers with the audio system, although at that time a separate active control unit was used from the audio system, since a number of different audio head units were offered on this vehicle. It is the full integration of the active control system with the audio system that would make this technology affordable on many vehicles", see (Elliott 2008).

This statement expresses the motivation for one trend in active control of interior noise, the combination of audio entertainment and ANC. Another trend is active sound design, see (Scheuren et al. 2002), that can be used to provide a choice of several different engine characteristics.

Examples for typical aircraft applications that use the tachometer signal as reference for the active control task, compare Fig. 1.10, were described e.g. in (Fuller et al. 1997) and (Johansson 2000). In the first reference the authors report that: "commercial systems have been installed in aircraft such as the Saab 340 or Saab 2000 and typically comprise of 24–36 optimally located loudspeakers in conjunction with 48–72 error microphones. The total weight is around 70 kg for a 10 dB attenuation", calculated from the changes in the sum of the squared sound pressure measured at all control microphones, "in the propeller fundamental frequency. Each speaker has a maximum power consumption of 25 Watts and approximately 1 km of wire is required in the Saab 2000. The total power consumption of the system is approximately 300 Watts."

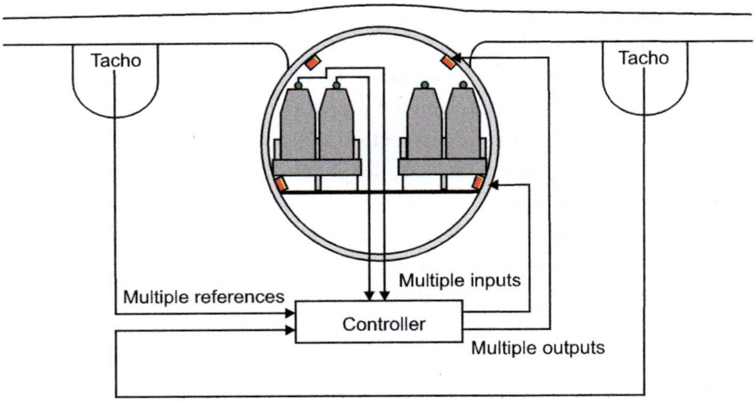

Fig. 1.10 Layout of ANC-system with multiple inputs and multiple outputs installed in a propeller driven aircraft (adapted from Johansson 2000)

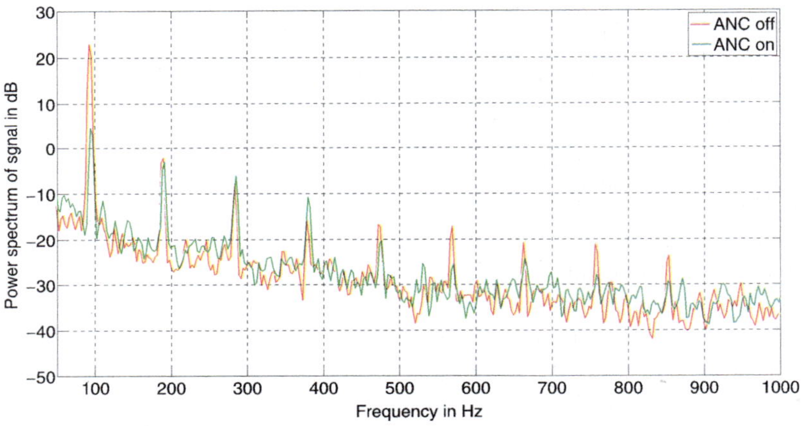

Fig. 1.11 Effectiveness of ANC applied to tonal noise in commercial propeller driven aircraft (Saab 2000). Original time recorded data were taken from (Ultra Electronics 2009). FFT-Analysis was performed to generate the power spectra of controlled and uncontrolled microphone signal

The effectiveness of the Saab 2000 ANC-system is shown in Fig. 1.11. The power spectra are obtained from a frequency analysis of time recorded demo data taken from the Internet presence of Ultra Electronics (Ultra Electronics 2009). The results indicate a significant noise reduction at $f = 97$ Hz that is the fundamental blade pass frequency (BPF) of the particular aircraft. However, the results have to be handled with care because of missing information such as flight conditions, place of measurement or averaging of measured data.

According to the manufacturer, see (Ultra Electronics 2010), the Ultra-system has up to now been adopted on the Bombardier Q100, Q200, Q300 and Q400, Bombardier Challenger 601 and 604, Airbus A400M, Lockheed Martin C-130, Beech

King Air 350, and Saab 340 and 2000. As reported in (Elliott Aviation 2010), an Ultra-system with 12 loudspeakers was successfully applied in a King Air 200. In (Hansen 2003) the author reports on an Ultra-system with 96 input channels (error- and reference sensors) and up to 48 output channels for actuators. The error sensors were mostly microphones, located just above the aircraft windows in the cabin lining as well as in the center of the ceiling lining and in the head racks. A performance of 10 dB, 7 dB and 3 dB was given for noise reduction (spatially averaged reduction determined at passenger head-level) of the fundamental frequency, and the first two harmonics.

Other examples of flight-proven systems are noise-canceling headphones with analogue feedback control. Because active headphones are able to act close to the error sensor, these systems provide a significant noise reduction. An active attenuation of 25 dB in the frequency range between 25 Hz and 500 Hz was reported for a closed headphone, see (Sennheiser 2010), and an attenuation of approximately 10 dB in a frequency range between 400 Hz and 1 kHz was reported for open headphones in the same reference.

1.4 Objective of Book

The previous sections have been (a) an introduction to the active control concept (Sect. 1.1) and (b) a motivation for the chosen acoustic approach to interior active noise control (Sect. 1.2). Using an illustrative example, it has been shown that the ANC-approach is capable of suppressing acoustic disturbances with relatively small (in relation to the excitation forces) control effort compared to that required by AVC or ASAC. Also a mention has been made that ANC can be applied, when reducing the primary excitation by passive noise treatments or by active control approaches based on AVC and/or ASAC is impossible. Off course ANC can also be used to simply support existing noise treatments. These are strong arguments to concentrate on ANC.

To indicate that some systems (applied to active control of sound and vibration) have already reached a high maturity level, textbooks and articles have been cited. Furthermore, examples have been given of existing commercial applications of ANC-systems which reduce interior noise in automobiles and aircraft cabins. This brings up the question: "Can this book contribute anything novel to ANC?"

This can only be answered with a convinced: "Yes!". Although, the "literature concerning the application of active noise control to airplane interior noise (as for automobiles, railed vehicles and other areas of applications; author's note) is extensive", as stated in (Wilby 1996), significant control profit is (except for noise canceling headphones) only obtained for (very) low excitation frequencies, and in "all the implementations reported, active noise control addressed the reduction of discrete frequency components associated with propeller noise or turbofan engine vibration", as also stated in (Wilby 1996). The state of the art can therefore be interpreted as follows:

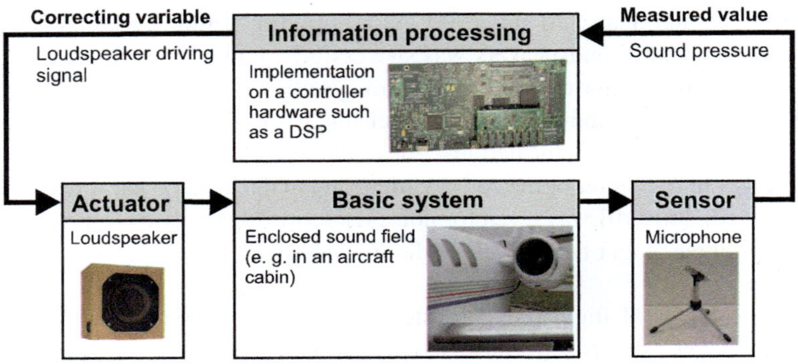

Fig. 1.12 Structure of ANC-system and internal information flow based on (VDI 2206 2004)

1. The theory of active control is well understood and has extensively been researched. However, many ANC-systems have not yet reached a satisfactory maturity level to be used in practice. This is—with active headphones as the exception—particularly true for broadband disturbances.
2. The design process of ANC-systems still pose a challenge, even if all components behave linear and are time-invariant. The fact that only few systems are in service supports this interpretation.

It is apparent that the development of ANC-systems as well as the actual trends in ANC-research are stimulated by the requirements of specific applications. Thus, there is a need for a guideline[2] describing both the ANC-system design tools and the ANC-system design process. This should be done on the basis of a well defined, unified theoretical approach covering the multidisciplinary fields of (a) engineering acoustics, (b) electro-acoustic and (c) adaptive signal processing.

The VDI has already published a general design methodology for mechatronic systems, see (VDI 2206 2004). According to (Harashima et al. 1996), the term *mechatronics* is (in this VDI guideline) specified as: "The synergetic integration of mechanical engineering with electronic and intelligent computer control in the design and manufacturing of industrial products and processes." In the same reference a *mechatronic system* is characterized as a system that consists of a *basic system*, *actuators*, *sensors*, and an *information processing* unit.

The interaction of engineering acoustics with electrical engineering and adaptive real-time computing is required in the design and manufacturing of ANC-systems. ANC can therefore be understood as a mechatronic approach, if engineering acoustics is seen as a sub-domain of engineering mechanics. This statement is supported by the fact that ANC-systems are structured in a manner similar to mechatronic systems, compare Fig. 1.12.

As shown in this figure, the basic system is given by the air filled volume (e.g., the cavity of an aircraft cabin). This elastic and compressible continuum is described

[2] An example of a guideline restricted to the explicitly defined topic of active vibration isolation is to be found in (VDI 2064 2008).

by the wave equation—a partial differential equation. Therefore, the plant that is to be controlled consists of a distributed parameter system. It behaves linear, if high sound pressure levels and also non-linear distortion of loudspeakers are avoided.

Acoustical sensors (e.g. microphones) are applied to measure the sound pressure that is used as error signal or simply to monitor the control profit. Predicated on this information, proper driving signals are determined during the information processing. This can be based on a fixed or on an adaptive control scheme. If a reference signal—that has to be well correlated with the disturbance—is additionally fed into the information processing, feed-forward control is realized. The loudspeaker driving signals are used as correcting variables. They control the canceling sources acting on the basic system. Therefore, ANC-systems are a special class of mechatronic systems.

Based on this statement it is possible to propose a unified and mechatronic approach to adaptive control of low frequency[3] interior noise that

1. Describes the required elements of system theory, engineering acoustics, electro-acoustics and adaptive signal processing in a consistent and systematic manner using a unified notation.
2. Includes a design methodology, explains its application and describes tools to be used for ANC-system design.

Developing an approach to ANC under these two premises is the objective of the present book. Something that has not yet been accomplished in existing (by all means, outstanding) textbooks, which focus on only one particular topic of ANC such as

- Engineering acoustics, see e.g. (Fahy 2003) and (Möser 2005),
- Electro-acoustics, see e.g. (Zwicker and Zollner 1984) and (Ballou 2009),
- The physical basis for active control of sound and structural vibrations, see e.g. (Nelson and Elliott 1992) and (Fahy and Gardonio 2007),
- Signal processing for ANC, see e.g. (Kuo and Morgan 1996) and (Elliott 2001).

Although all of these (sub-)domains are important for the understanding of ANC, the merging of several disciplines adds value to existing knowledge. Except for aforementioned (a) multidisciplinarity, this book not only distinguishes itself from former studies in its effort to (b) point out specific engineering applications for ANC, but also to (c) propose a mechatronic design methodology for ANC-systems, and to (d) discuss its practical use in view of five different projects of applied academic research.

Because signal processing is one key element of active control, a review and evaluation of noise control strategies is discussed. Furthermore, the book contains

---

[3]The determination of precisely what constitutes low frequency noise is not perfectly clear in terms of the lower and the upper limit (Berglund and Hassmén 1996). However, in this book the term low frequency noise means that the frequency of the disturbance f is less than the *Schröder* frequency f_{Sch}. The latter is determined by the reverberation time T_{60} and the size of the air filled volume V such as $f_{Sch} = 2000\sqrt{T_{60}/V}$, see (Nelson and Elliott 1992).

a unified description of optimal and adaptive control considering both multiple inputs and multiple outputs. To make the adaptive control algorithms "ready-to-use", algorithmic formulations are presented for different time-domain approaches. Such a consistent summary is either missing in existing textbooks on ANC,[4] see (Elliott 2001), or limited to examples considering one reference signal, two error signals and two canceling signals, see (Kuo and Morgan 1996).

In contrast to existing textbooks on ANC this book also contains a description of tools that can be used for ANC-system design.[5] Because the present work is focused on interior noise, the finite element method is discussed as a tool for numerical simulation. A novel approach—the inverse finite element method—is proposed for sound source localization in standing wave fields. Because it is necessary to evaluate the noise reduction potential of ANC-systems during the system design process, four tools for initial performance estimation are also discussed in this book. Although, the underlying equations of these performance estimators are not novel, their survey, presentation and exemplary use as ANC-system design tools however is.

Seen purely from the research point of view, the book presents new approaches to sound source localization in weakly damped interiors. One is based on the inverse finite element method, the other is based on a sound intensity probe with an active free field. Furthermore, a prototype of an ANC-system able to reach the physical limits of local (feed-forward) ANC is described. This is one example for applied research in ANC-system design. Other examples are given for (i) local ANC in a semi-enclosed subspace of an aircraft cargo hold and (ii) for the combination of audio entertainment with ANC.

Overview on (i) An engineering challenge for local ANC in a semi-enclosed sound field has been provided by the strong acoustic coupling between the cargo hold of a certain military aircraft that is (through an open surface) connected to a special working area. Because of the requirements, electro-acoustic equipment (e.g. microphones or loudspeakers) should be integrated in the operator's seat. Unfortunately this was impossible. Thus, it has been inevitable to realize local ANC around the operator's head (considering frequencies up to 300 Hz) using transducers remotely located to the control volume. The present book reports on this project. A comparably detailed and—due to the use of the proposed design methodology—systematic description of ANC-system design including (among others) transducer placement optimization, remote microphone control, and robust controller design is not to be found in any other publication on ANC.

[4]Please notice that excellent textbooks on adaptive filtering that naturally contain algorithmic formulations were of course already published, e.g. (Moschytz and Hofbauer 2000) and (Sayed 2003). However the "connection" to ANC, as established in (Kuo and Morgan 1996) and (Elliott 2001), is missing in these books.

[5]This statement is true except for (Fahy and Gardonio 2007). The main topic of this reference is however not ANC, but rather the physics of sound and structural vibration. For this reason it naturally contains an introduction to numerically based analyses of fluid-structure interaction.

Overview on (ii) As already mentioned, the combination of audio entertainment and ANC is an actual trend in applied research. The present book reports on a project in which this concept was applied to the interior noise field of a very light jet mock-up for the first time using time data recorded during in-flight measurements. What is also novel is the application of a special adaptive control algorithm that is, because of its structure, capable of separating the disturbing noise from the audio-signal. The level of noise reduction and the spatial extent of the resulting zone of quiet achieved in real-time experiments considering broadband disturbances up to 500 Hz are (except for noise canceling head phones) clearly above the control success reported for any other combination of audio entertainment and ANC.

The intention of this book is to present a unified approach to ANC as well as to show its application. It is not intended as a handbook such as (Havelock et al. 2008a) and (Havelock et al. 2008b). Therefore, the present work is restricted to adaptive feed-forward control of low frequency interior noise. Topics such as sound radiation in free field, fluid-structure coupling or feedback control have not been taken into account. Consequently this book is structured as follows: An introduction to the topic is given in Part I. Part II (to be understood as a monograph) is focused on the mechatronic background of interior active noise control. It starts with comments on signals and systems (Chap. 2) and continues with the acoustics of interior noise fields (Chap. 3). Leading sensing devices are discussed in Chap. 4. The dynamical behavior of some acoustic actuators is addressed in Chap. 5. Feed-forward control is discussed in Chap. 6. Part III is focused on design tools (Chap. 7) as well as on the design methodology (Chap. 8) for ANC-systems. Examples of systematic system design processes that fit into the frame of the proposed design methodology are discussed in Part IV (Chaps. 9–13). A comprehensive summary is given in Part V.

Part II
The Mechatronic Background of Feed-Forward Active Noise Control

Chapter 2
Comments on Signals and Systems

Abstract To prepare the mechatronic foundation of active noise control, this chapter contains some elements of system theory. Since we will be talking about active noise control *systems* and *signals*, it is necessary to define these terms. Furthermore, it is necessary to introduce values and functions that can be used to characterize signals and systems in time domain but also in frequency domain. However, it is not intended to present a compact summary of system theory that is in great detail presented in textbooks, such as in (Cadzov and van Landingham in Signals, systems, and transforms, Prentice Hall, New Jersey, 1985), (Fliege in Systemtheorie, Teubner, Stuttgart, 1991), (Girod et al. in Einführung in die Systemtheorie, Teubner, Stuttgart, 2005), (Johnson in Digitale Signalverarbeitung, Hanser, München in Cooperation with Prentice Hall International, London, 1991), (Oppenheim and Willsky in Signale und Systeme—Lehrbuch, VCH Verlagsgesellschaft, Weinheim, 1989), (Sundararajan in A practical approach to signals and systems, Wiley Eastern, Singapore, 2008), (Ziemer et al. in Signals and systems, continuous and discrete, Macmillan, New York, 1983). A description of stochastic signals and random vibrations of both linear and non-linear mechanical systems is to be found in (Lajos in Zufallsschwingungen und ihre Behandlung, Springer, Berlin, 1973), whereas digital audio signal processing is discussed in (Zölzer in Digital audio signal processing, Wiley, Chichester, 2008).

2.1 Comments on Signals

The concept of *signal* is central to the system theory philosophy. According to (Cadzov and van Landingham 1985) a signal connotes the process of conveying information in some format. In the present work, the expression signal is used to denote a measurement or observation that contains information describing some acoustic or electric phenomenon.

To define a mathematical structure, we designate signals by means of symbols such as x or y and refer them as *the signals* x and y. A signal in which the information characteristics fluctuate depends on other variables such as time, distance or frequency. For the purpose of this chapter, however, it is sufficient to use time as an independent variable.

T. Kletschkowski, *Adaptive Feed-Forward Control of Low Frequency Interior Noise*, Intelligent Systems, Control and Automation: Science and Engineering 56,
DOI 10.1007/978-94-007-2537-9_2, © Springer Science+Business Media B.V. 2012

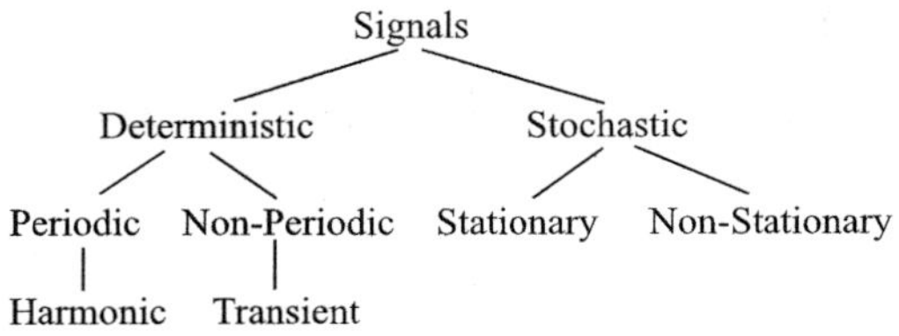

Fig. 2.1 Time-history classification of signals

2.1.1 Classification

A signal that can change at any instant of time is called *continuous-time signal* and will be denoted by $x(t)$ or $y(t)$. On the other hand, if a signal can only change its value at specific instants of time, it is called *discrete-time signal*. Such a signal can be denoted by $x(t_n)$ or $y(t_n)$, where $t_n = nT$ for $n = 0, \pm 1, \pm 2, \ldots$. T is a fixed time interval (e.g. the sample time) and n is the number of the time interval (e.g. the sample number). In the present work a discrete-time signal will be denoted by $x(n)$ or $y(n)$.

As shown in (DIN 1311-1 2000), signals can also be classified according to time history. An overview on time-history classification of signals discussed in this section is given in Fig. 2.1.

A signal is called *deterministic*, if the time history can be described by a function, e.g. $x(t) = f(t)$, that links the time t to the instantaneous value $x(t)$. A deterministic signal with a time history that is repeated after some period T such as

$$x(t) = x(t + T) \quad \forall t \tag{2.1}$$

is a *periodic signal*. Equation (2.1) also holds for $x(t) = x(t + nT)$ $\forall t$ with $n = 0, 1, 2, \ldots$. If a periodic signal can be described by one sine function (or one cosine function) such as

$$x(t) = \hat{x} \sin(\omega t + \varphi_0) \tag{2.2}$$

it is called a *harmonic signal*. Here $\omega = 2\pi/T$ is known as the angular frequency, φ_0 is known as the *zero phase angle*, and $\hat{x} > 0$ is known as the *amplitude*. Deterministic signals with time history

$$x(t) \neq x(t + T) \quad \forall t \tag{2.3}$$

are *non-periodic* signals. An important subclass of non-periodic signals are *transient signals* that describe the crossover from one condition to another one. Typical examples, see (DIN 1311-1 2000), are *quasi-harmonic signals* such as

$$x(t) = \hat{x} e^{-\sigma t} \cos(\omega t + \varphi_0) \quad \text{with decay coefficient } \sigma > 0 \tag{2.4}$$

and *sweep sine* signals such as

$$x(t) = \hat{x} \sin\big[\varphi(t)\big] \quad \text{with } \omega(t) = \frac{d\varphi(t)}{dt}. \tag{2.5}$$

A signal is called *stochastic* or *random*, if the time history cannot be described by a function. This means that predetermination of x is impossible for any point in time t. A stochastic signal can be seen as one realization of a stochastic process ξ that consists of an ensemble of realizations $x_i(t)$. The probability P_w that the realization $x_i(t)$ has a value between a and b at $t = t_j$ is obtained by integrating the probability density function of the stochastic process $f_{x(t)}(\xi)$ over the process such as

$$P_w\{a \leqslant x_i(t) \leqslant b\} = \int_{\xi=a}^{\xi=b} f_{x(t=t_j)}(\xi)d\xi. \tag{2.6}$$

If the probability P_w is independent of the observation time t_j for all values of a and b, the process is *stationary*, otherwise the process is *non-stationary*. Stochastic signals are characterized by their moments or *expectation values*. The most important moments are the *mean*

$$\bar{x} = E[x(t)] = \int_{\xi=-\infty}^{\xi=\infty} \xi \cdot f_x(\xi)d\xi \tag{2.7}$$

and the *variance*

$$\sigma_x^2 = E[x(t) - \bar{x}] = \int_{\xi=-\infty}^{\xi=\infty} (\xi - \bar{x})^2 \cdot f_x(\xi)d\xi. \tag{2.8}$$

If $\bar{x} = 0$ and σ_x^2 are independent of t_j, the signal is a weak (or wide-sense) *stationary* stochastic signal, otherwise the signal is a *non-stationary* stochastic signal. If the stochastic process is an *ergodic process*, the moments defined by (2.7) and (2.8) are identical with time averaged values obtained from analyzing one realization $x_i(t)$ of the stochastic process.

2.1.2 Characteristic Values and Functions

A signal x can be characterized by characteristic values and by characteristic functions. These values and functions will be introduced according to (Ziemer et al. 1983) and (Fliege 1991) but also considering (DIN 1311-1 2000).

Characteristic values that can directly be obtained from the time history of x measured in the time interval $t \in [t_1, t_2]$ are summarized in Table 2.1. In addition to these values other characteristic values are obtained by time averaging of continuous-time and discrete-time signals. Strictly speaking, time averaging of stochastic signals has to be performed on the i-th realization x_i of the stochastic process ξ. However, in what follows the i-th realization is identified with the signal x for convenience.

Continuous-Time Signals

If the continuous-time signal $x(t)$ is observed in the time interval $T = t_2 - t_1$, the *arithmetic mean* of $x(t)$ is given by

Table 2.1 Characteristic values directly obtained from time history of signal

Name and description	Definition
Maximum value: Maximum value of signal for $t \in [t_1, t_2]$.	$x_{\max} := \max(x)$
Minimum value: Minimum value of a signal for $t \in [t_1, t_2]$.	$x_{\min} := \min(x)$
Maximum absolute value: Minimum absolute value of a signal for $t \in [t_1, t_2]$.	$\lvert x_{\max}\rvert := \max(\lvert x(t)\rvert)$
Signal range: Difference between maximum and minimum value.	$x_h := x_{\max} - x_{\min}$

$$\bar{x} = E(x) = \lim_{T\to\infty} \frac{1}{T} \int_{-T/2}^{T/2} x(t)dt. \tag{2.9}$$

The *variance* of $x(t)$ is defined by

$$\sigma_x^2 = \lim_{T\to\infty} \frac{1}{T} \int_{-T/2}^{T/2} \left(x(t) - \bar{x}\right)^2 dt. \tag{2.10}$$

The *mean signal power*[1] of $x(t)$ can be defined by

$$\Pi_x = \lim_{T\to\infty} \frac{1}{T} \int_{-T/2}^{T/2} \left|x(t)\right|^2 dt. \tag{2.11}$$

It is obvious that the Π_x is given by σ_x^2, if the arithmetic mean is zero, i.e. $\bar{x} = 0$. The RMS[2] value of $x(t)$ is defined as the square root of Π_x

$$x_{RMS} = \sqrt{\lim_{T\to\infty} \frac{1}{T} \int_{-T/2}^{T/2} \left|x(t)\right|^2 dt.} \tag{2.12}$$

Furthermore, it is possible to define the *mean signal energy* for $x(t)$ by

$$E_x = \int_{-\infty}^{\infty} \left|x(t)\right|^2 dt. \tag{2.13}$$

The similarity between two continuous-time signals $x(t)$ and $y(t)$ can be characterized by the *cross correlation function* (CCF)

$$r_{xy}(\tau) = \overline{x(t)y(t+\tau)} = \lim_{T\to\infty} \frac{1}{T} \int_{-T/2}^{T/2} x(t)y(t+\tau)dt. \tag{2.14}$$

Consequently the *auto correlation function* (ACF) can be understood as a function that describes the change in time history of the signal $x(t)$. The ACF is defined by

[1] In system theory it is assumed that the signal x is normalized and dimensionless, see (Ziemer et al. 1983) and (Fliege 1991). Furthermore, it is assumed that the associated non-normalized signal has the dimension Volt and its electric energy is measured using a 1 Ohm resistance.

[2] The RMS of x is also used to define a logarithmic measure such as $L_x = 10\log_{10}(x_{RMS}^2/x_0^2)$, where x_0 is a normalization factor.

$$r_{xx}(\tau) = \overline{x(t)x(t+\tau)} = \lim_{T\to\infty} \frac{1}{T} \int_{-T/2}^{T/2} x(t)x(t+\tau)dt. \tag{2.15}$$

It is also possible to characterize continuous-time signals in frequency domain. The frequency domain representation of a continuous-time signal is obtained by applying the (continuous) *Fourier transform* which (according to Ziemer et al. 1983 and Fliege 1991) is defined by

$$x(t)\circ\!\!-\!\!\bullet X(j\omega) = \mathfrak{F}\big[x(t)\big] := \int_{-\infty}^{+\infty} x(t)e^{-j\omega t}dt. \tag{2.16}$$

Conditions for the existence of the Fourier transform are complicated to state in general (Champeney 1987), but it is sufficient for $x(t)$ to be *absolutely integrable*, i.e.,

$$\int_{-\infty}^{\infty} x(t)dt < \infty. \tag{2.17}$$

The inverse operation is defined by

$$X(j\omega)\bullet\!\!-\!\!\circ x(t) = \mathfrak{F}^{-1}\big[X(j\omega)\big] := \frac{1}{2\pi} \int_{-\infty}^{+\infty} X(j\omega)e^{j\omega t}d\omega. \tag{2.18}$$

Using (2.16) it is possible to introduce the frequency domain counterparts of the CCF and the ACF, known as the *cross spectral density* (CSD)

$$S_{xy}(j\omega) = \mathfrak{F}\big[r_{xy}(\tau)\big] \tag{2.19}$$

and the *auto spectral density* (ASD)

$$S_{xx}(j\omega) = \mathfrak{F}\big[r_{xx}(\tau)\big]. \tag{2.20}$$

For some applications it is advantageous to use the single-sided CSD or the single-sided ASD that are, according to (DIN 1311-1 2000), introduced as

$$G_{xy}(j\omega) = 2S_{xy}(j\omega) \quad \text{with } 0 \le \omega < \infty \tag{2.21}$$

and

$$G_{xx}(j\omega) = 2S_{xx}(j\omega) \quad \text{with } 0 \le \omega < \infty. \tag{2.22}$$

In general the CSD is a complex function of frequency, whereas the ASD is a real valued quantity. The complex nature of the CSD can be seen, if (2.19) is evaluated for two signals $x_T(t)$ and $y_T(t)$ that are observed in the time interval T. Using (2.14) the evaluation of (2.19) results in

$$\begin{aligned} S_{x_T y_T}(j\omega) &= \mathfrak{F}\big[r_{x_T y_T}(\tau)\big] \\ &= \int_{-\infty}^{\infty} \overline{x_T(t)y_T(t+\tau)}e^{-j\omega\tau}d\tau \\ &= \lim_{T\to\infty} \frac{1}{T} \int_{-T/2}^{-T/2} x_T(t)e^{j\omega t}\left(\int_{-\infty}^{\infty} y_T(t+\tau)e^{-j\omega(t+\tau)}d\tau\right)dt \end{aligned}$$

$$= \lim_{T\to\infty} \frac{1}{T} \int_{-T/2}^{-T/2} x_T(t)e^{j\omega t} Y_T(j\omega)dt$$

$$= \lim_{T\to\infty} \frac{1}{T}\left[X_T^*(j\omega)Y_T(j\omega)\right]. \tag{2.23}$$

Equation (2.23) also proves that $S_{x_T x_T}(j\omega) = \lim_{T\to\infty} \frac{1}{T}|X_T(j\omega)|^2$ is a real valued function.

Discrete-Time Signals

Discrete-time signals can be characterized similarly to continuous-time signals. However, instead of time integration, summation over all time steps is required for time averaging and the computing of both ACF and CCF.

The *arithmetic mean* of the discrete-time signal $x(n)$ is given by

$$\bar{x} = E(x) = \lim_{N\to\infty} \frac{1}{2N+1} \sum_{n=-N}^{n=N} x(n). \tag{2.24}$$

The *variance* of $x(n)$ is defined by

$$\sigma_x^2 = \lim_{N\to\infty} \frac{1}{2N+1} \sum_{n=-N}^{n=N} \left(x(n) - \bar{x}\right)^2. \tag{2.25}$$

The *mean signal power* of $x(n)$ can be defined by

$$\Pi_x = \lim_{N\to\infty} \frac{1}{2N+1} \sum_{n=-N}^{n=N} \left|x(n)\right|^2. \tag{2.26}$$

The RMS value of $x(n)$ is defined as the square root of Π_x

$$x_{RMS} = \sqrt{\lim_{N\to\infty} \frac{1}{2N+1} \sum_{n=-N}^{n=N} \left|x(n)\right|^2}. \tag{2.27}$$

The *Mean signal energy* of a discrete-time signal is calculated as

$$E_x = \sum_{-\infty}^{\infty} \left|x(n)\right|^2 (t_{n+1} - t_n). \tag{2.28}$$

The similarity between two discrete-time signals $x(n)$ and $y(n)$ can also be characterized by the CCF that is now defined as

$$r_{xy}(m) = \overline{x(n)y(n+m)} = \lim_{N\to\infty} \frac{1}{2N+1} \sum_{n=-N}^{n=N} x(n)y(n+m). \tag{2.29}$$

Consequently the discrete version of the ACF can be understood as a function that describes the change in time history of the signal $x(n)$. The ACF for the discrete-time signal $x(n)$ is defined as

$$r_{xx}(m)=\overline{x(n)x(n+m)}=\lim_{N\to\infty}\frac{1}{2N+1}\sum_{n=-N}^{n=N}x(n)x(n+m). \tag{2.30}$$

As with continuous-time signals it is also possible to characterize discrete-time signals in frequency domain. The (discrete) frequency domain representation of a discrete-time signal is obtained by applying the *Fourier transform of sampled signals* that is (according to Fliege 1991 and Nelson and Elliott 1992) defined by

$$x(n)\circ\!\!-\!\!\bullet X\left(e^{j\omega T}\right)=\mathfrak{F}_d\left[x(n)\right]:=\sum_{n=-\infty}^{\infty}x(n)e^{-j\omega nT}. \tag{2.31}$$

In direct analogy to (2.17) the existence of the Fourier transform for sampled signals is ensured, if

$$\sum_{-\infty}^{\infty}\left|x(n)\right|<\infty \tag{2.32}$$

holds for $x(n)$. The inverse operation is (according to Fliege 1991) defined by

$$X\left(e^{j\omega T}\right)\bullet\!\!-\!\!\circ x(n)=\mathfrak{F}_d^{-1}\left[X\left(e^{j\omega T}\right)\right]:=\frac{1}{2\pi}\int_{-\pi}^{+\pi}X\left(e^{j\omega T}\right)e^{j\omega nT}d(\omega T). \tag{2.33}$$

Using (2.31) it is possible to introduce the frequency domain counterparts of the CCR and the ACF for discrete-time signals. These are the CSD (of discrete-time signals)

$$S_{xy}\left(e^{j\omega T}\right)=\mathfrak{F}_d\left[r_{xy}(m)\right] \tag{2.34}$$

and the ASD (of discrete-time signals)

$$S_{xx}\left(e^{j\omega T}\right)=\mathfrak{F}_d\left[r_{xx}(m)\right]. \tag{2.35}$$

The *discrete Fourier transform* is used to calculate an approximate solution of the Fourier integral (2.16). According to (Fliege 1991) and (Nelson and Elliott 1992) it is defined by

$$x(n)\circ\!\!-\!\!\bullet X(k)=\mathfrak{F}_{DFT}\left[x(n)\right]:=\sum_{n=0}^{N-1}x(n)e^{-j\frac{2\pi nk}{N}}. \tag{2.36}$$

The associated inverse operation is defined by

$$X(k)\bullet\!\!-\!\!\circ x(n)=\mathfrak{F}_{DFT}^{-1}\left[X(k)\right]:=\frac{1}{N}\sum_{n=0}^{N-1}X(k)e^{j\frac{2\pi nk}{N}},\quad \text{with } 0\leqslant n\leqslant N-1, \tag{2.37}$$

where N is the number of time samples used to split the frequency range between zero and the sampling frequency into N values spaced at a frequency increment of $2\pi(NT)^{-1}$, and k is the index associated with the k-th discrete frequency line.

2.2 Comments on Systems

2.2.1 Definitions

In system theory the concept of a *system* is used alongside to that of signal. From the mathematical point of view a (single input/single output) system represents a transformation $\mathfrak{T}$ by which the input signal $x(t)$ is changed into the output signal $y(t)$, see (Cadzov and van Landingham 1985) and (Fliege 1991). A block diagram representing a system is shown in Fig. 2.2.

A system is called *linear*, if and only if

$$\mathfrak{T}\big[k_1x_1(t)+k_2x_2(t)\big]=k_1\mathfrak{T}\big[x_1(t)\big]+k_2\big[x_2(t)\big] \tag{2.38}$$

is fulfilled for arbitrary scalars k_1 and k_2 and arbitrary pairs of transformations $y_1(t)=k_1\mathfrak{T}[x_1(t)]$ and $y_2(t)=k_2\mathfrak{T}[x_2(t)]$.

An important subclass of linear systems are *linear time invariant systems* (LTI-systems). A (continuous) linear system with impulse response $h(t)$ caused by a Dirac impulse $\delta(t)$, compare Fig. 2.3, is a LTI-system if, and only if an arbitrary time shift τ in the input signal $\delta(t-\tau)$ causes the same time shift in the output signal such as

$$h(t-\tau)=\mathfrak{T}\big[\delta(t-\tau)\big]. \tag{2.39}$$

A (discrete) linear system with impulse response $h(n)$ caused by a Dirac impulse $\delta(n)$ is a LTI-system if, and only if an arbitrary time shift k in the input signal $\delta(n-k)$ causes the same time shift in the output signal such as

$$h(n-k)=\mathfrak{T}\big[\delta(n-k)\big]. \tag{2.40}$$

2.2.2 Transfer Behavior of LTI-Systems

The transfer behavior of a LTI-system can be described in both time domain and frequency domain. In time domain, the transformation or transfer behavior is described by the *impulse response* h, whereas in frequency domain the system behavior is described by its *transfer function* (or frequency response function (FRF)).

The transfer behavior of a continuous LTI-system is shown in Fig. 2.4. If the input signal $x(t)$ is a deterministic continuous-time signal, the output signal $y(t)$ can directly be obtained from the input by solving

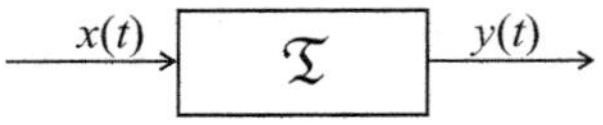

Fig. 2.2 Block diagram representing a system

Fig. 2.3 Linear system with Dirac impulse $\delta(t)$ as input and impulse response $h(t)$ as output

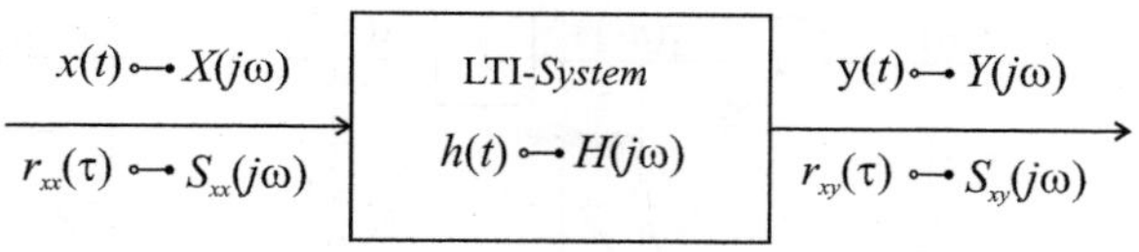

Fig. 2.4 Transmission of deterministic and stochastic signals through a continuous LTI-system

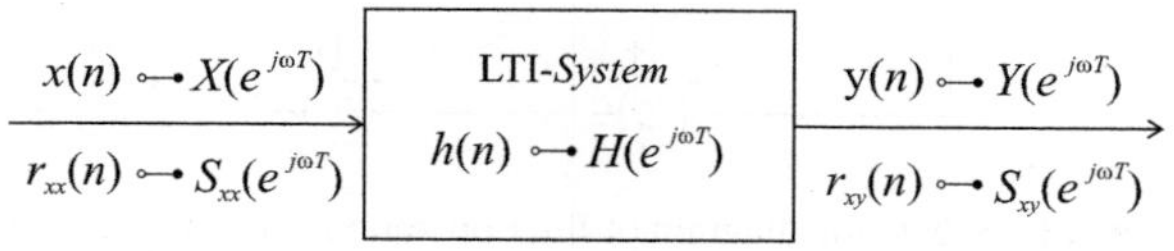

Fig. 2.5 Transmission of deterministic and stochastic signals through a discrete LTI-system

$$y(t) = x(t) * h(t) := \int_{-\infty}^{+\infty} x(\tau)h(t-\tau)d\tau. \tag{2.41}$$

If the calculation is performed in frequency domain using $X(j\omega) = \mathfrak{F}[x(t)]$ and $Y(j\omega) = \mathfrak{F}[y(t)]$, the convolution integral in (2.41) is replaced by a simple multiplication

$$Y(j\omega) = H(j\omega)X(j\omega). \tag{2.42}$$

Direct computation of the output signal is also possible for discrete LTI-systems with deterministic input

$$y(n) = x(n) * h(n) := \sum_{k=-\infty}^{+\infty} x(k)h(n-k). \tag{2.43}$$

The associated equation in frequency domain reads

$$Y\left(e^{j\omega T}\right) = H\left(e^{j\omega T}\right)X\left(e^{j\omega T}\right). \tag{2.44}$$

For stochastic input signals, see Fig. 2.5, it is only possible to analyze the similarity between the input signal $x(t)$ and the output signal $y(t)$ using the CCF (2.14) that is, for stationary continuous-time signals and continuous LTI-systems, given by

$$\begin{aligned} r_{xy}(\tau) &= \overline{x(t)y(t+\tau)} \\ &= \int_{-\infty}^{\infty} \overline{x(t)x(\lambda)}h(t+\tau-\lambda)d\lambda \\ &= \int_{-\infty}^{\infty} r_{xx}(\lambda-t)h(t+\tau-\lambda)d\lambda \\ &\quad \text{with } \lambda - t \to k \text{ and } d\lambda \to dk \\ &= \int_{-\infty}^{\infty} r_{xx}(k)h(\tau-k)dk =: r_{xx}(\tau) * h(\tau). \end{aligned} \tag{2.45}$$

The CCF (2.29) between a stationary discrete-time input signal $x(n)$ that is transformed into the stationary discrete-time output signal $y(n)$ by a discrete LTI-system is given by

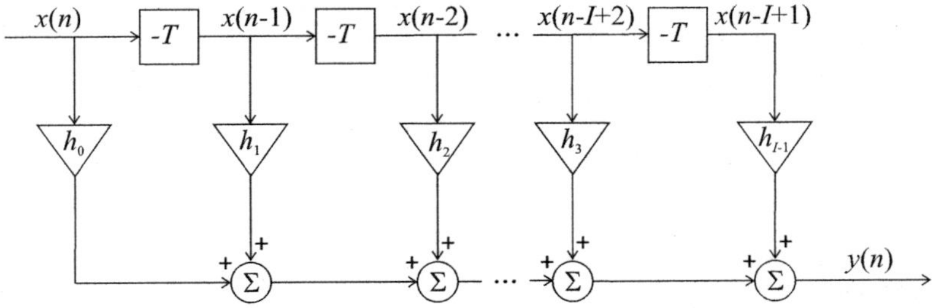

Fig. 2.6 Schematic diagram of filter operation using a finite impulse response filter of order I

$$
\begin{aligned}
r_{xy}(n) &= \overline{x(k)y(k+n)} \\
&= \sum_{m=-\infty}^{\infty} \overline{x(k)x(m)}h(k+n-m) \\
&= \sum_{m=-\infty}^{\infty} r_{xx}(m-k)h(k+n-m) \\
&\quad \text{with } m-k \to l \\
&= \sum_{l=-\infty}^{\infty} r_{xx}(l)h(n-l) =: r_{xx}(n) * h(n).
\end{aligned} \tag{2.46}
$$

The associated frequency domain equations are

$$S_{xy}(j\omega) = H(j\omega)S_{xx}(j\omega) \tag{2.47}$$

and

$$S_{xy}\big(e^{j\omega T}\big) = H\big(e^{j\omega T}\big)S_{xx}\big(e^{j\omega T}\big). \tag{2.48}$$

A convolution defined by (2.43) or (2.46) can be approximated by the inner product between a finite set of I coefficients $h_i(n)$ and a finite set of I input signals $x(n-i)$ such as

$$y(n) = x(n) * h(n) \approx \sum_{i=0}^{I-1} h_i(n)x(n-i) =: \mathbf{h}(n)^T\mathbf{x}(n), \tag{2.49}$$

where the $(I \times 1)$ column matrix $\mathbf{h}(n)$ is a *finite impulse response* (FIR) filter[3] used to approximate the impulse response $h(n)$ of the LTI-system at discrete time n.

The realization of the FIR filter operation defined by (2.49) is illustrated by Fig. 2.6 in which the I coefficients of $\mathbf{h}(n)$ can be interpreted as weights used to modify the actual as well as the previous values of the input signal x. The latter are obtained by successive delaying x with T. To get the output signal y, all weighted inputs are summed up, as also shown in Fig. 2.6.

[3] For details on FIR filters see for example (Kuo and Morgan 1996) or (Moschytz and Hofbauer 2000).

Chapter 3
Dynamics of Basic System

Abstract The present work is focused on adaptive feed-forward control to low-frequency interior noise. The basic system in which the noise is present is the air filled interior. To establish a control concept, it is important to know what field variables can be used to describe the transfer behavior of this basic system or to receive information about the systems state. For this reason, basic acoustic field variables such as the acoustic pressure, the acoustic velocity, and the change in density are introduced in this chapter. It will be shown that interior noise fields are described by the wave equation or, in case of harmonic excitation, by the Helmholtz equation considering the associated boundary conditions. These partial differential equations will be derived from basic field equations that are used to express the balance of linear momentum, the conservation of mass as well as the thermodynamic state. To give an introduction into the nature of standing waves, this chapter also contains analytical solutions for one-dimensional waveguides. However, this chapter is far away from being a comprehensive summary of (engineering) acoustics that is in great detail presented e.g. in (Baranek in Acoustics, McGraw-Hill, New York, 1954a), (Heckl and Müller in Taschenbuch der Technischen Akustik, Springer, Berlin, 1995), (Henn et al. in Ingenieurakustik, Vieweg, Wiesbaden, 1984), (Kuttfurff in Akustik: Eine Einführung, Hirzel, Stuttgart, 2004), (Morese and Ingard in Flügge (ed.) Encyclopedia of physics XI/1, acoustics 1, Springer, Berlin, 1961), (Morse and Ingard in Theoretical acoustics, McGraw-Hill, New York, 1968), (Möser in Technische Akustik, Springer, Berlin, 2005), (Skudrzyk in The foundations of acoustics, basic mathematics and basic acoustics, Springer, Wien, 1971), and in (Fahy in Foundations of engineering acoustics, Academic Press, Amsterdam, 2003). A compact summary of acoustics is given in (DEGA in Akustische Wellen und Felder; DEGA Deutsche Gesellschaft für Akustik e.V. http://www.dega-akustik.de/publikationen/online-publikationen. Cited 05 May 2010, 2006).

3.1 Basic Field Variables

The description of acoustic field is based on two independent variables. The first is the position vector $\vec{x}$ for any material point in the Euclidean space. The second independent variable is the time t.

T. Kletschkowski, *Adaptive Feed-Forward Control of Low Frequency Interior Noise*, Intelligent Systems, Control and Automation: Science and Engineering 56,
DOI 10.1007/978-94-007-2537-9_3,

The dependent variables are the *acoustic velocity* $\vec{v}(\vec{x},t)$ (also known as particle velocity), the *acoustic pressure* $p(\vec{x},t)$ (also known as sound pressure), the *change in density* $\rho(\vec{x},t)$ and the *change in temperature* $\theta(\vec{x},t)$. These dependent variables are introduced as (small) differences between the actual (or total) values and the equilibrium values such as

$$\begin{aligned} \vec{v}(\vec{x},t) &= \vec{v}_{tot}(\vec{x},t) - \vec{v}_{\infty} \\ p(\vec{x},t) &= p_{tot}(\vec{x},t) - p_{\infty} \\ \rho(\vec{x},t) &= \rho_{tot}(\vec{x},t) - \rho_{\infty} \\ \theta(\vec{x},t) &= \theta_{tot}(\vec{x},t) - \theta_{\infty}. \end{aligned} \tag{3.1}$$

The equilibrium values for (the atmospheric) pressure $p_{\infty} = 0.101325$ MPa, density $\rho_0 = 1.292\ \mathrm{kg\,m^{-3}}$ and temperature $\theta_{\infty} = 273.15$ K defined for air in the standard state are to be found in (DIN 1314 1990). In the present work the acoustic medium air is seen as fluid in rest, hence $\vec{v}_{\infty} = \vec{0}$.

An example for the order of magnitudes of the acoustic field variables is given in (Müller 1994). It clarifies that a (relatively high) sound pressure amplitude of 20 Pa $\approx 0.0002 p_{\infty}$ is related to a velocity amplitude of $0.0064\ \mathrm{m\,s^{-1}}$. The corresponding changes in density and temperature are $2.4 \cdot 10^{-4}\ \mathrm{kg\,m^{-3}}$ and 0.022 K.

3.2 Acoustic Field Equations

The field variables introduced in Sect. 3.1 are usually small enough to obtain reasonably accurate equations, involving these variables to the first order, in terms of the equilibrium values (not necessarily to the first order). Neglecting the change in temperature we therefore use the acoustic equations for a fluid in rest, compare (Morese and Ingard 1961), to describe the basic system by the following set of partial differential equations (PDE)

$$\operatorname{grad} p(\vec{x},t) + \rho_{\infty} \frac{\partial \vec{v}(\vec{x},t)}{\partial t} = \vec{0} \tag{3.2}$$

$$\frac{\partial \rho(\vec{x},t)}{\partial t} + \rho_{\infty} \operatorname{div} \vec{v}(\vec{x},t) = 0 \tag{3.3}$$

$$p(\vec{x},t) - c^2 \rho(\vec{x},t) = 0. \tag{3.4}$$

The linearized *Euler equation* (3.2) expresses *Newton's* second law for an infinitesimal volume element. Equation (3.2) also clarifies that the normal component of the pressure gradient is linked to the normal component of the acoustic velocity such as

$$\vec{n}(\vec{x}) \cdot \operatorname{grad} p(\vec{x},t) = -\vec{n}(\vec{x}) \cdot \rho_{\infty} \frac{\partial \vec{v}(\vec{x},t)}{\partial t} =: -\rho_{\infty} \frac{\partial v(\vec{x},t)}{\partial t}. \tag{3.5}$$

The principle of conservation of mass is expressed by the linearized *continuity equation* (3.3), which connects the change in density $\rho(\vec{x},t)$ with the acoustic velocity $\vec{v}(\vec{x},t)$. The state equation (3.4) is the first order approximation of

$$p_{tot}(\vec{x},t) = p_\infty \left(\frac{\rho_{tot}(\vec{x},t)}{\rho_0} \right)^{C_p/C_V}$$
$$\Rightarrow d\left[p(\vec{x},t) - \frac{C_p}{C_V} \frac{p_\infty}{\rho_\infty} \rho(\vec{x},t) \right] = 0 \tag{3.6}$$

which is used to describe isentropic compression of an ideal gas. C_p is the *specific heat* for constant pressure, and C_V is the specific heat for constant volume. A comparison of (3.4) and (3.6) clarifies that the constant c that is known as the *speed of sound* is given by

$$c = \sqrt{\frac{C_p}{C_V} \frac{p_\infty}{\rho_\infty}}. \tag{3.7}$$

Using the equilibrium values for pressure p_∞ and density ρ_∞ specified in Sect. 3.1 and considering that air is mixture of two-atomic gases with $C_p/C_V = 1.4$, the speed of sound is given by $c \approx 331\ \mathrm{m\,s^{-1}}$. It is also possible to define the *bulk modulus* of air. Replacing the acoustic density in (3.3) using (3.4) yields

$$\frac{\partial p(\vec{x},t)}{\partial t} = -\rho_\infty c^2 \operatorname{div} \vec{v}(\vec{x},t) = -\rho_\infty c^2 \frac{\partial \varepsilon_{vol}(\vec{x},t)}{\partial t} =: -E \dot{\varepsilon}_{vol}(\vec{x},t), \tag{3.8}$$

where $E = \rho_\infty c^2$ is the bulk modulus and $\dot{\varepsilon}_{vol}(\vec{x},t)$ is the time derivative of the volume compression $\varepsilon_{vol}(\vec{x},t) = \Delta V/V$. Using $\rho_\infty = 1.292\ \mathrm{kg\,m^{-3}}$ and considering $c \approx 331\ \mathrm{m\,s^{-1}}$, the bulk modulus of air is given by $E \approx 0.141$ MPa.

Instead of solving three PDEs for three dependent variables, it is advantageous to eliminate two field variables using (3.2)–(3.4). As shown before, the time derivative of the acoustic density in (3.3) can be replaced by the time derivative of the acoustic pressure using the constitutive relation (3.4). Partial differentiation of the resulting equation (3.8) with respect to time yields in combination with the equation that results from taking the divergence of (3.2) in the *homogeneous wave equation* for the acoustic pressure

$$\Box p(\vec{x},t) := \operatorname{div}\left[\operatorname{grad} p(\vec{x},t) \right] - \frac{1}{c^2} \cdot \frac{\partial^2 p(\vec{x},t)}{\partial t^2} = 0. \tag{3.9}$$

This PDE describes the state of a linear, homogeneous, loss-less and source-free medium at rest. Solution of (3.9) requires *boundary conditions* (BC) such as the:

- *Dirichlet* BC, where a prescribed sound pressure p_{R_p} is imposed on the pressure boundary R_p at the position $\vec{x}_r$

$$p(\vec{x}_r,t) = p_{R_p}(\vec{x}_r,t). \tag{3.10}$$

- *Neumann* BC, where the normal component of the acoustic velocity[1] v_{R_v} is specified on the velocity boundary R_v

$$v(\vec{x}_r,t) = v_{R_v}(\vec{x}_r,t). \tag{3.11}$$

[1] It is also possible to specify $\dot{v}(\vec{x}_r,t) = \dot{v}_{R_u}(\vec{x}_r,t) =: a_{R_v}(\vec{x}_r,t)$ on R_v.

- *Robin* BC, where a linear combination consisting of the time derivative of the acoustic pressure and the normal component of the pressure gradient is (using the ratio between sound pressure and normal acoustic velocity at a point on the boundary, called impedance Z_{R_Z}) specified on R_Z

$$\frac{\partial p(\vec{x}_r, t)}{\partial t} + \frac{Z_{R_Z}}{\rho_\infty} \vec{n}(\vec{x}_r) \cdot \operatorname{grad} p(\vec{x}_r, t) = 0. \tag{3.12}$$

It is also possible to formulate alternative wave equations such as the *vector wave equation* for the acoustic velocity or the wave equation for the *velocity potential.*

The velocity potential $\phi(\vec{x}, t)$, compare (Bathe 1996) and (Fahy and Gardonio 2007), is introduced as a scalar quantity that is connected to the acoustic variables by two PDEs such as

$$p(\vec{x}, t) = \rho_\infty \cdot \frac{\partial \phi(\vec{x}, t)}{\partial t} \tag{3.13}$$

and

$$\vec{v}(\vec{x}, t) = -\operatorname{grad} \phi(\vec{x}, t). \tag{3.14}$$

The linearized *Euler equation* (3.2) is satisfied by (3.13) and (3.14). Equating the acoustic variables in the *continuity equation* (3.3) using these PDEs results in a scalar wave equation for the velocity potential

$$\Box \phi(\vec{x}, t) := \operatorname{div}\left[\operatorname{grad} \phi(\vec{x}, t)\right] - \frac{1}{c^2} \cdot \frac{\partial^2 \phi(\vec{x}, t)}{\partial t^2} = 0. \tag{3.15}$$

To derive the vector wave equation for the acoustic velocity, we have again to replace the time derivative of the acoustic density in (3.3) by the time derivative of the acoustic pressure using the constitutive relation (3.4). By taking the spatial gradient of the resulting equation (3.8) we obtain

$$\frac{1}{c^2} \cdot \frac{\partial \operatorname{grad} p(\vec{x}, t)}{\partial t} + \rho_\infty \cdot \operatorname{grad}\left[\operatorname{div} \vec{v}(\vec{x}, t)\right] = \vec{0}. \tag{3.16}$$

Partial differentiation of (3.2) with respect to time and multiplication with c^{-2} yields

$$\frac{1}{c^2} \cdot \frac{\partial \operatorname{grad} p(\vec{x}, t)}{\partial t} + \frac{\rho_\infty}{c^2} \cdot \frac{\partial^2 \vec{v}(\vec{x}, t)}{\partial t^2} = \vec{0}. \tag{3.17}$$

The vector wave equation for the acoustic velocity is obtained as the difference between (3.17) and (3.16), if this difference is multiplied by ρ_∞^{-1}. It is given by

$$\vec{\Box} \vec{v}(\vec{x}, t) := \operatorname{grad}\left[\operatorname{div} \vec{v}(\vec{x}, t)\right] - \frac{1}{c^2} \cdot \frac{\partial^2 \vec{v}(\vec{x}, t)}{\partial t^2} = \vec{0}. \tag{3.18}$$

In the special case in which the interior noise field is caused by harmonic disturbances, it is advantageous to solve the acoustic field equations in frequency domain. If the acoustic pressure $p(\vec{x}, t)$ is seen as the real part of the *complex acoustic pressure*

$$p(\vec{x},t) = \text{Re}\{P(\vec{x},j\omega)e^{j\omega t}\}, \tag{3.19}$$

and, if furthermore, the acoustic velocity $\vec{v}(\vec{x},j\omega)$ is seen as the real part of the *complex acoustic velocity*

$$\vec{v}(\vec{x},t) = \text{Re}\{\vec{V}(\vec{x},j\omega)e^{j\omega t}\}, \tag{3.20}$$

it is straightforward to show that evaluation of (3.2) using (3.19) and (3.20) results in a relation that links the amplitude of the complex pressure and the amplitude of the complex velocity such as

$$\vec{V}(\vec{x},j\omega) = \frac{j}{\rho_\infty \omega} \,\text{grad}\, P(\vec{x},j\omega). \tag{3.21}$$

Inserting (3.19) into (3.9) results in a PDE that is independent of time

$$\text{div}\big[\text{grad}\, P(\vec{x},j\omega)\big] + k^2 P(\vec{x},j\omega) = 0. \tag{3.22}$$

This equation in which $k = \omega/c$ is the *wave number*, is the *homogeneous Helmholtz equation*. The associated BC are the

- *Dirichlet* BC with prescribed pressure P_{R_P} defined on R_P

$$P(\vec{x}_r,j\omega) = P_{R_P}(\vec{x}_r,j\omega). \tag{3.23}$$

- *Neumann* BC with prescribed velocity V_{R_V} defined on R_V

$$V(\vec{x}_r,j\omega) = V_{R_V}(\vec{x}_r,j\omega) \quad \text{with } V(\vec{x},j\omega) = \vec{n}(\vec{x}) \cdot \frac{j}{\rho_\infty \omega} \,\text{grad}\, P(\vec{x},j\omega) \tag{3.24}$$

- *Robin* BC with prescribed impedance Z_{R_Z} specified at R_Z

$$\frac{P(\vec{x}_r,j\omega)}{V(\vec{x}_r,j\omega)} = Z_{R_Z}(\vec{x}_r). \tag{3.25}$$

Special BCs are the *sound hard boundary condition*: $V(\vec{x}_r,j\omega) = 0$ on R_V, the *sound soft boundary condition*: $P(\vec{x}_r,j\omega) = 0$ on R_P, and the *free field boundary condition*: $P(\vec{x}_r,j\omega) = \rho_\infty c V(\vec{x}_r,j\omega)$ on R_Z.

3.3 Energy Density and Sound Intensity

Based on the results of the previous sections it is possible to introduce field variables that can be used to describe the transport of energy and power in acoustic systems. The instantaneous *acoustic energy density* $e(\vec{x},t)$ at every material point is given by the sum of the instantaneous *acoustic kinetic energy density* and the instantaneous *acoustic potential energy density*. These two field variables are defined by

$$e_{kin}(\vec{x},t) = \frac{1}{2}\rho_\infty \big|\vec{v}(\vec{x},t)\big|^2 \tag{3.26}$$

and

$$e_{pot}(\vec{x},t) = \frac{1}{2\rho_\infty c^2}\left|p(\vec{x},t)\right|^2. \tag{3.27}$$

Using (3.2) and (3.8) it is possible to introduce the instantaneous *sound intensity* $\vec{i}(\vec{x},t)$ via the time derivative of $e(\vec{x},t)$ such as

$$\begin{aligned}
\frac{\partial e(\vec{x},t)}{\partial t} &= \frac{\partial}{\partial t}\left[\frac{1}{2}\rho_\infty\left|\vec{v}(\vec{x},t)\right|^2 + \frac{1}{2\rho_\infty c^2}\left|p(\vec{x},t)\right|^2\right] \\
&= \frac{\partial}{\partial t}\left[\frac{1}{2}\rho_\infty \vec{v}(\vec{x},t)\cdot\vec{v}(\vec{x},t) + \frac{1}{2\rho_\infty c^2}p^2(\vec{x},t)\right] \\
&= -\frac{\partial}{\partial t}\left[\vec{v}(\vec{x},t)\cdot \operatorname{grad} p(\vec{x},t) + p(\vec{x},t)\operatorname{div}\vec{v}(\vec{x},t)\right] \\
&= -\frac{\partial}{\partial t}\left[p(\vec{x},t)\vec{v}(\vec{x},t)\right] \\
&=: -\operatorname{div}\vec{i}(\vec{x},t).
\end{aligned} \tag{3.28}$$

Equation (3.28) states that the change in energy density depends on the spatial distribution of the energy flux through the boundaries of the enclosed sound field. "If the sound field is temporally stationary, the time integral of (3.28) will converge to zero as the integration time extends beyond the period of lowest frequency component present." as stated in (Fahy 1989).

The normal component of the instantaneous sound intensity is obtained by multiplying the instantaneous sound pressure with the normal component of the instantaneous acoustic velocity

$$i(\vec{x},t) = p(\vec{x},t)v(\vec{x},t) = p(\vec{x},t)\left[\vec{n}(\vec{x})\cdot\vec{v}(\vec{x},t)\right]. \tag{3.29}$$

The arithmetic mean of $i(\vec{x},t)$ is for stationary signals given by

$$\bar{i}(\vec{x},t) = \overline{p(\vec{x},t)v(\vec{x},t)} = r_{pv}(\vec{x},\tau=0). \tag{3.30}$$

Taking into account that the cross correlation $r_{pv}(\vec{x},\tau)$ is the inverse Fourier transform of the associated cross spectral density, compare Sect. 2.1.2,

$$r_{pv}(\vec{x},\tau) = \mathfrak{F}^{-1}\left[S_{pv}(\vec{x},j\omega)\right] \tag{3.31}$$

and considering that $\mathrm{Re}[S_{pv}(\vec{x},j\omega)]$ is even and $\mathrm{Im}[S_{pv}(\vec{x},j\omega)]$ is odd, see (Fahy 1989), the real valued normal component of the *mean intensity* (also known as *active intensity*) can be calculated from the cross spectrum $S_{pv}(\vec{x},j\omega)$ such as

$$\bar{i}(\vec{x},t) = r_{pv}(\vec{x},0) = \frac{1}{2\pi}\int_{-\infty}^{\infty} S_{pv}(\vec{x},j\omega)d\omega = \frac{1}{2\pi}\int_{-\infty}^{\infty}\mathrm{Re}\left[S_{pv}(\vec{x},j\omega)\right]d\omega. \tag{3.32}$$

In this sense the real part of $S_{pv}(\vec{x},j\omega)$ represents the frequency distribution of the mean intensity and describes the energy transport. The imaginary part of $S_{pv}(\vec{x},j\omega)$ is proportional to the magnitude of the *reactive intensity*. The latter does not represent the frequency distribution of a time-averaged quantity, because the mean reactive intensity is zero at all frequencies, compare (3.32). For harmonic signals

$$p(\vec{x},t)=\mathrm{Re}\big[P(\vec{x},j\omega)e^{j\omega t}\big] \quad \text{and} \quad v(\vec{x},t)=\mathrm{Re}\big[V(\vec{x},j\omega)e^{j\omega t}\big] \tag{3.33}$$

(3.32) reduces to

$$\begin{aligned}
\bar{i} &= \lim_{T\to\infty}\frac{1}{T}\int_{-T/2}^{-T/2} p(\vec{x},t)v(\vec{x},t)dt \\
&= \lim_{T\to\infty}\frac{1}{T}\int_{-T/2}^{-T/2} \mathrm{Re}\big[P(\vec{x},j\omega)e^{j\omega t}\big]\mathrm{Re}\big[V(\vec{x},j\omega)e^{j\omega t}\big]dt \\
&= \lim_{T\to\infty}\frac{1}{4T}\int_{-T/2}^{-T/2} \big[P^* e^{-j\omega t}+Pe^{j\omega t}\big]\big[V^* e^{-j\omega t}+Ve^{j\omega t}\big]dt \\
&= \lim_{T\to\infty}\frac{1}{2T}\int_{-T/2}^{-T/2} \mathrm{Re}\big[PV^*+PVe^{2j\omega t}\big]dt \\
&= \frac{1}{2}\mathrm{Re}\big[P(\vec{x},j\omega)V^*(\vec{x},j\omega)\big] =: \bar{I}.
\end{aligned} \tag{3.34}$$

We finally introduce the instantaneous *sound power*

$$\Pi(t)=\int_S \vec{i}(t)\cdot d\vec{S} \tag{3.35}$$

as the power that is transmitted through the area S, where $d\vec{S}=\vec{n}dS$ is the vectorial surface element that is at any point perpendicular to S.

3.4 One-Dimensional Enclosed Sound Fields

To get familiar with interior noise problems, this section is focused on one-dimensional enclosed sound fields for which free as well as forced vibrations will be analyzed analytically in frequency domain.

3.4.1 Free Vibrations in One-Dimensional Sound Fields

A representative example of a one-dimensional enclosure is a straight duct with length L and cross section S. The acoustic field in this interior is described using *Cartesian coordinates*, assuming that the x-axis falls together with the axis of the duct. The dimensions in cross direction are small compared to L. The uniaxial acoustic velocity $\vec{V}(x,j\omega)=V_x(x,j\omega)\vec{e}_x$ is therefore identified with $V_x(x,j\omega)$. In the following we will distinguish between three sets of BCs for the left ($x=0$) and the right ($x=L$) end of the duct. These are the

Sound hard set: $V_x(x=0,j\omega)=V_x(x=L,j\omega)=0$

Sound soft set: $P(x=0,j\omega)=P(x=L,j\omega)=0$

Mixed set: $V_x(x=0,j\omega)=0$ and $P(x=L,j\omega)=0.$

Table 3.1 Eigenfrequencies and eigenmode shapes of a finite duct with length L

Name of set	Eigenfrequency	Pressure mode shape	Velocity mode shape
Sound soft	$f_n = \frac{nc}{2L}$	$P_n(x, j\omega) \sim \sin(\frac{n\pi x}{L})$	$V_{nx}(x, j\omega) \sim \cos(\frac{n\pi x}{L})$
Sound hard	$f_n = \frac{nc}{2L}$	$P_n(x, j\omega) \sim \cos(\frac{n\pi x}{L})$	$V_{nx}(x, j\omega) \sim \sin(\frac{n\pi x}{L})$
Mixed	$f_n = \frac{(2n-1)c}{4L}$	$P_n(x, j\omega) \sim \cos(\frac{\pi(2n-1)x}{2L})$	$V_{nx}(x, j\omega) \sim \sin(\frac{\pi(2n-1)x}{2L})$

It is straightforward to show that the solution of the one-dimensional Helmholtz equation (3.22) is for the n-th eigenfrequency f_n given by a function that consists of two waves traveling in opposite direction

$$P_n(x, j\omega) = A_n e^{-jk_n x} + B_n e^{jk_n x} \quad \text{with } k_n = \frac{2\pi f_n}{c}. \tag{3.36}$$

Equation (3.36) is also known as *traveling wave model*. Because of (3.21) the associated acoustic velocity is given by

$$V_{nx}(x, j\omega) = \frac{1}{\rho_\infty c}\left(A_n e^{-jk_n x} - B_n e^{jk_n x}\right). \tag{3.37}$$

The eigenfrequencies and eigenfunctions (also known as *eigenmode shapes*) are obtained from evaluating the BC-sets using (3.36) and (3.37). These *modal parameters* are listed in Table 3.1.

Summation over all modes results in

$$P(x, j\omega) = \sum_{n=1}^{\infty}\left(A_n e^{-jk_n x} + B_n e^{jk_n x}\right) \tag{3.38}$$

and

$$V_x(x, j\omega) = \frac{1}{\rho_\infty c}\sum_{n=1}^{\infty}\left(A_n e^{-jk_n x} - B_n e^{jk_n x}\right). \tag{3.39}$$

The magnitude of the acoustic potential energy density of the one-dimensional enclosure is, according to (3.27), for the n-th eigenmode given by

$$E_{npot}(x, j\omega) = \frac{1}{2\rho_\infty c^2}\left|P_n(x, j\omega)\right|^2. \tag{3.40}$$

In case of free vibrations this quantity will in engineering applications decay with time because of internal losses. This decay can be described using an exponential function that relates the actual value of the acoustic potential energy density to its magnitude, defined in (3.40), using a *decay coefficient* σ_n for each eigenmode such as

$$e_{npot}^{\sigma_n}(x, t) = \frac{1}{2\rho_\infty c^2}\left|P_n(x, j\omega)\right|^2 e^{-2\sigma_n t}. \tag{3.41}$$

The time span T_{n60} in which the acoustic potential energy density is reduced to a millionth of its initial value, as specified by

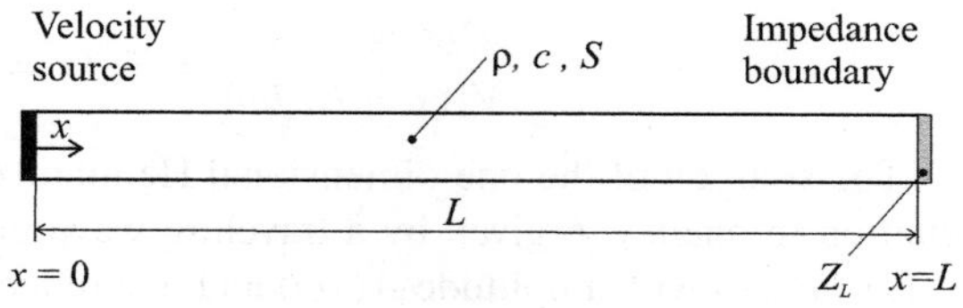

Fig. 3.1 One-dimensional acoustic system with velocity source located at $x = 0$ and impedance boundary at $x = L$

$$e_{npot}^{\sigma_n}(x, T_{n60}) = \frac{1}{2\rho_\infty c^2} \left| P_n(x, j\omega) \right|^2 e^{-2\sigma_n T_{n60}}$$
$$= 10^{-6} \frac{1}{2\rho_\infty c^2} \left| P_n(x, j\omega) \right|^2, \tag{3.42}$$

is known as the *reverberation time* of the n-th eigenmode. Using (3.42) the associated decay parameter can be approximated by

$$\sigma_n = -\frac{\ln(10^{-6})}{2T_{n60}} \approx \frac{6.9}{T_{n60}}. \tag{3.43}$$

If $1 \leq N < \infty$ eigenmodes are present in the frequency band of interest, the decay of acoustic potential energy density in this frequency band can be described by an average decay coefficient $\langle\sigma\rangle$ such as

$$e_{npot}^{\langle\sigma\rangle}(x, t) = \frac{1}{2\rho_\infty c^2} e^{-2\langle\sigma\rangle t} \sum_{n=1}^{N} \left| P_n(x, j\omega) \right|^2. \tag{3.44}$$

The time span $\langle T_{60}\rangle$ in which $e_{npot}^{\langle\sigma\rangle}(x, t)$ is reduced to a millionth of its initial value can analogue to (3.43) be used to approximate $\langle\sigma\rangle$ by

$$\langle\sigma\rangle \approx \frac{6.9}{\langle T_{60}\rangle}. \tag{3.45}$$

3.4.2 Forced Vibrations in One-Dimensional Sound Fields

Forced vibrations in an one-dimensional waveguide will also be studied by the example of a straight duct with length L and cross section S as shown in Fig. 3.1. The acoustic field in this interior is again described using Cartesian coordinates, assuming that the x-axis falls together with the axis of the duct, see Fig. 3.1. The dimensions in cross direction may again be small compared to L. The velocity field $\vec{V}(x, j\omega) = V_x(x, j\omega)\vec{e}_x$ is therefore identified with $V_x(x, j\omega)$ as in the previous subsection.

At the left end, the duct is excited by a velocity source with *volume velocity* $Q(x = 0, j\omega) = V_0(j\omega)S$. The acoustic velocity at $x = 0$ is therefore prescribed by

$$V_x(x = 0, j\omega) = V_0(j\omega). \tag{3.46}$$

At the right end, the duct is terminated by an impedance boundary such as

$$\frac{P(x=L,j\omega)}{V_x(x=L,j\omega)} = Z_L(x=L,j\omega). \tag{3.47}$$

The solution of the one-dimensional Helmholtz equation (3.22) is for every excitation frequency f given by a traveling wave model that consists of an incident wave $P_i(j\omega)$ with amplitude $A(j\omega)$ and of a reflected wave $P_r(j\omega)$ with amplitude $B(j\omega)$ traveling in opposite direction

$$P(x,j\omega) = P_i + P_r = Ae^{-jkx} + Be^{jkx}. \tag{3.48}$$

The associated acoustic velocity is because of (3.21) given by

$$V_x(x,j\omega) = \frac{1}{\rho_\infty c}\big(Ae^{-jkx} - Be^{jkx}\big). \tag{3.49}$$

Evaluation of the BCs specified by (3.46) and (3.47) yields

$$A(j\omega) = \frac{\rho_\infty c e^{jkL}}{e^{jkl} - \frac{Z_L-\rho_\infty c}{Z_L+\rho_\infty c}e^{jkL}} V_0 \quad \text{and} \quad B(j\omega) = \frac{\rho_\infty c \frac{Z_L-\rho_\infty c}{Z_L+\rho_\infty c}e^{-jkL}}{e^{jkl} - \frac{Z_L-\rho_\infty c}{Z_L+\rho_\infty c}e^{jkL}} V_0. \tag{3.50}$$

The ratio between the reflected wave component P_r and the incident wave component P_i is known as the complex *reflection coefficient*. This quantity is, according to (3.48) and (3.50) for $x = L$ given by

$$\begin{aligned} R(j\omega) := \big|R(x=L,j\omega)\big|e^{j\chi(x=L,\omega)} &= \frac{P_i(x=L,j\omega)}{P_r(x=L,j\omega)} \\ &= \frac{Z_L(j\omega) - \rho_\infty c}{Z_L(j\omega) + \rho_\infty c}. \end{aligned} \tag{3.51}$$

Equation (3.51) clarifies that the complex reflection coefficient can be used to describe an impedance boundary. Furthermore, it can be used to introduce the *absorption coefficient* $\alpha(j\omega)$ at the right termination of the duct such as

$$\alpha(j\omega) := \frac{|P_i(L,j\omega)|^2 - |P_r(L,j\omega)|^2}{|P_i(L,j\omega)|^2} = 1 - \big|R(j\omega)\big|^2. \tag{3.52}$$

Based on (3.51) it is also possible to introduce a complex pressure amplitude

$$\hat{P}(j\omega) = \frac{\rho_\infty c}{e^{jkL} - Re^{-jkL}} V_0(j\omega) \tag{3.53}$$

that can be used to present the solution of the investigated problem in a more compact manner such as

$$P(x,j\omega) = \hat{P}(j\omega)\big(e^{-jk(x-L)} + R(j\omega)e^{jk(x-L)}\big). \tag{3.54}$$

Because of (3.21), the associated acoustic velocity is given by

$$V(x,j\omega) = \frac{\hat{P}(j\omega)}{\rho_\infty c}\big(e^{-jk(x-L)} - R(j\omega)e^{jk(x-L)}\big). \tag{3.55}$$

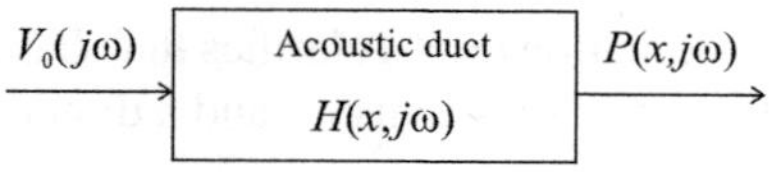

Fig. 3.2 Transfer behavior of an one-dimensional wave-guide for harmonic excitation with $V_0(j\omega)$

If Z_L is unknown, (3.54) allows for direct calculation of the complex reflection coefficient using two measurements performed at x and $x - \Delta x$. Evaluating the ratio of the associated pressure values

$$\frac{P(x, j\omega)}{P(x - \Delta x, j\omega)} = \frac{e^{-jk(x-L)} + R(j\omega)e^{jk(x-L)}}{e^{-jk(x-L-\Delta x)} + R(j\omega)e^{jk(x-L-\Delta x)}} \tag{3.56}$$

yields[2]

$$R(j\omega) = -e^{-2jk(x-L)} \frac{P(x - \Delta x, j\omega) - P(x, j\omega)e^{jk\Delta x}}{P(x - \Delta x, j\omega) - P(x, j\omega)e^{-jk\Delta x}}. \tag{3.57}$$

It should also be noticed that the solution (3.54) can be interpreted as the superposition of a standing wave with amplitude $\hat{P}_s := \hat{P}R$ and a traveling wave with amplitude $\hat{P}_p := \hat{P}(1 - R)$ such as

$$P(x, j\omega) = \hat{P}_s(j\omega)\big(e^{-jk(x-L)} + e^{jk(x-L)}\big) + \hat{P}_p(j\omega)e^{-jk(x-L)}. \tag{3.58}$$

The standing wave represents the acoustic potential energy stored in the system and is known as the *reactive sound field*, while the traveling wave describes the energy transport through the duct. Therefore, this part is called *active sound field*.

The meaning of these terms can be understood, if the energy transport in the investigated one-dimensional waveguide is calculated. Considering (3.52), (3.54), and (3.55), the mean intensity can—based on (3.34)—be expressed in terms of the absorption coefficient

$$\bar{I} = \frac{1}{2}\frac{|\hat{P}|^2}{\rho_\infty c}\alpha. \tag{3.59}$$

Equation (3.59) clarifies that a purely active sound field is present, if $\alpha = 1$. This means that we have no reflection at $x = L$ and a perfect transmission of energy through the system. According to (3.59), a purely reactive sound field (in which $\alpha = 0$), will disable any energy transport.

A system representation of the finite acoustic duct is shown in Fig. 3.2. It clarifies that the system output $P(x, j\omega)$, compare (3.53) and (3.54), caused by the system input $V_0(j\omega)$ is at any position x determined by the space and frequency dependent transfer function

$$\begin{aligned} H(x, j\omega) &:= \frac{P(x, j\omega)}{V_0(j\omega)} \\ &= \frac{\rho_\infty c}{e^{jkL} - R(j\omega)e^{-jkL}}\big(e^{-jk(x-L)} + R(j\omega)e^{jk(x-L)}\big) \\ &=: \kappa \tilde{H}(x, j\omega). \end{aligned} \tag{3.60}$$

[2]Further reading: (Möser 2010; EN ISO 10534-1 2001), and (EN ISO 10534-2 2001).

Equation (3.60) clarifies that $H(x, j\omega)$ can be decomposed into a proportionality factor (or sensitivity) κ and a dimensionless transfer function $\tilde{H}(x, j\omega)$ such as

$$\kappa = \rho_\infty c \quad \text{and} \quad \tilde{H}(x, j\omega) = \kappa^{-1} H(x, j\omega). \tag{3.61}$$

The maximum absolute value of $H(x, j\omega)$ can be found by analyzing the magnitude of the acoustic potential energy density as defined by (3.40). Combining (3.53) and (3.54), this quantity is at the left boundary given by

$$E_{pot}(x = 0, j\omega) = \frac{1}{2}\rho_\infty \frac{1 + |R|^2 + 2|R|\cos(2kL - \chi(\omega))}{1 + |R|^2 - 2|R|\cos(2kL - \chi(\omega))} \left|V_0(j\omega)\right|^2. \tag{3.62}$$

$E_{pot}(x = 0, j\omega)$ peaks for

$$f_{nR} = \frac{\omega_{nR}}{2\pi} = \frac{c}{2L}\left(n + \frac{\chi(\omega_{nR})}{2\pi}\right) \tag{3.63}$$

that is the n-th *resonance frequency*. The associated pressure distribution is the n-th *resonance mode*. f_{nR} differs from the n-th eigenfrequency of a sound hard terminated duct, compare Table 3.1. This difference is specified by

$$f_{nR} = \left(1 + \frac{\chi(\omega_{nR})}{2\pi n}\right) f_n. \tag{3.64}$$

Equation (3.62) also clarifies that the losses introduced by the impedance boundary cause a limitation of $\hat{P}$ in case of resonance in which $R(j\omega_{nR})$ can be used to motivate *modal damping* via a *complex wave number*

$$k_{nR} = \frac{\omega_{nR}}{c}(1 - j\xi_{nR}). \tag{3.65}$$

ξ_{nR} is the *modal damping ratio* of the n-th resonance. This quantity can be determined by solving (compare denominator of (3.53))

$$e^{jk_{nR}L} - R(j\omega)e^{-jk_{nR}L} = 0 \tag{3.66}$$

that yields

$$\xi_{nR} = -\frac{c}{2L}\frac{\ln|R(j\omega_{nR})|}{\omega_{nR}}. \tag{3.67}$$

If both the impedance Z_L and the reflection coefficient R are unknown, ξ_{nR} can be obtained from the *half-width* Δf_{nR} of n-th resonance frequency. This frequency span is determined by two frequencies $f_{nR} \pm \Delta f_{nR}$ with $\Delta f_{nR} > 0$ and corresponds with a decay of the sound pressure maximum that is determined by the factor $1/\sqrt{2}$. Using this approach, compare (Kuttruff 2004), the modal damping ratio can be approximated such as

$$\xi_{nR} \approx \frac{\Delta f_{nR}}{f_{nR}} \quad \Leftrightarrow \quad (f_{nR} + \Delta f_{nR}) - (f_{nR} - \Delta f_{nR}) \approx 2\xi_{nR} f_{nR}. \tag{3.68}$$

Finally we will motivate a connection between σ_n, compare (3.43), that has been introduced as the decay coefficient of the n-th eigenmode $P_n(x, j\omega)$ and ξ_{nR}, the damping ratio of the n-th (damped) resonance mode $P_{nR}(x, j\omega, \xi_{nR}(|R|))$. Strictly speaking we have to evaluate

$$\left|P_n(x, j\omega)\right|^2 e^{-2\sigma_n T_{n60}} - \left|P_{nR}\left(x, j\omega, \xi_{nR}\left(|R|\right)\right)\right|^2 = 0. \tag{3.69}$$

Equation (3.69) postulates that the acoustic potential energy density of the n-th resonance mode equals a millionth of the acoustic potential energy density of the n-th eigenmode. This yields a relation between the modal damping parameter ξ_{nR} and the decay coefficient σ_n. However, instead of solving (3.69), we will use a simplified approach to motivate this relation.

For this purpose we replace the magnitude of the reflection coefficient in (3.67) by the associated absorption coefficient according to (3.52). This results in

$$\xi_{nR} = -\frac{c}{2L}\frac{\ln|\sqrt{1-\alpha_{nR}}|}{\omega_{nR}}. \tag{3.70}$$

For weakly damped interiors we can use a linear approximation of the logarithm in (3.70).[3] This yields

$$\xi_{nR} \approx \frac{c}{2L}\frac{\alpha_{nR}}{\omega_{nR}}. \tag{3.71}$$

α_{nR} can be replaced by an average *absorption coefficient* such as

$$\bar{\alpha}_{nR} := \frac{\sum_{i=1}^{N} \alpha_{nR}^{i} S^i}{\sum_{i=1}^{N} S^i}, \tag{3.72}$$

where S^i is the i-th absorbing surface and α_{nR}^i is the absorption coefficient of the i-th absorbing surface for the n-the resonance frequency. The average absorption coefficient can now be approximated using the *Sabine equation*

$$\bar{\alpha}_{nR} \approx 55.2 \frac{V}{\sum_{i=1}^{N} S^i}\frac{1}{cT_{nR60}}, \tag{3.73}$$

see e.g. (Möser 2005), in which T_{nR60} is the reverberation time of n-th resonance mode. Because T_{nR60} will in engineering systems be equal to T_{n60}, the reverberation time associated with the n-th eigenfrequency, compare (3.43), (3.73) provides the link between the n-th decay coefficient σ_n and ξ_{nR}, the damping ratio of the n-th resonance. Taking into account that the volume of the one-dimensional waveguide shown in Fig. 3.1 is given by $V = S \cdot L$ and considering (3.43), (3.71) as well as (3.73), this link is for $T_{nR60} = T_{n60}$ given by

$$\xi_{nR} \approx 4 \cdot \frac{S}{\sum_{i=1}^{N} S^i} \cdot \frac{\sigma_n}{\omega_{nR}}. \tag{3.74}$$

If the dependence of the average decay coefficient on frequency can be described such as $\sigma_n = \alpha_\sigma \langle\sigma\rangle\omega_{nR}$, where α_σ is a proportionality factor with unit seconds, it is possible to use the same modal damping ratio for each resonance frequency.

---

[3] The approximation error is less than 5%, if the absorption coefficient α_{nR} is less than 0.1.

This yields, compare (3.65), to a *complex wave number* that is based on a frequency independent damping ratio ξ such as

$$k = \frac{\omega}{c}(1 - j\xi). \tag{3.75}$$

Chapter 4
Sensors for Active Noise Control

Abstract In order to establish an adaptive control approach, it is necessary to collect information about the systems state. The most common way of monitoring the state of interior noise fields is based on the observation of the acoustic pressure. This field variable is detected by microphones. Vectorial field variables such as the sound intensity can be approximated by sound pressure measurements, using two (or more) microphones. However, in some applications it is also useful to obtain a non-acoustic (reference) signal such as the acceleration of a vibrating boundary. This is particularly true, if a signal must not be altered by acoustical sources. Therefore, the upcoming chapter contains an introduction into both acoustical and structural sensing for active noise control. It is however limited to the description of condenser microphones, microphone based sound intensity probes, and inertial sensors, because (from the viewpoint of the author) these transducer types are of most practical importance. The equations of motion will be derived from simplified electro-mechanical transducer models. The particular transducer behavior is discussed in frequency domain. More details on acoustic transducers can be found in (Harris in Handbook of noise control, McGraw-Hill, New York, 1975), (Havelock et al. in Handbook of signal processing in acoustics, vol 2, Springer, New York, 2008b), (Heckl and Nüller in Taschenbuch der Technischen Akustik, Springer, Berlin, 1995), (Möser in Technische Akustik, Springer, Berlin, 2005), (Möser in Messtechnik der Akustik, Springer, Heidelberg, 2010), (Zwicker and Zollner in Elektroakustik, Springer, Berlin, 1984), (B&K in Microphone handbook, vol. 1: theory, Bruel & Kjaer, Nærum, http://www.bksv.com/. Cited 10 May 2010, 1996) and also in (DIN EN 60268-4 in Elektroakustische Geräte. Teil 4: Mikrofone (IEC 60268-4:2004); Deutsche Fassung EN 60268-4:2004, DIN Deutsches Institut für Normung e.V., Germany, 2004). The fundamentals of sound intensity measurement are described in (Fahy in Sound intensity, Elsevier, London, 1989). Structural sensors are discussed in (Cremer et al. in Structure-borne sound, Springer, Berlin, 1995), (Fahy and Gardonio in Sound and structural vibration, Amsterdam, Elsevier, 2007) and (Havelock et al. in Handbook of signal processing in acoustics, vol 2, Springer, New York, 2008b).

T. Kletschkowski, *Adaptive Feed-Forward Control of Low Frequency Interior Noise*, Intelligent Systems, Control and Automation: Science and Engineering 56, DOI 10.1007/978-94-007-2537-9_4,

4.1 Acoustical Sensing by Condenser Type Microphones

As illustrated by Fig. 4.1, a microphone can be interpreted as a linear system. The input quantity is given by the acoustic pressure $P(j\omega)$, and the output quantity by the voltage $U(j\omega)$. The system response is determined by its frequency response $H_{Mic}(j\omega)$. *Condenser type microphones* are widely used as sensing devices for the acoustic pressure.

Such a microphone consists of a metal housing inside which an electrical insulator with a backplate is mounted behind a thin membrane. This membrane is attached in a way which makes it parallel to the backplate. As stated in (B&K 1996), the membrane thickness may vary from about 1.5 to 8 μm depending on the microphone type. The backplate is relatively heavy and massive, compared to the membrane. The distance between both elements is typically 20 μm.

The membrane and the front of the backplate form the plates of an active capacitor whose working principle can be described as follows. The (external) sound pressure (input quantity) that is acting on the membrane causes a displacement. This displacement reduces the distance between membrane and backplate, and therefore the capacity of the condenser. The latter determines the transducer voltage, also known as electromagnetic force, see (DIN EN 60268-4 2004), that is measured as the output quantity.

A schematic diagram of a condenser microphone is shown in Fig. 4.2, where $u(t)$ is the output voltage, u_0 is the voltage of a direct current (DC) source, $i(t)$ is the electric current, and $u_c(t)$ is the voltage of the condenser with capacity $c_C(t)$. R_a is the microphone resistance. Applying *Kirchhoff's* second law, the output voltage is given by

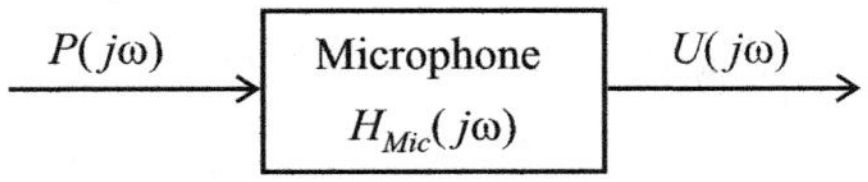

Fig. 4.1 Microphone interpreted as linear system with acoustic pressure as input quantity and transducer voltage as output quantity

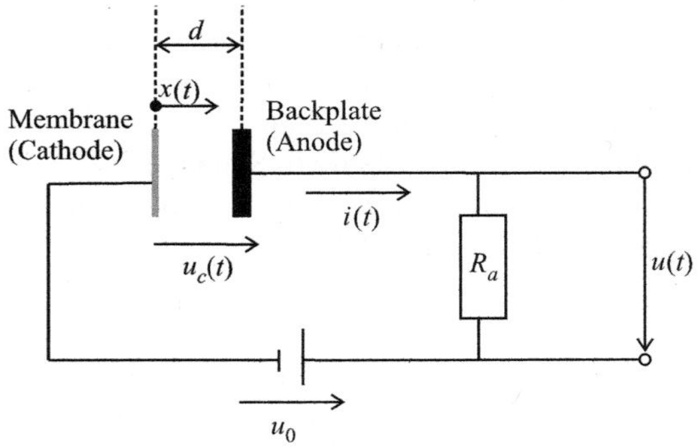

Fig. 4.2 Principle of a condenser microphone (adapted from Möser 2005). Cathode is usually made of a thin membrane. Compared to this, the anode is relatively massive. DC power supply u_0 is needed to polarize the condenser. Membrane displacement $x(t)$ is small compared to the distance between the condenser plates d

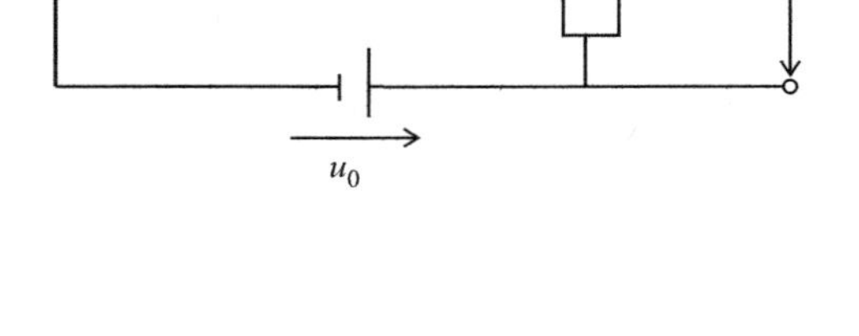

$$u(t) = u_0 - u_c(t) \quad \text{with } u_c(t) = \frac{q(t)}{c_c(t)}. \tag{4.1}$$

Assuming that the displacement of the membrane is small compared to the distance between the plates of the condenser $x \ll d$, and assuming small fluctuations of the electric charge $q(t) = q_0 + q_f(t)$ with $q_f(t) \ll q_0$, compare (Möser 2005), the output voltage is described by

$$u(t) = u_0 \frac{x(t)}{d} - \frac{q_f(t)}{C_0}. \tag{4.2}$$

Taking the time derivative of (4.2), and equating $\dot{q}_f(t) = i(t)$ by *Ohm's* law, the equation of motion for the electrical circuit of the condenser microphone is

$$\dot{u}(t) + \frac{1}{RC_0} u(t) = \frac{u_0}{d} \dot{x}(t), \tag{4.3}$$

where $\dot{x}(t)$ is the velocity of the cathode. For frequencies that are much below the fundamental frequency of the membrane, the equation of motion for the mechanical part of the transducer is given by

$$\ddot{x}(t) + 2D\omega_0 \dot{x}(t) + \omega_0^2 x(t) = p(t)S, \tag{4.4}$$

where S is the cross section area of the membrane, ω_0^2 is the ratio of the membrane stiffness k and the mass of the membrane m. $D = r/2\sqrt{km}$ is a dimensionless damping ratio, based on the membrane damping coefficient r.

Equations (4.3) and (4.4) form a set of two coupled ODEs. Coupling between the mechanical and the electrical circuit is provided by the velocity of the microphone membrane $\dot{x}(t)$. Assuming harmonic fluctuation of the external sound pressure $p(t)$, membrane displacement $x(t)$ and output voltage $u(t)$, the frequency response of the condenser microphone calculated as the ratio between output and input signal

$$H_{Mic}(j\omega) = \frac{U(j\omega)}{P(j\omega)} \tag{4.5}$$

is given by

$$\begin{aligned} H_{Mic}(j\omega) &= \kappa_{Mic} \frac{1}{[1 - j\frac{\omega_k}{\omega_0}(\frac{\omega}{\omega_0})^{-1}][1 - (\frac{\omega}{\omega_0})^2 + j2D\frac{\omega}{\omega_0}]} \\ &=: \kappa_{Mic} \tilde{H}_{Mic}(j\omega), \end{aligned} \tag{4.6}$$

where

$$\kappa_{Mic} = \frac{U_0}{d} \frac{S}{k}, \qquad \omega_k = \frac{1}{R_a C_0}. \tag{4.7}$$

Equation (4.6) clarifies that the frequency response can be split into the proportional factor κ_{Mic} with units V Pa^{-1}, known as the sensitivity, and the dimensionless, but frequency dependent transfer function $H_{Mic}(j\omega)$.

Equation (4.7) shows that the output voltage is proportional to the cross section area of the membrane. For this reason, the diameter S must be increased, to obtain

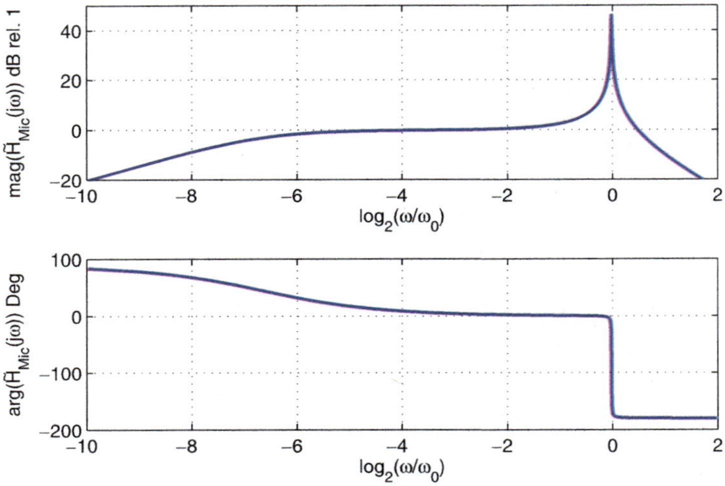

Fig. 4.3 Magnitude (*top*) and phase (*below*) of the dimensionless transfer function of the condenser microphone. Magnitude is almost constant within four octave-bands below the natural frequency. Output voltage is in phase with input pressure within two octave-bands

a high sensitivity at (very) low frequencies.[1] The theoretical frequency response of the condenser microphone is shown in Fig. 4.3.

The ratio between the electric bend frequency $\omega_k = (R_a C_0)^{-1}$ and the natural frequency of the membrane ω_0 was set to 0.01. Furthermore, a damping ratio of $D = 0.0025$ was taken into account. It can be seen that the magnitude response is nearly constant within four octave-bands below the natural frequency. The output voltage is in phase with the input pressure within two octave-bands below ω_0. As described in (Möser 2005), and illustrated by Fig. 4.3 (top), the condenser microphone acts as a band-pass filter. Its lower cut-off frequency ω_k is determined by the electrical circuit, the upper cut-off frequency is given by ω_0 that depends on the mechanical parts of the membrane. A short comment on *A-weighting* of microphone signals is to be found in Sect. 9.4.2.

4.2 Sound Intensity Measurement

4.2.1 Functional Principle of Sound Intensity Probes

To quantify the energy transport in acoustic systems or to locate acoustic hot spots on the interior boundaries, it can be necessary to perform sound intensity measurements. This can be done in time as well as in frequency domain, compare

[1] As shown in (Möser 2005), a cross section diameter of up to 23.8 mm is realized for condenser microphones with a working range between 4 Hz and 18 kHz.

(Fahy 1989), (Möser 2005), and (Möser 2010) using a pressure-velocity sound intensity probe (*p-v probe*) or a pressure-pressure sound intensity probe (*p-p probe*).

A p-v probe combines a microphone with a sensor designed to measure the acoustic velocity. An established example is the Microflown particle velocity sensor, see (Jacobsen and de Bree 2005), (Havelock et al. 2008b) and (Microflown 2009). This transducer is a micromachined hot wire anemometer that is in contrast to a classical one, compare (Baker 1955) and (Hamann 1980), based on two short, thin and closely spaced wires of silicon nitride coated with platinum. These wires are heated by a DC current to about 300 °C. The electrical resistance of the wires is changed by the acoustic velocity component that is perpendicular to the wires.

Because the upstream wire is cooled more than the downstream wire, it is possible to measure the resistance difference between the wires using a bridge that provides a signal proportional to the particle velocity. However, as stated in (Havelock et al. 2008b), the dynamic range of this transducer is limited to about 125 dB rel. 50 nm/s. Furthermore, standardized calibration procedures have not yet been established. For this reasons the upcoming section is limited to sound intensity measurements using p-p probes.

A p-p probe consists of two closely spaced pressure microphones usually mounted "face to face". The microphone pair is used to determine the average value of sound pressure as well as to approximate the pressure gradient at $\Delta x/2$ that is one half of the spacer distance between the microphones. "The success of this method can be explained by the fact that condenser microphones are more stable and reliable than any other acoustic transducer." as stated in (Havelock et al. 2008b).

Considering harmonic sound waves in an one-dimensional waveguide, the functional principle as well as the associated measurement errors will be discussed for such an intensity probe. For this purpose we also assume that the x-axis of the Cartesian coordinate system falls together with the axis of the p-p probe.

To calculate the mean intensity it is necessary to approximate the pressure gradient in x-direction by a finite difference schema that is used to discretize (3.21) such as

$$V(x, j\omega) = \frac{j}{\rho_\infty \omega} \operatorname{grad} P(x, j\omega) \approx \frac{j}{\rho_\infty \omega} \frac{P(x+\Delta x, j\omega) - P(x, j\omega)}{\Delta x}. \tag{4.8}$$

It is strait forward to show that evaluation of (3.34) using (4.8) yields to an approximation $\bar{I}_M(x, j\omega)$ of the true intensity $\bar{I}_T(x, j\omega)$ that is given by

$$\begin{aligned} \bar{I}_T &:= \bar{I}_T(x + \Delta x/2, j\omega) \\ &\approx \frac{1}{2\rho_\infty \omega \Delta x} \operatorname{Im}\big[P(x, j\omega) P^*(x+\Delta x, j\omega)\big] =: \bar{I}_M. \end{aligned} \tag{4.9}$$

In practice, however, the measured mean intensity $\bar{I}_M$ is determined by the output signals of the microphones. If the output of microphone 1 is given by $U_1 = \kappa_1 \tilde{H}_1 P_1$ with $P_1 = P(x, j\omega)$, and if the output of microphone 2 is given by $U_2 = \kappa_2 \tilde{H}_2 P_2$ with $P_2 = P(x + \Delta x, j\omega)$, compare (4.5), $\bar{I}_M$ is calculated as

$$\bar{I}_M = \frac{1}{2\rho_\infty \omega \Delta x} \operatorname{Im}\big[(U_1/\kappa_1)(U_2/\kappa_2)^*\big]$$

$$
\begin{aligned}
&= \frac{1}{2\rho_\infty \omega \Delta x} \operatorname{Im}\left[(\tilde{H}_1 P_1)(\tilde{H}_2 P_2)^*\right] \\
&= \frac{1}{2\rho_\infty \omega \Delta x} |\tilde{H}_1||\tilde{H}_2||P_1||P_2| \operatorname{Im}\left[e^{j\Delta\varphi_H} e^{j\Delta\varphi_p}\right] \\
&= \frac{1}{2\rho_\infty \omega \Delta x} |\tilde{H}_1||\tilde{H}_2||P_1||P_2| \sin(\Delta\varphi_H + \Delta\varphi_p),
\end{aligned} \tag{4.10}
$$

where $\Delta\varphi_H$ is the phase difference due to the difference between the transducers and $\Delta\varphi_p$ is the phase difference between P_1 and P_2. Equation (4.10) clarifies that the measured mean intensity $\bar{I}_M$ is for every frequency determined by the

- Magnitudes of the transducer transfer functions $\tilde{H}_1$ and $\tilde{H}_2$,
- Phase difference between the transducers $\Delta\varphi_H$,
- Phase difference between the two sound pressures $\Delta\varphi_p$,
- Microphone separation distance Δx.

Without any phase difference, the measured signal will be zero. However, in a perfect (or perfectly calibrated) p-p probe the phase difference should only be given by $\Delta\varphi_p$.

4.2.2 Errors in Sound Intensity Measurements

The approximation (4.9) limits the frequency range in which a p-p probe can be used. On the one hand the finite difference approximation (4.8) will fail, if the wave length becomes small compared to the microphone separation distance. On the other hand, a certain separation distance is required to enable the measurement of a phase difference between the microphone signals that is essential for the determination of the mean intensity. Hence, the ratio of spacer length and wave length will determine both a high and a low frequency limit for sound intensity measurements.

High Frequency Error of p-p Probes

The high frequency limit of p-p probes will be derived from the ratio between the measured intensity $\bar{I}_M$ and the true intensity $\bar{I}_T$ assuming one-dimensional wave propagation in a finite acoustic duct with perfect acoustic absorption $\alpha = 1$ at both ends.

According to (3.59) the true intensity is under these BC given by

$$
\bar{I}_T = \frac{1}{2} \frac{|\hat{P}|^2}{\rho_\infty c}, \tag{4.11}
$$

and phase difference between the complex acoustic pressure measured at the microphones of the p-p probe is determined by the ratio between spacer distance Δx and wave length λ such as

$$\Delta\varphi_p = k\Delta x = \frac{2\pi f}{c}\Delta x = 2\pi\frac{\Delta x}{\lambda}. \tag{4.12}$$

Considering (4.10), (4.11), (4.12) and $|P_1| = |P_2| = |\hat{P}|$, the ratio between measured and true intensity results for a perfect p-p probe ($\Delta\varphi_H = 0$) in

$$\frac{\bar{I}_M}{\bar{I}_T} = |\tilde{H}_1||\tilde{H}_2|\frac{\sin(k\Delta x)}{k\Delta x}. \tag{4.13}$$

If the microphones are used in the their desired frequency range in which $|H_1| = 1$ and $|H_2| = 1$, compare Fig. 4.3, the *high frequency error* also known as *finite difference approximation error* is given by

$$E_{HF} := 10\log_{10}\left(\frac{\bar{I}_M}{\bar{I}_T}\right) = 10\log_{10}\left(\frac{\sin(k\Delta x)}{k\Delta x}\right). \tag{4.14}$$

E_{HF} causes an underestimation of the true intensity. To guarantee that this error is limited to -1 dB, the normalized spacer distance must be bounded by

$$\frac{\Delta x}{\lambda} < 0.18. \tag{4.15}$$

The *high frequency limit* for a perfect p-p probe can therefore be specified by

$$f < \frac{1}{5}\frac{c}{\Delta x}. \tag{4.16}$$

Low Frequency Error of p-p Probes

To motivate the low frequency limit of p-p probes, we assume one-dimensional wave propagation in a finite acoustic duct with velocity source at $x = 0$ and impedance boundary, specified by the complex reflection coefficient R, at $x = L$ as discussed in Sect. 3.4.2.

Without loss of generality it is possible to approximate the intensity by the values of the acoustic pressure measured at $x = L - \Delta x$ and $x = L$ that are according to (3.54) given by

$$P_1 = P(x = L - \Delta x, j\omega) = \hat{P}\left(e^{jk\Delta x} + Re^{-jk\Delta x}\right) \tag{4.17}$$

and

$$P_2 = P(x = L, j\omega) = \hat{P}(1 + R). \tag{4.18}$$

Because we are interested in the low frequency limit, where $\Delta x/\lambda$ is small, it is possible to approximate P_1 using a first order approximation of the exponential functions such as $e^x \approx 1 + x$ by

$$P_1 \approx \hat{P}\left[1 + jk\Delta x + R(1 - jk\Delta x)\right]. \tag{4.19}$$

Based on (4.19) the cross spectral density between P_2 and P_1 results in

$$P_2 P_1^* = \left(1 + R^*\right)\left[1 + jk\Delta x + R(1 - jk\Delta x)\right]. \tag{4.20}$$

The mean intensity measured with a perfect p-p probe with $|\tilde{H}_1| = 1$ and $|\tilde{H}_2| = 1$ and $\Delta\varphi_H = 0$ is therefore given by

$$\begin{aligned}\bar{I}_M &= \frac{1}{2\rho_\infty \omega \Delta x}\,\mathrm{Im}\big(P_1 P_2^*\big) = -\frac{1}{2\rho_\infty \omega \Delta x}\,\mathrm{Im}\big(P_2 P_1^*\big) \\ &\approx -\frac{1}{2\rho_\infty \omega \Delta x}|\hat{P}|^2\big(1-|R|^2\big)k\Delta x. \end{aligned} \tag{4.21}$$

If $|\tilde{H}_1| = 1$ and $|\tilde{H}_2| = 1$ but a phase mismatch $\Delta\varphi_H \neq 0$ is present, the measured mean intensity can be approximated by

$$\begin{aligned}\bar{I}_M^{\Delta\varphi_H} &= \frac{1}{2\rho_\infty \omega \Delta x}\,\mathrm{Im}\big(P_1 P_2^* e^{-j\Delta\varphi_H}\big) = -\frac{1}{2\rho_\infty \omega \Delta x}\,\mathrm{Im}\big(P_2 P_1^* e^{j\Delta\varphi_H}\big) \\ &\approx -\frac{1}{2\rho_\infty \omega \Delta x}|\hat{P}|^2 \frac{k\Delta x(1-|R|^2) + |1+R|^2\Delta\varphi_H}{2\rho_\infty \omega \Delta x}, \end{aligned} \tag{4.22}$$

if terms of higher order such as $\Delta\varphi_H k\Delta x$ can be neglected[2] in the low frequency range, where $k\Delta x \ll 1$. The ratio between $\bar{I}_M^{\Delta\varphi_H}$ and $\bar{I}_T$ is bounded by

$$\frac{\bar{I}_M^{\Delta\varphi_H}}{\bar{I}_T} = 1 + \frac{\Delta\varphi_H}{k\Delta x}\frac{|1+R|^2}{1-|R|^2} < 1 + \frac{\Delta\varphi_H}{k\Delta x}\frac{1+|R|^2}{1-|R|^2}. \tag{4.23}$$

Using (3.52) it is therefore possible to introduce the *low frequency error* as a function of the acoustic absorption coefficient

$$E_{LF} := 10\log_{10}\left(\frac{\bar{I}_M^{\Delta\varphi_H}}{\bar{I}_T}\right) = 10\log_{10}\left(1 + \frac{\Delta\varphi_H}{k\Delta x}\cdot\frac{2-\alpha}{\alpha}\right). \tag{4.24}$$

Equation (4.24) clarifies that the influence of $\Delta\varphi_H$ on the measurement results increases, if the sound field consists of an active and a reactive part. The low frequency limit in which the overestimation error E_{LF} is less than 1 dB can be derived from (4.24) and is for $\alpha = 0.1$, $\Delta x = 5$ mm and $\Delta\varphi_H = 0.3°$ determined by

$$0.08\frac{c}{\Delta x} < f. \tag{4.25}$$

Near Field Error of p-p Probes

Another error occurs, if the intensity changes along the p-p probe. As mentioned in (Crocker and Arenas 2003): "This can happen if the source is comprised of multipoles of high order (e.g., a rotating propeller) or when the sound radiating surface vibrates with a small wavelength (e.g., thin plates)." This inaccuracy is referred to as the *near field error* of p-p probes.

As shown in (Gade 1982) it is possible to quantify this error for a spherical wave that is emitted by a monopole source with source strength Q, if, for simplicity, the line joining the centers of the p-p probe microphones passes through the monopole source. This arrangement is shown in Fig. 4.4.

---

[2] According to (Möser 2005), the phase mismatch $\Delta\varphi_H$ is for a good p-p probe less than 0.3°.

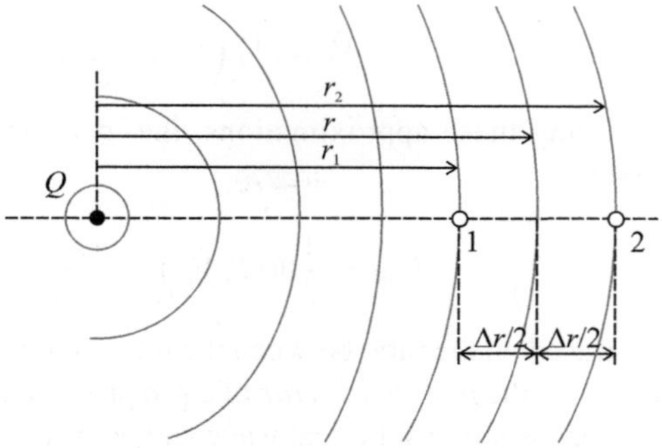

Fig. 4.4 Orientation of a p-p probe in a spherical propagating wave (adapted from Gade 1982)

The velocity potential, compare Sect. 3.2, of the sound field generated by the monopole source is according to (Gade 1982) given by

$$\Phi = \frac{Q}{4\pi r} \cdot e^{j(\omega t - kr)}. \tag{4.26}$$

Thus, the sound pressure along the center line is because (3.13) given by

$$P_r = \rho_\infty \frac{\partial \Phi}{\partial t} = jk\rho c \frac{Q}{4\pi r} e^{j(\omega t - kr)}. \tag{4.27}$$

The associated radial component of the acoustic velocity is because of (3.14) determined by

$$V_r = -\frac{\partial \Phi}{\partial r} = \left(\frac{1}{r} + jk\right) \frac{Q}{4\pi r} e^{j(\omega t - kr)}. \tag{4.28}$$

Considering (3.34), and neglecting the term $e^{j\omega t}$, the radial component of the true intensity at a distance r from the source can be specified as

$$\bar{I}_{r,T} = \frac{1}{2} \operatorname{Re}(P_r V_r^*) = \frac{1}{2} \omega k \rho_\infty \left(\frac{Q}{4\pi r}\right)^2 \frac{1}{r^2}. \tag{4.29}$$

Let us now calculate the measured mean sound intensity at a distance $r + \Delta r/2$ from the source. The sound pressure at this point can be approximated as

$$\bar{P}_r := P_r\left(r_1 + \frac{\Delta r}{2}\right) \approx \frac{1}{2}(P_1 + P_2), \tag{4.30}$$

where

$$P_1 = jk\rho c \frac{Q}{4\pi r_1} e^{-jkr_1}, \tag{4.31}$$

is the sound pressure picked up at a distance r_1 from the source, and

$$P_2 = jk\rho c \frac{Q}{4\pi r_2} e^{-jkr_2} \tag{4.32}$$

is the sound pressure measured at a distance r_2 from the source. The estimate of the radial component of the particle velocity at a distance $r + \Delta r/2$ from the source can according to (4.8) be approximated by

$$\bar{V}_r := V_r\left(r_1 + \frac{\Delta r}{2}\right) \approx \frac{j}{\rho_\infty \omega}(P_2 - P_1). \tag{4.33}$$

Using these approximations, the mean intensity measured with a p-p probe is given by

$$\bar{I}_{r,M} = \frac{1}{2}\Re\left(\bar{P}_r \bar{V}_r^*\right) = \frac{\omega\rho_\infty}{2\Delta r}\frac{1}{r_1 r_2}\left(\frac{Q}{4\pi}\right)^2 \sin(k\Delta r). \tag{4.34}$$

Based on the ratio between the measured and the true mean intensity, it is possible to define the *near field error* of a p-p probe used to measure the mean sound intensity of a propagating spherical wave and pointing directly at the center of the source such as

$$E_{NF} := 10\log_{10}\left(\frac{\bar{I}_M^{\Delta\varphi_H}}{\bar{I}_T}\right) = 10\log_{10}\left[\frac{\sin(k\Delta r)}{k\Delta r}\cdot\left(1 - \frac{1}{4}\frac{\Delta r^2}{r^2}\right)^{-1}\right]. \tag{4.35}$$

From (4.35) we observe that the error is not only a function of $k\Delta r$ but also of the ratio between the microphone spacing and the radial distance to the source $\Delta r/r$. This error can be neglected, if $r > 2\Delta r$, while for $r = \Delta r$ the error becomes an overestimation of approximately 1 dB as shown in (Gade 1982).

This indicates that sound intensity measurement can be performed in the near field of a source without introducing significant errors, if the microphone spacing is "adjusted" to the frequency range of interest. An overview on the frequency range of various microphone and spacer configurations can also be found in (Gade 1982).

4.3 Structural Sensing by Accelerometers

An accelerometer, also known as inertial sensor (Fahy and Gardonio 2007), is a sensor that is often used to measure structural vibrations at a specific point. For ANC applications this non-acoustic transducer type is of particular interest, if the required signal (e.g. the reference signal) must not be altered by acoustical sources such as canceling loudspeakers.

As illustrated by Fig. 4.5, an accelerometer can be interpreted as a linear system with input quantity $A(j\omega)$ (structural acceleration) and output quantity by $U_a(j\omega)$ (the transducer voltage). The system response is determined by its frequency response $H_{Acc}(j\omega)$.

A schematic diagram is shown in Fig. 4.6. As described in (Fahy and Gardonio 2007), an accelerometer compromises a seismic mass m_a connected to a vibrating

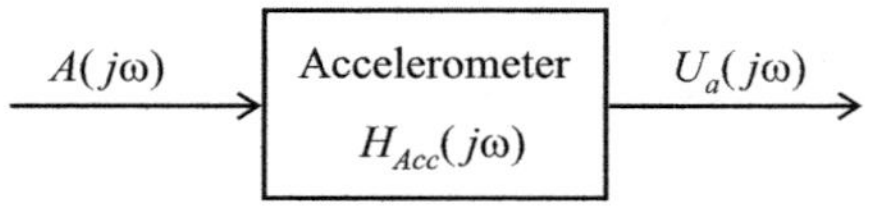

Fig. 4.5 Accelerometer interpreted as linear system with structural acceleration as input quantity and transducer voltage as output quantity

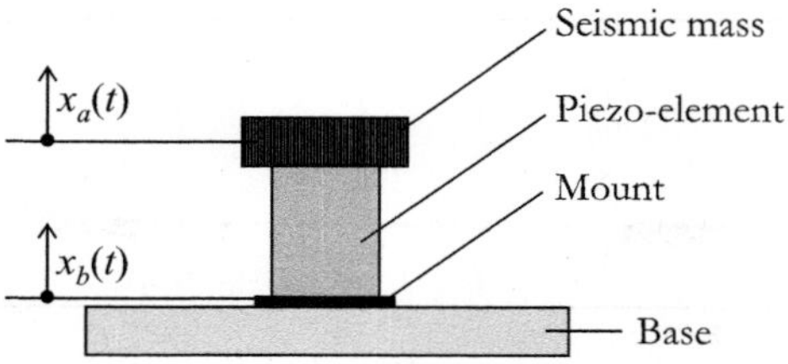

Fig. 4.6 Accelerometer consisting of a seismic mass and a piezoelectric element mounted on a base (adapted from Fahy and Gardonio 2007). $x_a(t)$ is the relative displacement between seismic mass and base. $x_b(t)$ is the base displacement

base via a spring of stiffness k_a and damping coefficient r_a (usually realized by a piezoelectric element). The voltage generated by this device is proportional to the relative displacement between the seismic mass, described by $x_a(t)$ and the base $x_b(t)$. This is expressed by

$$u_a(t) = \kappa_{Acc} x(t) \quad \text{with } x(t) = x_a(t) - x_b(t), \tag{4.36}$$

where κ_{Acc} is the sensitivity of the accelerometer with units V m^{-1}. The equation of motion for the seismic mass is given by

$$m_a \ddot{x}(t) + r_a \dot{x}(t) + k_a x(t) = m_a \ddot{x}_b(t). \tag{4.37}$$

Assuming harmonic motion of the base, the transfer behavior of the accelerometer is described by the ratio between the output voltage of the accelerometer and the displacement of the base

$$\frac{U_a(j\omega)}{X_b(j\omega)} = \kappa_{Acc} \frac{(\frac{\omega}{\omega_a})^2}{1 - (\frac{\omega}{\omega_a})^2 + j2D_a \frac{\omega}{\omega_a}}, \tag{4.38}$$

where $\omega_a = \sqrt{k_a m_a^{-1}}$, and $D_a = r_a / 2\sqrt{k_a m_a}$. The complex output voltage with reference to the complex base acceleration is given by

$$\frac{U_a(j\omega)}{-\omega^2 X_b(j\omega)} = \frac{\kappa_{Acc}}{\omega_a^2} \tilde{H}_{Acc}(j\omega) =: H_{Acc}(j\omega), \tag{4.39}$$

with

$$\tilde{H}_{Acc}(j\omega) = \frac{-1}{1 - (\frac{\omega}{\omega_a})^2 + j2D_a \frac{\omega}{\omega_a}}. \tag{4.40}$$

Equation (4.40) clarifies that the output voltage of the accelerometer is inversely proportional to the square of its natural frequency ω_a. The resulting sensitivity $\kappa_{Acc}\omega_a^2$ has the units $\mathrm{V}/(\mathrm{m\,s}^{-2})$. Therefore, as concluded in (Fahy and Gardonio 2007), the resonance frequency should be kept as close as possible to the (upper) range of measurement, to obtain a high signal-to-noise ratio. The frequency response of the accelerometer is shown in Fig. 4.7. To draw the curve, a damping ratio of $D_a = 0.0025$ was taken into account. It can be seen that the magnitude response is nearly constant within eight octave-bands below the natural frequency ω_a.

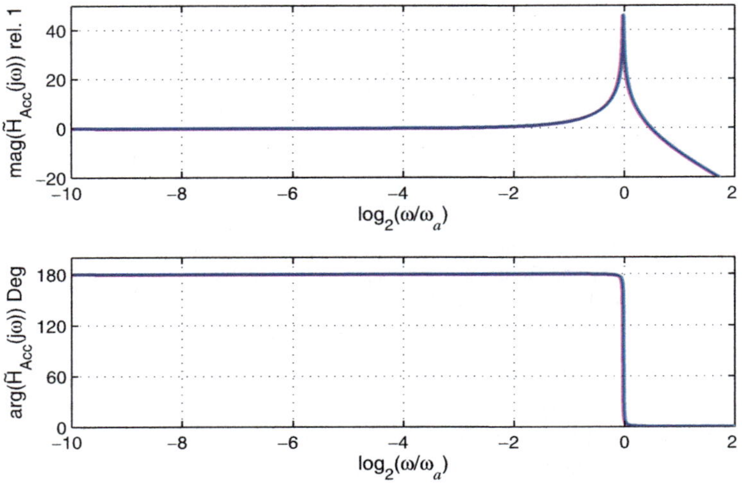

Fig. 4.7 Magnitude (*top*) and phase (*below*) of the dimensionless transfer function of the accelerometer. Magnitude is constant within eight octave-bands below the natural frequency. Phase difference between output voltage and acceleration of base is 180° within these octave-bands

The phase difference between output voltage and acceleration of base is 180° within these octave-bands. The accelerometer acts as a low-pass filter. Its (upper) cut-off frequency is given by ω_a.

Chapter 5
Actuators for Active Noise Control

Abstract To establish an active control system, it is necessary to integrate actuators. Therefore, this chapter gives a short introduction to acoustical actuation for active noise control that is limited to electro-dynamical actuation by loudspeakers and electro-dynamical exciters. The first actuator type is widely used as secondary sound source. The second one is of special interest, if an elastic structure is to be used as an acoustical source. The equation of motions will be derived from simplified electro-mechanical actuator models and the particular actuator behavior is discussed in frequency domain. A general overview on loudspeakers is given e.g. in (Baranek in J. Acoust. Soc. Am. 26(5):618–629, 1954b) and (Havelock et al. in Handbook of signal processing in acoustics, vol 1, Springer, New York, 2008a). Comments on loudspeaker design and performance evaluation can also be found in (Havelock et al. in Handbook of signal processing in acoustics, vol 1, Springer, New York, 2008a). A detailed description of audio amplifiers is given in (Shea in Amplifier handbook, McGraw-Hill, New York, 1966) and the transfer behavior of electro-dynamical loudspeakers is analyzed e.g. in (Zwicker and Zollner in Elektroakustik, Springer, Berlin, 1984) and (Möser in Technische Akustik, Springer, Berlin, 2005). Sound radiation from vibrating structures has been studied extensively in (Cremer et al. in Structure-borne sound, Springer, Berlin, 1995) and (Fahy and Gardonio in Sound and structural vibration, Elsevier, Amsterdam, 2007).

5.1 Electro-Dynamical Loudspeakers

Especially at low frequencies acoustical actuation is realized by electro-dynamical loudspeakers. As shown in Fig. 5.1 such a sound source can be interpreted as a linear electro-vibro-acoustical system that consists of an electro-mechanical actuator with transfer function $H_{Lsp-EM}(j\omega)$ that acts in an acoustic interior. The dynamical behavior of this interior is described by the transfer function $H_{Int}(j\omega)$.

The input quantity is the applied voltage $U(j\omega)$ that results in a structural velocity $U_{Lsp}(j\omega)$ at the loudspeakers membrane. Because the air is compressed as a consequence of the membrane-motion, an acoustic pressure $P(j\omega)$ is generated. The overall response of the electro-vibro-acoustical system is therefore determined by the transfer functions of both subsystems.

T. Kletschkowski, *Adaptive Feed-Forward Control of Low Frequency Interior Noise*, Intelligent Systems, Control and Automation: Science and Engineering 56, DOI 10.1007/978-94-007-2537-9_5, © Springer Science+Business Media B.V. 2012

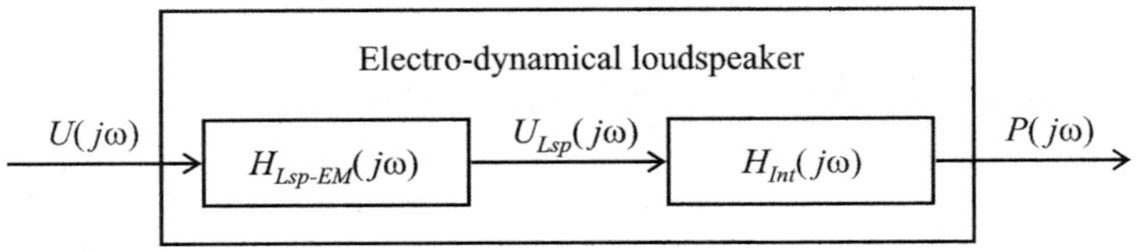

Fig. 5.1 Electro-dynamical loudspeaker (acting in an acoustic field) interpreted as linear system with applied voltage as input quantity and acoustic pressure as output quantity

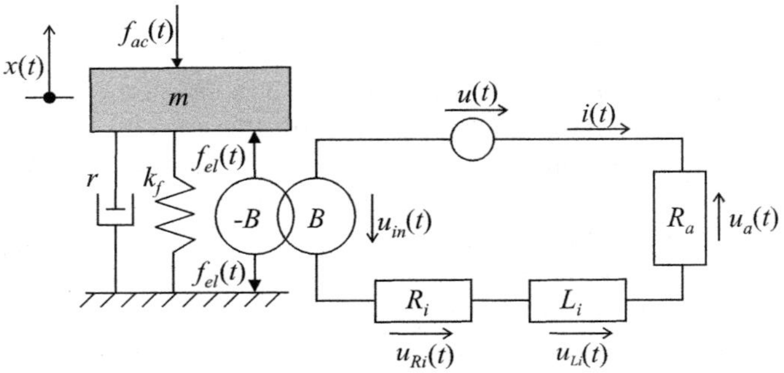

Fig. 5.2 Schematic diagram of an electro-dynamical loudspeaker. The electrical circuit is modeled by an inner resistance R_i and an outer resistance R_a as well as an inner inductance L_i. B is the electromagnetic induction. Mechanical circuit is modeled as an one-degree of freedom oscillator with mass m, stiffness k_f and damping r. f_{ac} represents the force caused by the external sound pressure, and f_{el} is the electrical force (adapted from Fahy and Gardonio 2007)

An electro-dynamical loudspeaker consists of an electrical and a mechanical circuit. A schematic diagram is shown in Fig. 5.2.

Because of the applied voltage $U(t)$, the electrical force $f_{el}(t)$ is generated by the voice-coil. This force causes the motion of the loudspeaker membrane that is described by the coordinate $x(t)$. However, due to the (external) sound pressure that acts on the membrane, the voltage $U_{in}(t)$ is induced in the voice-coil at the same time. Therefore, an electro-dynamical loudspeaker can be used as an electro-dynamical microphone as well.[1]

Applying *Kirchhoff's* second law, the equation of motion for the electrical circuit is given by

$$u(t) = (R_i + R_a)i(t) + L_i\left[di(t)/dt\right] + Bl\dot{x}(t) \quad \text{with } \dot{x}(t) = u_{Lsp}(t), \tag{5.1}$$

where $i(t)$ is the electric current, l is the length of the wire in the voice-coil, and B is the electromagnetic induction. The inner resistance of the loudspeaker is given by R_i, and the outer resistance by R_a. L_i is the inner inductance of the actuator. The equation of motion for the mechanical part of the actuator is given by

[1] A detailed description of electro-dynamical microphones is given e.g. in (Möser 2005) and (B&K 1996).

$$m\ddot{x}(t) + r\dot{x}(t) + k_f x(t) = Bli(t) - p(t)S, \tag{5.2}$$

where $p(t)$ is the external sound pressure acting on the loudspeakers membrane with cross section S. Equations (5.1) and (5.2) form a set of coupled ODEs. Coupling between the mechanical and the electrical circuit is provided by the velocity of the loudspeaker membrane $\dot{x}(t) = u_{Lsp}(t)$ as well as by the electrical current $i(t)$. The magnitude of the acoustical force $f_{ac} = -Sp(t)$ is (in general) small compared to the magnitude of the electrical force $f_{el} = Bli(t)$. The same holds for the relation between the applied voltage $u(t)$ and the voltage $u_{in}(t) = Bl\dot{x}(t)$ that is induced by the velocity of the membrane.

Neglecting $Bl\dot{x}(t)$ in (5.1) and $-p(t)S$ in (5.2), and assuming harmonic fluctuation for all remaining quantities, the frequency response of the electro-mechanical part of the loudspeaker is given by

$$H_{Lsp-EM}(j\omega) := \kappa_{Lsp-EM}\tilde{H}_{Lsp-EM}(j\omega) = \frac{U_{Lsp}(j\omega)}{U(j\omega)} \tag{5.3}$$

with

$$\tilde{H}_{Lsp-EM}(j\omega) = \frac{1}{[1 - j\frac{\omega_k}{\omega_0}(\frac{\omega}{\omega_0})^{-1}][1 - (\frac{\omega}{\omega_0})^2 + j2D\frac{\omega}{\omega_0}]} \tag{5.4}$$

and

$$\kappa_{Lsp-EM} = \frac{Bl}{k_f L_i}, \qquad \omega_k = \frac{R_i + R_a}{L_i}, \qquad \omega_0^2 = \frac{k_f}{m}. \tag{5.5}$$

Equation (5.3) clarifies that the frequency response can be split into a proportional factor κ_{Lsp-EM} with units $\mathrm{m\,s^{-1}V}$, the sensitivity of the loudspeaker, and a dimensionless, but frequency dependent transfer function $\tilde{H}_{Lsp-EM}(j\omega)$.

This transfer behavior can also be described by the ratio between the acceleration of the loudspeaker membrane $A_{Lsp}(j\omega) = j\omega U_{Lsp}(j\omega)$ and the applied voltage

$$\frac{A_{Lsp}(j\omega)}{U(j\omega)} = \frac{j\omega U_{Lsp}(j\omega)}{U(j\omega)} = j\omega H_{Lsp-EM}(j\omega). \tag{5.6}$$

For acoustical applications however, not the structural acceleration $A_{Lsp}(j\omega)$ of the loudspeakers membrane, but rather the sound pressure generated by the loudspeaker is the quantity of interest. If, according to Fig. 5.1, the acoustic interior is seen as a linear system, the acoustic pressure is (at any point $\vec{x}$ in the interior) determined by $H_{Int}(\vec{x}, j\omega)$, the transfer function of the interior. This is expressed by

$$\begin{aligned} P(\vec{x}, j\omega) &= H_{Int}(\vec{x}, j\omega)A_{Lsp}(j\omega) \\ &= H_{Int}(\vec{x}, j\omega)j\omega H_{Lsp-EM}(j\omega)U_{Lsp}(j\omega). \end{aligned} \tag{5.7}$$

The acoustic transfer function $H_{Int}(\vec{x}, j\omega)$ can also be decomposed into a sensitivity that in general depends on frequency, compare Sect. 3.4.2 (3.60), and a dimensionless transfer function such as

$$H_{Int}(\vec{x} j\omega) = \kappa_{Int}(j\omega)\tilde{H}_{Int}(j\omega). \tag{5.8}$$

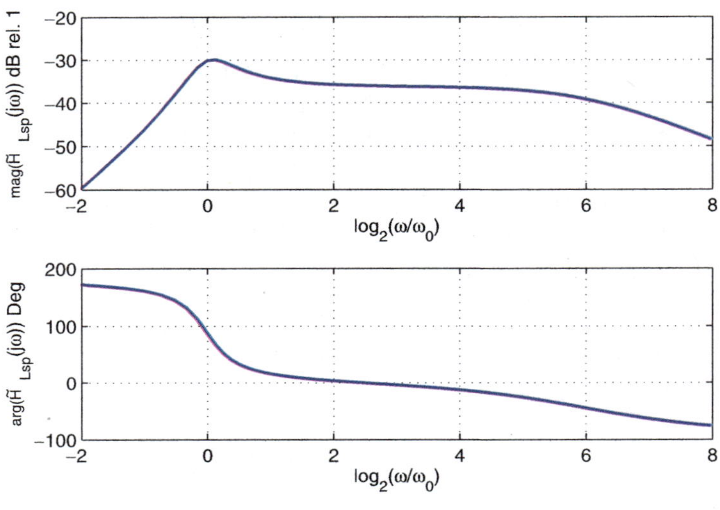

Fig. 5.3 Magnitude (*top*) and phase (*below*) of the dimensionless transfer function of an electro-dynamical loudspeaker. Magnitude is nearly constant within two octave-bands above the natural frequency of the mechanical components. Sound pressure is (nearly) in phase with the input voltage in these two octave-bands

The transfer function of the overall system is therefore given by

$$\begin{aligned} H_{eva}(\vec{x}, j\omega) &:= \frac{P(\vec{x}, j\omega)}{U(j\omega)} \\ &= \frac{P(\vec{x}, j\omega)}{A_{Lsp}(j\omega)} \cdot \frac{j\omega U_{Lsp}(j\omega)}{U(j\omega)} \\ &= H_{Int}(\vec{x}, j\omega)\big[j\omega H_{Lsp-EM}(j\omega)\big] \\ &= \omega_0 \cdot \kappa_{Int} \cdot \kappa_{Lsp-EM} \cdot \tilde{H}_{Int}(\vec{x}, j\omega) \cdot \frac{j\omega}{\omega_0} \tilde{H}_{Lsp-EM}(j\omega) \\ &=: \omega_0 \cdot \kappa_{Int} \cdot \kappa_{Lsp-EM} \cdot \tilde{H}_{Lsp}(j\omega) \cdot \tilde{H}_{int}(\vec{x}, j\omega) \end{aligned} \tag{5.9}$$

with

$$\tilde{H}_{Lsp}(j\omega) = \frac{j\frac{\omega}{\omega_0}}{[1 - j\frac{\omega_k}{\omega_0}(\frac{\omega}{\omega_0})^{-1}][1 - (\frac{\omega}{\omega_0})^2 + j2D\frac{\omega}{\omega_0}]}. \tag{5.10}$$

A graphical representation of (5.10) is shown in Fig. 5.3. The ratio between the electric bend frequency ω_k and the natural frequency of the mechanical circuit ω_0 was set to 64. Furthermore, a damping ratio of $D = 0.25$ was taken into account for the mechanical circuit. It can be seen that the magnitude response is nearly constant within two octave-bands above ω_0. The sound pressure is more or less in phase with the input voltage within these two octave-bands.

As described in (Möser 2005), and illustrated by Fig. 5.3 (top), the electro-dynamical loudspeaker acts as a band-pass filter. Its (lower) cut-off frequency ω_0 is determined by the mechanical circuit, the (upper) cut-off frequency is given by ω_k that depends on the electrical components of the loudspeaker.

5.2 Electro-Dynamical Panel Speakers

Effective sound generation is also possible, if an electro-dynamical actuator is used to drive a panel that (because of its motion) radiates sound into the interior. Such a sound source is known as electro-dynamical *panel speaker*.

Typical examples of this technique are trim panel speakers (QinetiQ 2007) that have been found to be very effective in the mid (500 Hz $< f <$ 1 kHz) and high frequency range ($f >$ 1 kHz).

The main idea of this technique that is based on the theory of the electro-dynamic linear motor (Fahy and Gardonio 2007) is reviewed in this section. At first we will analyze the situation in which the electro-dynamical exciter is attached to a rigid base. Then we will use a very simple one-degree of freedom model to motivate the sound generation of a panel that is driven by such an electro-dynamical exciter.

As illustrated by Fig. 5.4, an electro-dynamical exciter (the reciprocal to an inertial sensor) can be interpreted as a linear system with applied voltage $U(j\omega)$ as input quantity and $F_t(j\omega)$—the force transmitted into the base—as output quantity. The system response is determined by its frequency response $H_{Iac}(j\omega)$.

Figure 5.5 shows the schematic diagram of an electro-dynamic exciter that is also known as *inertial actuator*. This diagram is similar to the schematic diagram of an electro-dynamical loudspeaker, compare Fig. 5.2, but, in contrast to the loudspeaker, $f_t(t)$ the force transmitted from the inertial actuator into the base is the relevant output quantity. Assuming that such an exciter is mounted on a rigid base, compare (Fahy and Gardonio 2007), the equation of motion for the electrical and the mechanical circuits are given by

$$u(t) = R_i i(t) + L_i\big[di(t)/dt\big] + Bl\dot{x}(t), \tag{5.11}$$

and

$$m\ddot{x}(t) + r\dot{x}(t) + k_f x(t) = Bli(t). \tag{5.12}$$

R_i is the resistance, and L_i is the inductance of the actuator. The mechanical properties are represented by a mass m, a damping coefficient r, and a stiffness k_f.

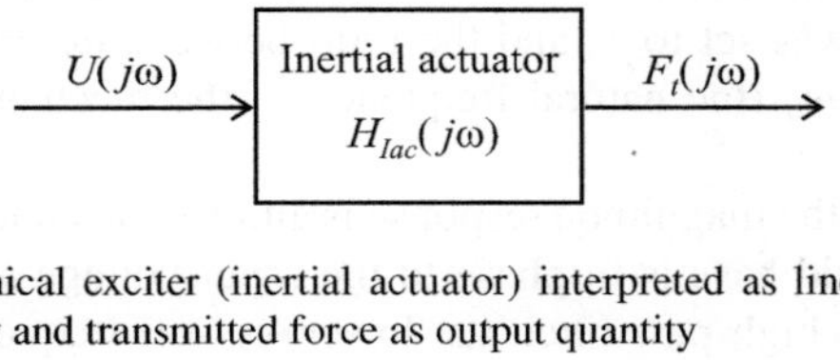

Fig. 5.4 Electro-dynamical exciter (inertial actuator) interpreted as linear system with applied voltage as input quantity and transmitted force as output quantity

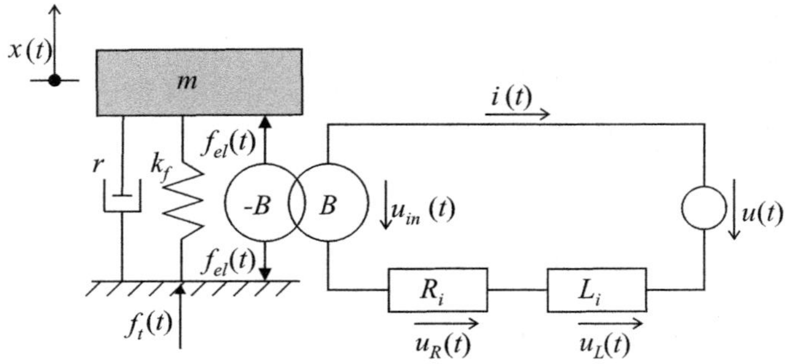

Fig. 5.5 Schematic diagram of an electro-dynamical exciter. Electrical circuit is modeled by a resistance R_i and an inductance L_i. B is the frequency-independent voice-coil coefficient. Mechanical circuit is modeled as a one-degree of freedom oscillator with mass m, stiffness k_f and damping r. f_t represents the force transmitted into the base (adapted from Fahy and Gardonio 2007)

The coupling between the mechanical and electrical circuit is determined by the voltage $U_{in}(t) = Bl\dot{x}(t)$ induced by the motion of the mass m, and the electrical actuator force $f_{el}(t) = Bli(t)$. The force transmitted into the base is given by

$$f_t(t) = k_f x(t) + r\dot{x}(t) - Bli(t). \tag{5.13}$$

Assuming harmonic fluctuation of all time-dependent quantities, the transfer behavior of an electro-dynamical exciter is described by

$$H_{Iac}(j\omega) := \kappa_{Iac}\tilde{H}_{Iac}(j\omega) = \frac{F_t(j\omega)}{U(j\omega)}, \tag{5.14}$$

where

$$\tilde{H}_{Iac}(j\omega) = \frac{-j(\frac{\omega}{\omega_0})}{1 + \frac{L_i k}{B^2 l^2}[1 - j\frac{\omega_k}{\omega_0}(\frac{\omega}{\omega_0})^{-1}][1 - (\frac{\omega}{\omega_0})^2 + j2D\frac{\omega}{\omega_0}]} \tag{5.15}$$

and

$$\kappa_{Iac} = \frac{\sqrt{mk_f}}{Bl}, \qquad \omega_k = \frac{R_i}{L_i}, \qquad \omega_0^2 = \frac{k_f}{m}. \tag{5.16}$$

κ_{Iac} is the sensitivity of the exciter with units $\mathrm{N\,V}^{-1}$ and $L_i k_f (Bl)^{-2}$ is a dimensionless parameter that is determined by the electro-mechanical properties of the actuator.

The frequency response of the inertial actuator is shown in Fig. 5.6. To draw the curve, a damping ratio of $D = 0.025$ was taken into account for the mechanical circuit. $L_i k_f (Bl)^{-2}$ was set to 1, and the ratio between $\omega_k = R_i L_i^{-1}$ (the electric bent frequency) and ω_0 (the natural frequency of the mechanical part) was set to 5333.

It can be seen that the magnitude response is almost constant above ω_0. The force transmitted into a rigid base is in phase to the input voltage above this frequency. The actuator acts as a high-pass filter. Its (lower) cut-off frequency is given by ω_0.

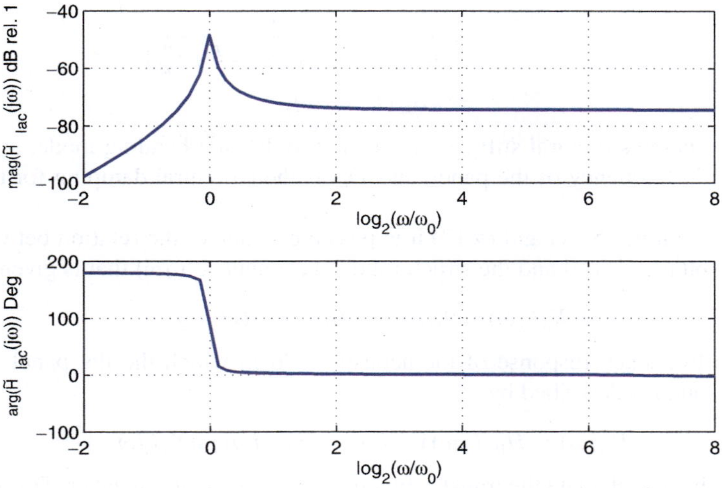

Fig. 5.6 Magnitude (*top*) and phase (*below*) of the dimensionless transfer function of the inertial actuator. Magnitude is constant above the natural frequency. Force transmitted into rigid base is in phase to the input voltage above this frequency

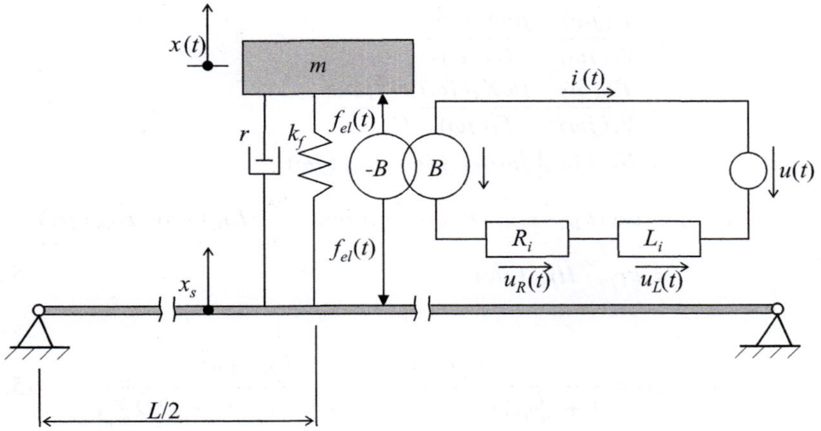

Fig. 5.7 Schematic diagram of an electro-dynamical exciter that is attached to an elastic structure

An example of structural actuation that is used for sound generation is a flat panel speaker. Such an actuator consists of an elastic panel with at least one electro-dynamical exciter as illustrated by Fig. 5.7. If the structural response $x_s(t)$ of the elastic panel is only analyzed for the first bending mode, its mechanical behavior may be represented by the behavior of a harmonic oscillator with transfer function

$$H_{Osz}(j\omega) = \frac{X_s(j\omega)}{F_t(j\omega)}$$

$$
\begin{aligned}
&= \frac{1}{k_m} \frac{1}{[1 - (\frac{\omega}{\omega_m})^2 + j2D_m \frac{\omega}{\omega_m}]} \\
&=: \kappa_{Osz} \tilde{H}_{Osz}(j\omega),
\end{aligned} \tag{5.17}
$$

where k_m is the structural stiffness associated to the first bending mode, ω_m is the first natural frequency of the panel, and D_m is the structural damping for this frequency.

By combining (5.14) and (5.17) it is possible to derive the relation between the applied voltage $U(j\omega)$ and the structural displacement $X_s(j\omega)$ that is given by

$$
X_s(j\omega) = H_{Osz}(j\omega) H_{Iac}(j\omega) U(j\omega). \tag{5.18}
$$

If the frequency response of the acoustic field in which the flat panel speaker radiates sound is described by

$$
P(j\omega) = H_{Int}(j\omega) V_s(j\omega) = H_{Int}(j\omega) j\omega X_s(j\omega) \tag{5.19}
$$

it is possible to calculate the transfer function of the flat panel speaker. This transfer function characterizes the relation between the acoustic pressure $P(j\omega)$ and the applied voltage $U(j\omega)$. It is given by

$$
\begin{aligned}
H_{Fps}(j\omega) &:= \frac{P(j\omega)}{U(j\omega)} \\
&= \frac{P(j\omega)}{V_s(j\omega)} \cdot \frac{j\omega X_s(j\omega)}{U(j\omega)} \\
&= \frac{P(j\omega)}{V_s(j\omega)} \cdot \frac{j\omega X_s(j\omega)}{F_t(j\omega)} \frac{F_t(j\omega)}{U(j\omega)} \\
&= H_{Int}(j\omega) \big[j\omega H_{Osz}(j\omega) \big] H_{Iac}(j\omega) \\
&= \omega_0 \cdot \kappa_{Int} \cdot \kappa_{Osz} \cdot \kappa_{Iac} \cdot \tilde{H}_{Int}(j\omega) \cdot \frac{j\omega}{\omega_0} \tilde{H}_{Osz}(j\omega) \cdot \tilde{H}_{Iac}(j\omega) \\
&=: \kappa_{Fps} \cdot \tilde{H}_{Fps}(j\omega)
\end{aligned} \tag{5.20}
$$

with

$$
\tilde{H}_{Fps}(j\omega) = \frac{(\frac{\omega}{\omega_0})^2 \cdot \tilde{H}_{Int}(j\omega) \cdot \tilde{H}_{Osz}(j\omega)}{1 + \frac{Lk}{B^2 l^2} [1 - j \frac{\omega_k}{\omega_0} (\frac{\omega}{\omega_0})^{-1}] [1 - (\frac{\omega}{\omega_0})^2 + j2D \frac{\omega}{\omega_0}]} \tag{5.21}
$$

and

$$
\kappa_{Fps} = \omega_0 \cdot \kappa_{Int} \cdot \kappa_{Osz} \cdot \kappa_{Iac}. \tag{5.22}
$$

The theoretical frequency response of a flat panel speaker is shown in Fig. 5.8. To draw the curve a damping ratio of $D = 0.025$ was taken into account for the mechanical part of the exciter. D_m was set to 0.05. $L_i k_f (Bl)^{-2}$ was set to 1, and the ratio between $\omega_k = RL^{-1}$ and ω_0 was set to 5333. The ratio between ω_0 and ω_m was set to 1/3.

It can be seen that the magnitude is almost constant in a half-octave band between the natural frequency of the exciter and the frequency that belongs to the first bending mode of the panel. Therefore, a flat panel speaker acts as a band-pass filter. Its

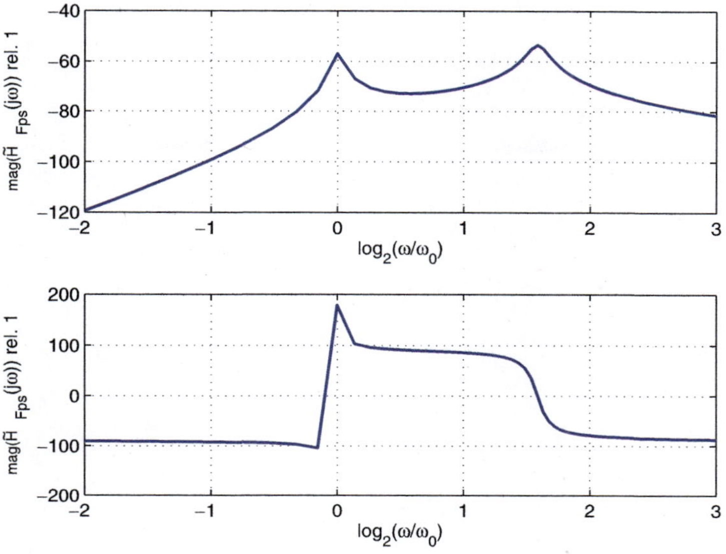

Fig. 5.8 Magnitude (*top*) and phase (*below*) of the dimensionless transfer function of a flat panel speaker. Magnitude is nearly constant in a half-octave band between the natural frequency of the exciter and the first bending mode of the panel

(lower) cut-off frequency is given by ω_0. Its (upper) cut-off frequency is determined by ω_m that indicates the first structural bending mode of the panel.

Chapter 6
Active Control of Tonal and Broadband Noise

Abstract Active reduction of disturbing noise requires the generation of appropriate control signals in order to drive the canceling sources. For this purpose we have to organize the *signal processing* of data provided by sensors. These data contain information about the disturbance as well as about the systems state. Because we assume that we are able to work with reference signals that are well correlated to the disturbing noise, the upcoming chapter is restricted to signal processing in *feed-forward control systems*. It starts with a mathematical preparation, because we will use some concepts of matrix calculus, see (Zurmühl in Matrizen und ihre technischen Anwendungen, Springer, Berlin, 1964) or (Strang in Introduction to linear algebra, Cambridge University Press, Wellesley, 1993). It is continued with an introduction to feed-forward control signal processing, before we discuss active control of both harmonic excitations and stochastic disturbances. Besides the topic of *optimal control*, attention is paid to *adaptive control* that enables the controller to track down non-stationary effects, e.g. the change of the operational speed of a turbo-machinery, or changes in the system response that can be caused by a change of environmental conditions such as ambient pressure, air temperature or humidity as well as by the warming-up of the electro-acoustic equipment. Adaptive feed-forward control of harmonic disturbances is discussed in frequency domain, whereas we will present four time domain approaches for adaptive feed-forward control of stochastic disturbances. However, this chapter is far away from presenting the basics of digital signal processing and digital filters as described in (Antoniou in Digital filters: analysis and design, McGraw-Hill, New York, 1979), (Bose in Digital filters—theory and applications, Elsevier, New York, 1985), (Diniz in Adaptive filtering—algorithms and practical implementations, Springer, New York, 2008), (Haykin in Adaptive filter theory, Prentice Hall, London, 1996), (Hess in Digitale Filter—Eine Einführung, Teubner, Stuttgart, 1989), (Johnson in Digitale Signalverarbeitung, Hanser, München in Cooperation with Prentice Hall International, London, 1991), and (Lücker in Grundlagen digitaler Filter—Eine Einführung in die Theorie linearer zeitdiskreter Systeme und Netzwerke, 1980) to review the theory of adaptive filtering as presented in (Farhang-Boroujeny in Adaptive filters—theory and applications, Wiley, New York, 1998), (Honig and Messerschmidt in Adaptive filters—structure, algorithms and applications, Kluwer Academic, Boston, 1984), (Sayed in Fundamentals of adaptive filtering, Wiley, Hoboken, 2003) and (Widrow and Stearns in Adaptive signal processing, Prentice Hall International,

T. Kletschkowski, *Adaptive Feed-Forward Control of Low Frequency Interior Noise*,
Intelligent Systems, Control and Automation: Science and Engineering 56,
DOI 10.1007/978-94-007-2537-9_6, © Springer Science+Business Media B.V. 2012

London, 1985) or to discuss DSP implementations, see (Akpan et al. in Active noise and vibration control literature survey: controller technologies. DREA-CR-1999-177, Contractor Report, Defence Research Establishment Atlantic, Dartmouth NS (CAN); MARTEC Ltd, Halifax NS (CAN); Sherbrooke Univ, Sherbrooke QUE (CAN), 1999) and (Chassaing in Digital signal processing and applications with the C6713 and C6416 DSK, Wiley, Canada, 2005). Its intension is to provide a basis for adaptive feed-forward control of low frequency interior noise that also includes algorithmic formulations of the control schemes. The content of the upcoming chapter is therefore oriented on (Kuo and Morgan in Active noise control systems—algorithms and DSP implementations, Wiley, Canada, 1996) and (Elliott in Signal processing for active noise control, Academic Press, London, 2001)—two fundamental books about signal processing for ANC that are (without detailed algorithmic formulations) summarized in (Kuo and Morgan in Proc. IEEE 87(6):943–973, 1999) and (Elliott in Tokhi and Veres (eds.) Active sound and vibration control, Institution of Electrical Engineers, London, pp. 57–72, 2002).

6.1 Mathematical Preparation

Signal processing for active control systems is focused on the minimization of real (time domain) or complex (frequency domain) matrix quadratic forms that can be established on the input and output signals of the basic system. Minimization of quadratic forms can be performed analytically by *quadratic optimization* and numerically using the *steepest decent algorithm*. The latter is the basis for the most common adaptive algorithms. For this reason, the upcoming section contains a mathematical preparation that is essential for active feed-forward control of interior noise.

6.1.1 Hermitian Matrices

Let $\mathbf{A}$ be a quadratic matrix consisting of complex elements a_{ij} that can be split into a real $\mathbf{A}_R$ with elements $a_{Rij} = \mathrm{Re}[a_{ij}]$ and into an imaginary part $\mathbf{A}_I$ with elements $a_{Iij} = \mathrm{Im}[a_{ij}]$ such as

$$\mathbf{A} = \mathbf{A}_R + j\mathbf{A}_I. \tag{6.1}$$

The conjugate complex and transposed form known as the *hermitian* of $\mathbf{A}$ is defined by

$$\mathbf{A}^H := (\mathbf{A}^*)^T = \mathbf{A}_R^T - j\mathbf{A}_I^T. \tag{6.2}$$

If $\mathbf{A} = \mathbf{A}^T$, the sum $\mathbf{A} + (\mathbf{A}^*)^T$ equals $2\,\mathrm{Re}[\mathbf{A}] = 2\mathbf{A}_R$. The definition of a *hermitian matrix* (the complex generalization of a symmetric matrix), is, compare (Zurmühl 1964) and (Strang 1993), given by

$$\mathbf{A}^H = \mathbf{A}. \tag{6.3}$$

Equation (6.3) implies that a hermitian matrix $\mathbf{A}$ can be decomposed into a symmetric real part, and into a skew-symmetric imaginary part such as

$$\mathbf{A}_R = \mathbf{A}_R^T, \qquad \mathbf{A}_I = -\mathbf{A}_I^T, \quad \text{i.e. } a_{ij} = a_{ji}^*. \tag{6.4}$$

Every eigenvalue of a hermitian matrix is real. The eigenvectors of a hermitian matrix are orthogonal, provided they correspond to different eigenvalues. The *hermitian quadratic form* $\mathbf{x}^H\mathbf{A}\mathbf{x}$, where $\mathbf{A} = \mathbf{A}^H$ is a hermitian matrix with dimension $(N \times N)$, and $\mathbf{x}$ is an $(N \times 1)$ column matrix, is always a real valued quantity. $\mathbf{A}$ is positive definite, if

$$\mathbf{x}^H\mathbf{A}\mathbf{x} > 0 \quad \text{holds for all } \mathbf{x} \neq 0. \tag{6.5}$$

For real matrices with $a_{Iij} = 0$, that can be seen as a special case of complex matrices, we can replace the superscript H by the superscript T.

6.1.2 Quadratic Optimization

Let us assume that we have to search for the minimum of a real valued function that is based on the *complex quadratic form* given by

$$J_R = \text{Re}\{J(\mathbf{x})\} \quad \text{with } J(\mathbf{x}) = \mathbf{x}^H\mathbf{A}\mathbf{x} + \mathbf{x}^H\mathbf{b}_1 + \mathbf{b}_2^H\mathbf{x} + c, \tag{6.6}$$

where $\mathbf{A}$ is a quadratic matrix with dimension $(N \times N)$, $\mathbf{b}_1$, $\mathbf{b}_2$, $\mathbf{x}$ are $(N \times 1)$ column matrices, and c is a scalar. These quantities are also known as *cost function parameters*. The total increment of (6.6) is given by

$$dJ_R = \text{Re}\big[dJ(\mathbf{x})\big] = \text{Re}\big[\mathbf{x}^H(\mathbf{A}^H + \mathbf{A})d\mathbf{x} + \mathbf{b}_1^H d\mathbf{x} + \mathbf{b}_2^H d\mathbf{x}\big] =: \text{Re}\big[\mathbf{g}^H d\mathbf{x}\big], \tag{6.7}$$

where the $(N \times 1)$ column matrix

$$\mathbf{g}(n) := \big(\mathbf{A} + \mathbf{A}^H\big)\mathbf{x}(n) + \mathbf{b}_1 + \mathbf{b}_2, \tag{6.8}$$

is known as the *gradient matrix*, compare (Elliott 2001). The column matrix $\mathbf{x}_{opt}$ that minimizes J_R is obtained by setting dJ_R to zero. Assuming that $\mathbf{A} + \mathbf{A}^H$ can be inverted yields

$$\mathbf{x}_{opt} = -\big(\mathbf{A}^H + \mathbf{A}\big)^{-1}(\mathbf{b}_1 + \mathbf{b}_2). \tag{6.9}$$

The minimum value of the quadratic form is obtained by substituting (6.9) into (6.6), this gives

$$J_{\min} = c + \big[(\mathbf{b}_1 + \mathbf{b}_2)^H\big[\big(\mathbf{A}^H + \mathbf{A}\big)^{-1}\big]^H\mathbf{A} - \mathbf{b}_1^H - \mathbf{b}_2^H\big]\big(\mathbf{A}^H + \mathbf{A}\big)^{-1}(\mathbf{b}_1 + \mathbf{b}_2). \tag{6.10}$$

If $\mathbf{A} = \mathbf{A}^T$, the optimal solution (6.9) reduces to

$$\mathbf{x}_{opt} = -\big(2\,\mathrm{Re}[\mathbf{A}]\big)^{-1}(\mathbf{b}_1 + \mathbf{b}_2), \tag{6.11}$$

and the corresponding minimum is given by

$$J_{\min} = c + \big[(\mathbf{b}_1 + \mathbf{b}_2)^H \big(2\,\mathrm{Re}[\mathbf{A}]\big)^{-T}\mathbf{A} - \mathbf{b}_1^H - \mathbf{b}_2^H\big]\big(2\,\mathrm{Re}[\mathbf{A}]\big)^{-1}(\mathbf{b}_1 + \mathbf{b}_2). \tag{6.12}$$

If $\mathbf{A} = \mathbf{A}^H$, and if $\mathbf{b} = \mathbf{b}_1 = \mathbf{b}_2$, the optimal solution (6.11) reduces further to

$$\mathbf{x}_{opt} = -\mathbf{A}^{-1}\mathbf{b}, \tag{6.13}$$

and the corresponding minimum is given by

$$J_{\min} = c - \mathbf{b}^H \mathbf{A}^{-1}\mathbf{b}. \tag{6.14}$$

As in the previous subsection, we can replace the superscript H by the superscript T, if we deal with real valued quantities, i.e. if $\mathrm{Im}[\mathbf{x}] = \mathbf{0}$, $\mathrm{Im}[\mathbf{A}] = \mathbf{0}$, $\mathrm{Im}[\mathbf{b}_1] = \mathbf{0}$, $\mathrm{Im}[\mathbf{b}_2] = \mathbf{0}$, and $\mathrm{Im}[c] = 0$.

6.1.3 Steepest-Decent Algorithm

We can also use an iterative procedure that is known as *steepest decent algorithm* to find the minimum of the complex vector quadratic form (6.6). Therefore we assume that the evolution of $\mathbf{x}$ between the time steps n and $n+1$ can be described by the direction of the steepest-decent such as

$$\mathbf{x}(n+1) = \mathbf{x}(n) - \frac{\mu}{2}\mathbf{g}, \tag{6.15}$$

where μ is a convergence parameter (also known as *step size*). If we use the gradient matrix as defined in (6.8), the iterative mapping (6.15) becomes

$$\mathbf{x}(n+1) = \mathbf{x}(n) - \frac{\mu}{2}\big[(\mathbf{A} + \mathbf{A}^H)\mathbf{x}(n) + \mathbf{b}_1 + \mathbf{b}_2\big]. \tag{6.16}$$

If we subtract $\mathbf{x}_{opt}$ as specified in (6.9) from both sides of (6.16) and, if we further replace $\mathbf{b}_1 + \mathbf{b}_2$ with $-(\mathbf{A} + \mathbf{A}^H)\mathbf{x}_{opt}$ also using (6.9), we obtain the following iterative mapping

$$\Delta\mathbf{x}(n+1) = \left[\mathbf{I} - \frac{\mu}{2}\big(\mathbf{A} + \mathbf{A}^H\big)\right]\Delta\mathbf{x}(n), \tag{6.17}$$

where

$$\Delta\mathbf{x}(n+1) := \mathbf{x}(n+1) - \mathbf{x}_{opt} \quad \text{and} \quad \Delta\mathbf{x}(n) := \mathbf{x}(n) - \mathbf{x}_{opt}. \tag{6.18}$$

If all eigenvalues of $\mathbf{A} + \mathbf{A}^H$ are distinct, so that the eigenvectors are linearly independent, we can use the spectral representation

$$\mathbf{A} = \mathbf{Q}\boldsymbol{\Lambda}\mathbf{Q}^{-1}, \tag{6.19}$$

where $\mathbf{\Lambda}$ contains the N distinct eigenvalues of $\mathbf{A}$

$$\mathbf{\Lambda} = \begin{bmatrix} \lambda_1 & 0 & \dots & 0 \\ 0 & \lambda_2 & & \\ \vdots & & & \\ 0 & & & \lambda_N \end{bmatrix}, \tag{6.20}$$

and the columns of $\mathbf{Q}$ are expressed by the associated eigenvectors

$$\mathbf{Q} = [\mathbf{q}_1\ \mathbf{q}_2 \dots \mathbf{q}_N] \tag{6.21}$$

to introduce principle coordinates

$$\Delta\mathbf{v}(n+1) := \mathbf{Q}^{-1}\Delta\mathbf{v}(n+1) \quad \text{and} \quad \Delta\mathbf{v}(n) := \mathbf{Q}^{-1}\Delta\mathbf{v}(n). \tag{6.22}$$

Considering (6.19) and (6.22), the iterative mapping (6.17) reads

$$\Delta\mathbf{v}(n+1) = \left[\mathbf{I} - \frac{\mu}{2}\left(\mathbf{\Lambda} + \mathbf{\Lambda}^H\right)\right]\Delta\mathbf{v}(n). \tag{6.23}$$

Stability of (6.23) is ensured, if

$$\left|1 - \frac{\mu}{2}\left(\lambda_m + \lambda_m^*\right)\right| = \left|1 - \mu\,\mathrm{Re}(\lambda_m)\right| < 1 \tag{6.24}$$

holds for all eigenvalues λ_m. For stable iteration, the convergence coefficient must therefore be bounded such as

$$0 < \mu < \frac{2}{\mathrm{Re}(\lambda_m)}. \tag{6.25}$$

Equation (6.25) clarifies that the real part of each eigenvalue λ_m must be positive, such as $\mathrm{Re}(\lambda_m) > 0$, to obtain a stable iteration schema. Because

$$\frac{2\,\mathrm{Re}(\lambda_m)}{|\lambda_m|^2} \leq \frac{2}{\mathrm{Re}(\lambda_m)}, \tag{6.26}$$

it is possible to replace (6.25) by[1]

$$0 < \mu < \frac{2\,\mathrm{Re}(\lambda_m)}{|\lambda_m|^2}. \tag{6.27}$$

For a hermitian matrix, compare (6.3) and (6.4), the iterative mapping (6.17) reduces to

$$\Delta\mathbf{x}(n+1) = [\mathbf{I} - \mu\mathbf{A}]\Delta\mathbf{x}(n). \tag{6.28}$$

[1] In some textbooks, e.g. (Elliott 2001), the stability criteria (6.26) is motivated by the argument that the iterative mapping (6.28) can also performed with a non-hermitian matrix $\mathbf{A}$. The stability criteria for the in this case complex eigenvalues λ_m is given by

$$|1 - \mu\lambda_m| < 1 \quad \text{i.e. } 0 < \mu < \frac{2\,\mathrm{Re}(\lambda_m)}{|\lambda_m|^2}.$$

As shown in the present subsection, it is not necessary to argue this way.

Because all eigenvalues of a hermitian are real, as noticed in Sect. 6.1.1, the convergence coefficient must in this case for all eigenvalues λ_m of the matrix $\mathbf{A}$ be bounded by

$$|1 - \mu\lambda_m| < 1 \quad \text{i.e. } 0 < \mu < \frac{2}{\lambda_m}. \tag{6.29}$$

Equation (6.29) is fulfilled, if the hermitian matrix $\mathbf{A}$ is positive definite, because all eigenvalues of a positive definite hermitian matrix are positive, as mentioned in Sect. 6.1.1.

6.2 Terms and Structure of Feed-Forward Control Approaches

Before we introduce adaptive algorithms for feed-forward control, it is necessary to introduce both terms and structures used to describe this control strategy.

The objective of all control schemes is the reduction or complete cancellation of a *disturbance* that is in acoustic interiors given by the *primary noise* (short: noise). Therefore, a *canceling signal* is superimposed on the disturbance. The latter is in acoustic interiors given by the *secondary noise* (short: anti-noise). The superposition of noise and anti-noise gives the *error signal.*

In feed-forward control, as shown in Fig. 6.1, we have the possibility to incorporate a *reference signal* that should be well correlated to the disturbance. This signal passes through a system called *primary path.* This gives us the disturbance that is determined by the transfer behavior of the primary path. To generate the canceling signal, the reference signal is passed through the *controller*.

The controller output is known as *control signal.* It has to pass the *secondary path,* a system that describes the transfer behavior between an actuator and an error sensor, and is hereby transformed into the canceling signal. If the controller is not adjusted by an adaptive algorithm, *non-adaptive* control is performed.

Figure 6.1 also clarifies that the time delay caused by the secondary path must be smaller that the delay caused by the primary path to enable the implementation of a *causal controller*. The difference between these two time spans is the time period that is left for active control.

The controller can be adjusted, if *adaptive feed-forward control,* compare Fig. 6.2, is applied. This is advantageous, if it is necessary to compensate for a change in magnitude, phase or frequency of the disturbance, but also to compensate

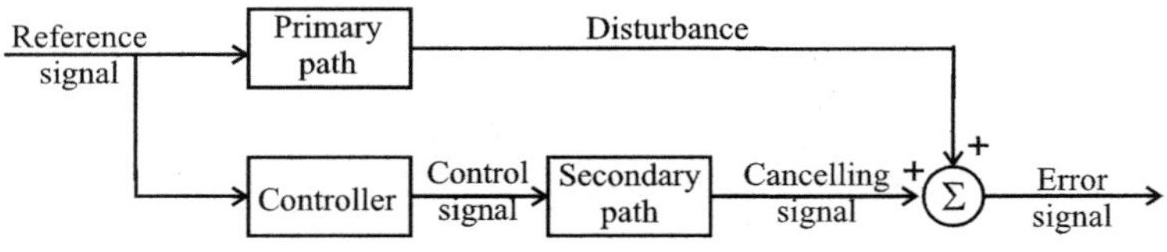

Fig. 6.1 Block diagram representation of a non-adaptive feed-forward control schema illustrating the information flow between three systems (primary path, controller and secondary path)

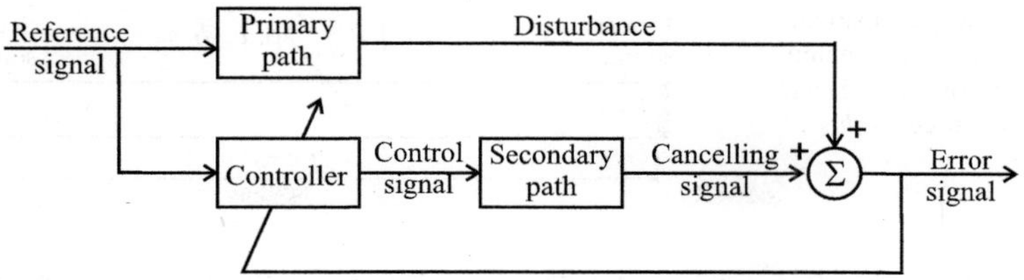

Fig. 6.2 Block diagram representation of an adaptive feed-forward controller illustrating the information flow between three systems (primary path, controller and secondary path) including a feed back of error signal

a variation in the transfer behavior of the secondary path. For this purpose, the error signal is traced back to the controller and used to adapt the control signal.

6.3 Review and Evaluation of Control Strategies

In this section we will discuss the physical basis for active noise control. For this purpose we will consider pure tone sound fields and a frequency range that includes the first five acoustic resonances.[2]

The underlying physical concepts range from controlling the acoustic potential energy density (that is in most applications approximated by the squared sound pressure measured with a finite number of microphones, see Nelson et al. 1987a, 1987b, and 1987c) to more sophisticated methods such as the minimization of the active intensity, see (Kang and Kim 1997), but also (Böhme et al. 2009) and (Sachau et al. 2008c). Active control can also be established on the minimization of the total power input (Elliott et al. 1991), or the minimization of the acoustic energy density (Parkins et al. 2000). Furthermore it is possible to increase the dissipation by maximizing the absorption power (Elliott et al. 1991) or the absorption coefficient (Guicking and Karcher 1984). In the present work we will consider:

1. Global control of acoustic potential energy (*Epot_global*),
2. Local control of acoustic potential energy density (*Epot_local*),
3. Active control of total power input (*TPI*),
4. Local control of acoustic energy density (*E_local*),
5. Active control of power absorption (*PowerAbs*),
6. Active control of acoustic absorption (*AcousticAbs*).

To evaluate these strategies, the acoustic field in a weakly damped finite duct will be used as a simple benchmark. A close form solution will be derived in frequency

---

[2]Active cancellation of sound pressure at a point in a pure tone diffuse sound field is discussed in (Elliott et al. 1988), whereas (Elliott and Garcia-Bonito 1995) reports on active cancellation of pressure and pressure gradient in a diffuse sound field. Active control of stationary random sound fields is analyzed in (Nelson et al. 1990).

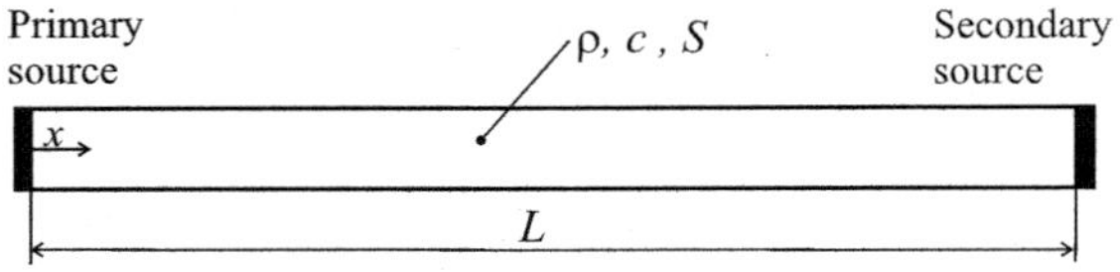

Fig. 6.3 One-dimensional acoustic system terminated by primary source (*left*) and secondary source (*right*)

domain using a traveling wave model. The advantages and disadvantages of different control strategies will be discussed with respect to the reduction of the acoustic potential energy of the benchmark system.

Curtis et al. (1987)[3] used a similar model to illustrate the effects of the control strategies 2, 3 and 6. It turned out that these strategies are capable of suppressing acoustic resonances. Only the approaches 2 and 6 result in causally acting controller, as also noticed in (Curtis et al. 1987).

However, the amount of noise reduction that can be realized by a certain control strategy in and between acoustic resonances as well as the influence of passive damping on the control profit has not yet been discussed. This will be done in the upcoming section.

6.3.1 Description of Benchmark System and Close Form Solution

To keep the calculations simple, the benchmark system is given by an air filled finite acoustic duct with length L and cross section S that is at both ends terminated by loudspeakers. These loudspeakers are seen as ideal volume velocity sources with source strengths Q_p at $x = 0$, the position of the primary source, and Q_s at $x = L$, the position of the canceling source. (See Fig. 6.3.)

The choice of this benchmark system can be motivated as follows: "It is obvious that by placing a single secondary ring source in coincidence with a single primary ring source that complete cancellation of sound field will occur, if the secondary source is driven in exact antiphase to the primary source. Since the secondary source is moved away from the primary source the complete cancellation can never occur.", see (Curtis et al. 1987). This means we will discuss a worst case scenario and the lower limits of active noise reduction.

Assuming harmonic fluctuation of the source strengths, the problem is described by the Helmholtz equation (3.22) and the Neumann BC, compare (3.24), at $x = 0$ and $x = L$. Applying a traveling wave model, compare (3.48), for both the acoustic pressure

$$P(x, j\omega) = A(j\omega)e^{-jkx} + B(j\omega)e^{+jkx} =: P_i(x, j\omega) + P_r(x, j\omega) \quad (6.30)$$

and the volume velocity

[3] A modified approach, based on input power, that includes damping effects, unfortunately hidden behind real parts of transfer impedances, was presented in (Elliott 2001).

$$Q(x, j\omega) = \frac{S}{\rho_\infty c}\left[A(j\omega)e^{-jkx} - B(j\omega)e^{+jkx}\right] \tag{6.31}$$

where $k = (1 - j\xi)\omega c^{-1}$ is the complex wave number, compare (3.75). $A(j\omega)$ and $B(j\omega)$ are complex coefficients that determine magnitude and phase of the incident wave component and the reflected wave component. This approach yields the following solution

$$P(x, j\omega) = \frac{\rho_\infty c}{S}\frac{(e^{jkL}Q_p + Q_s)e^{-jkx} + (e^{-jkL}Q_p + Q_s)e^{jkx}}{e^{jkL} - e^{-jkL}}. \tag{6.32}$$

To prepare a discussion that is based on a manageable analytical solution, the damping terms are approximated by a first order series expansion of the exponential function such as

$$\begin{aligned} e^{\xi kx} &\approx 1 + \xi kx, & e^{-\xi kx} &\approx 1 - \xi kx, \\ e^{\xi kL} &\approx 1 + \xi kL, & e^{-\xi kL} &\approx 1 - \xi kL. \end{aligned} \tag{6.33}$$

Using the approximations defined by (6.34) combined with the definition of the *normalized frequency* $\zeta := L/\lambda$ and the definition of the *normalized position* $\tilde{x} := x/L$ it is possible to rewrite (6.32) as follows

$$P(\tilde{x}, j\omega) = \frac{\rho_\infty c}{S}\left(\frac{A_1 + A_2\xi}{B_1 + B_2\xi}\right) \quad \Leftrightarrow \quad \tilde{P}(\tilde{x}) = \frac{S}{\rho_\infty c}\frac{P(\tilde{x}, j\omega)}{Q_p}. \tag{6.34}$$

$\tilde{P}(\tilde{x})$ is the normalized sound pressure. All other variables are defined as follows

$$\begin{aligned} A_1 &= \cos\left[2\pi\zeta(\tilde{x} - 1)\right]Q_p + \cos[2\pi\zeta\tilde{x}]Q_s \\ A_2 &= j2\pi\zeta(\tilde{x} - 1)\sin\left[2\pi\zeta(\tilde{x} - 1)\right]Q_p + j\tilde{x}\sin[2\pi\zeta\tilde{x}]Q_s \\ B_1 &= j\sin[2\pi\zeta] \\ B_2 &= 2\pi\zeta\cos[2\pi\zeta]. \end{aligned} \tag{6.35}$$

Equation (6.34) clarifies that both numerator and denominator of the solution depend linearly on the dimensionless damping ratio ξ. In the limit case $\xi = 0$ the numerator would reduce to a pure real valued quantity and the denominator to a pure imaginary number. This indicates that real valued source strengths Q_p and Q_s would cause a 90° phase shift in the pressure response. The resonances of the undamped system are given by $\zeta_{nR} = n/2$, compare Table 3.1, whereas the anti-resonances are determined by $\zeta_{nA} = n/4$.

The numerator depends on both the normalized position $\tilde{x}$ and the normalized frequency ζ. The denominator is determined by the normalized frequency only. It can bee seen that every damping ratio $\xi \neq 0$ causes an additional imaginary part A_2 in the numerator, and an additional real part B_2 in the denominator. The real parts of both numerator and denominator are linked with a cos-function.

The imaginary parts are linked with the corresponding sin-equivalent. Therefore, the magnitudes of numerator and denominator are nonzero for every damping ratio $\xi \neq 0$.

To calculate the acoustic potential energy of the enclosure, the acoustic potential energy density, compare (3.27), has to be integrated over the volume such as

$$E_{pot}^{V} = \frac{1}{2\rho_\infty c^2}\frac{L}{S}\int_{\tilde{x}=0}^{\tilde{x}=1} \left|P(\tilde{x}, j\omega)\right|^2 d\tilde{x}. \tag{6.36}$$

This leads to

$$E_{pot}^{V} = \frac{\rho_\infty L}{S}\frac{(a_{01}+a_{11}\xi^2)(|Q_p|^2+|Q_s|^2)+(a_{02}+a_{22}\xi^2)\,\mathrm{Re}\{Q_p^* Q_s\}}{(a_{03}+a_{33}\xi^2)} \tag{6.37}$$

with

$$\begin{aligned}
a_{01} &= 24\pi\zeta + 6\sin[4\pi\zeta]\\
a_{11} &= 32\pi^3\zeta^3 - 12\pi\zeta\cos[4\pi\zeta] + \left(3 - 24\pi^2\zeta^2\right)\sin[4\pi\zeta]\\
a_{02} &= 48\pi\zeta\cos[2\pi\zeta] + 24\sin[2\pi\zeta]\\
a_{22} &= -8\pi\zeta\left(3 + 4\pi^2\zeta^2\right)\cos[2\pi\zeta] + 12\sin[2\pi\zeta]\\
a_{03} &= 96\pi\zeta\sin^2[2\pi\zeta]\\
a_{33} &= 384\pi^3\zeta^3\cos^2[2\pi\zeta].
\end{aligned} \tag{6.38}$$

Equation (6.37) shows that the acoustic potential energy depends on the magnitudes of the primary and secondary source strengths as well as on the phase angle between these two quantities. Furthermore, both numerator and denominator in (6.37) depend on ξ^2. In the following (6.37) will be used to calculate the acoustic potential energy for both the uncontrolled and the controlled system.

6.3.2 Analysis of Specific Control Strategies

Passive Noise Treatments

The noise reduction (NR) that is obtained by an increase of passive damping is evaluated as reference for the control profit that can be obtained by ANC. Considering only a primary excitation ($Q_p \neq 0$ and $Q_s = 0$) in the undamped case $\xi = 0$, (6.37) reduces to

$$E_{pot}^{V,ini}\Big|_{\xi\to 0} = \frac{\rho L}{S}\frac{a_{01}}{a_{03}}|Q_p|^2. \tag{6.39}$$

If a certain amount of damping $\xi \neq 0$ is taken into account, (6.37) takes the following form

$$E_{pot}^{V,ini} = \frac{\rho L}{S}\frac{(a_{01}+a_{11}\xi^2)}{(a_{03}+a_{33}\xi^2)}|Q_p|^2. \tag{6.40}$$

The ratio between these two quantities

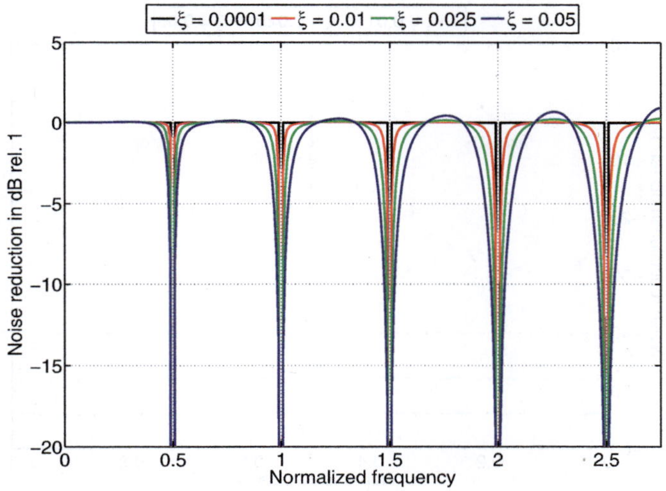

Fig. 6.4 Noise reduction obtained by passive noise control for a damping ratio of $\xi = 0.0001$ (*black*), $\xi = 0.01$ (*red*), $\xi = 0.025$ (*green*), and $\xi = 0.05$ (*blue*)

$$f_{pass} = \frac{1 + \frac{a_{11}}{a_{01}}\xi^2}{1 + \frac{a_{33}}{a_{03}}\xi^2} \tag{6.41}$$

can be used to quantify the NR obtained by passive noise treatments

$$NR_{pass} = 10 \log_{10}[f_{pass}]. \tag{6.42}$$

The global effect of passive damping is shown in Fig. 6.4 in which NR_{pass} is plotted against the normalized frequency ζ. It can be seen that a small damping ratio $\xi = 0.0001$ results in global NR in narrow bands around the acoustic resonances $\zeta_{nR} = 0.5, 1.0, 1.5, 2.0, 2.5$. An increase of passive damping enlarges these frequency bands. For higher damping ratios the potential acoustic energy of the enclosure can be amplified at the anti-resonances $\zeta_{nA} = 0.25, 0.75, 1.25, 1.75, 2.25$, as also shown in Fig. 6.4.

The local noise reduction obtained by passive damping applied to the first acoustic mode ($\zeta_{1R} = 0.5$) is shown in Fig. 6.5. The NR is calculated using the results obtained for a damping ratio of $\xi = 0.001$ as reference. The curves shown in Fig. 6.5 prove that the NR obtained by an increase of viscous damping is for this frequency outside the nodal point at $\tilde{x} = 0.5$ constant with respect to ζ. An increase of ξ from $\xi_1 = 0.01$ to $\xi_2 = 0.025$ results in an additional NR of approximately -8 dB. If the damping ratio is further increased form $\xi_2 = 0.025$ to $\xi_3 = 0.05$, the NR is increased by -6 dB, as shown in Fig. 6.5.

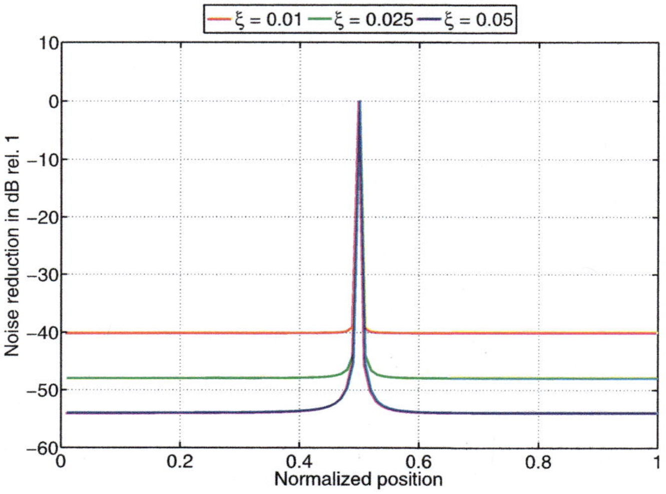

Fig. 6.5 Local noise reduction shown for the first acoustic mode ($\zeta = 0.5$) and damping ratios of $\xi = 0.01$ (*red*), $\xi = 0.025$ (*green*), and $\xi = 0.05$ (*blue*). Noise reduction is calculated using the normalized sound pressure with $\xi = 0.001$ as reference

Global Control of Acoustic Potential Energy

If it is possible to obtain a measure for the potential acoustic energy of the whole enclosure, this quantity can be used to calculate optimal secondary source strengths that effectively suppress acoustic resonances, but also prevent amplification at anti-resonances. This control strategy is known as *global control of acoustic potential energy*. The cost function of this control strategy, given by (6.37), reads

$$J_{Epot\_global} = E_{pot}^{V}(Q_p, Q_s). \tag{6.43}$$

Minimizing of (6.43) with respect to Q_s leads to the following relation between the primary and the optimal source strength

$$Q_{Epot\_global}^{opt} = \frac{n_{00} + n_{22}\xi^2}{r_{00} + r_{22}\xi^2} Q_p =: H_{Epot\_global}(\zeta) Q_p \tag{6.44}$$

with

$$\begin{aligned} n_{00} &= -12\big(2\pi\zeta \cos[2\pi\zeta] + \sin[2\pi\zeta]\big) \\ n_{22} &= 2\big[2\pi\zeta\big(3 + 4\pi^2\zeta^2\big)\cos[2\pi\zeta] - 3\sin[2\pi\zeta]\big] \\ r_{00} &= 6\big(4\pi\zeta + \sin[4\pi\zeta]\big) \\ r_{22} &= 32\pi^3\zeta^3 - 12\pi\zeta\cos[4\pi\zeta] + 3\big(1 - 8\pi^2\zeta^2\big)\sin[4\pi\zeta]. \end{aligned} \tag{6.45}$$

Equation (6.44) clarifies that both numerator and denominator of $Q_{Epot\_global}^{opt}$ have a quadratic dependence on the damping ratio ξ. Hence, only a significant

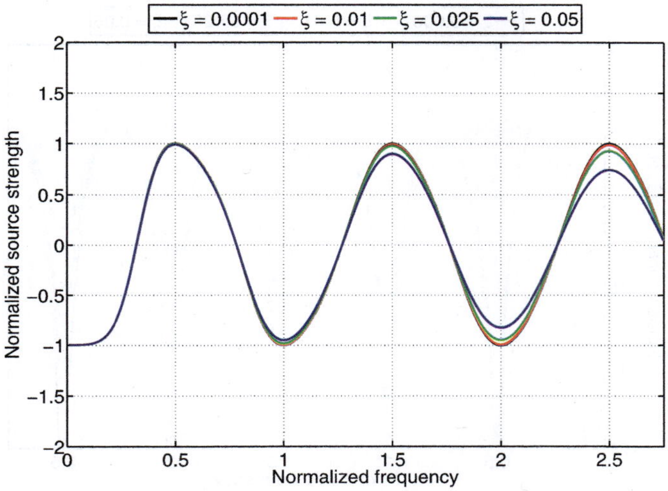

Fig. 6.6 Influence of viscous damping on optimal source strength, if active control of potential acoustic energy is applied in systems with different damping ratios

amount of passive damping will have an effect on the optimal solution. Furthermore, it can be seen that the frequency dependent controller transfer function $H_{Epot\_global}(\zeta)$ is a real valued quantity. Therefore, a real valued primary source strength will cause a real valued secondary source strength.

The effect of viscous damping on the optimal secondary source strength is shown in Fig. 6.6. It can be seen that an increase in viscous damping lowers the magnitude of $Q^{opt}_{Epot\_global}$ and therefore the costs of this control strategy. If the effect of viscous damping can be neglected, (6.44) reduces to

$$\begin{aligned} Q^{opt}_{Epot\_global} &= -\left(\cos[2\pi\zeta] + \frac{2\sin^3[2\pi\zeta]}{4\pi\zeta + \sin[4\pi\zeta]}\right) Q_p \\ &=: H_{Epot\_global}(\zeta, \xi = 0)\, Q_p, \end{aligned} \tag{6.46}$$

compare (Curtis et al. 1987). Equation (6.46) makes clear that this control concept is based on a 180° phase shift between primary excitation and canceling signal in combination with an amplitude-modulation of the excitation signal. The second summand in the bracket of (6.46) represents the difference to a control strategy that is based on the total acoustic energy of the enclosure (including the acoustic kinetic energy). The influence of this term is reduced with an increase in frequency.

The magnitude of $H_{Epot\_global}(\zeta, \xi = 0)$ is limited. The maxima are found at the acoustic resonances $\zeta_{nR} = 0.5, 1.0, 1.5, 2.0, 2.5$. At the anti-resonances $\zeta_{nA} = 0.25, 0.75, 1.25, 1.75, 2.25$, the secondary source strength is zero, because the system is already in a state of minimum acoustic potential energy. For the undamped system $\xi = 0$, amplification in magnitude is limited to one. Therefore, overamplification cannot be caused by this control strategy as also mentioned in Part I of this book.

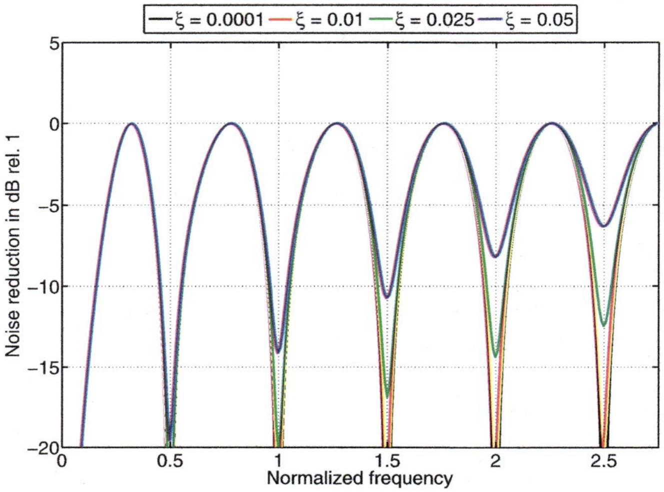

Fig. 6.7 Noise reduction by global control of acoustic potential energy for a damping ratio of $\xi = 0.0001$ (*black*), $\xi = 0.01$ (*red*), $\xi = 0.025$ (*green*), and $\xi = 0.05$ (*blue*)

Active control based on (6.46) results in a secondary source acting non-causally, because the inverse Fourier transform of $-\cos[2\pi\zeta]$ is given by

$$h(t) = -1/2\big[\delta(t + L/c) + \delta(t - L/c)\big]. \tag{6.47}$$

According to (Curtis et al. 1987) this impulse response may be interpreted as follows. The primary source emits a pulse of peak amplitude r. However, before this happens, the secondary source emits a pulse of amplitude $-r/2$. This secondary pulse is emitted at a time L/c before the primary pulse. Therefore, it arrives in front of the primary source, then the primary pulse is emitted. The resulting pulse of amplitude $r/2$ travels to the secondary source and is completely canceled by a second secondary pulse of amplitude $-r/2$. Therefore, an a-priory measure of the primary source strength, as in the case of periodic excitation, is required, if this control strategy shall be applied as an optimal controller.

The ratio between the acoustic potential energy of the uncontrolled system (6.40) and the acoustic potential energy in case of optimal control based on (6.44) is defined by

$$f_{Epot\_global} = \frac{E^{V}_{pot}(Q_p, Q^{opt}_{Epot\_global})}{E^{V,ini}_{pot}}. \tag{6.48}$$

The NR obtained by controlling the acoustic potential acoustic energy is

$$NR_{Epot\_global} = 10\log_{10}[f_{Epot\_global}]. \tag{6.49}$$

The global effect of active control applied to the potential acoustic energy is shown in Fig. 6.7 in which NR_{Epot_global} is plotted against the normalized frequency ζ. Effect of active control means that the noise reduction obtained by ANC is compared to the uncontrolled system with the same damping ratio.

It can be seen that an increase in viscous damping results in a decrease of NR that can be realized in the acoustic resonances $\zeta_{nR} = 0.5, 1.0, 1.5, 2.0, 2.5$. At the anti-resonances $\zeta_{nA} = 0.25, 0.75, 1.25, 1.75, 2.25$ the potential acoustic energy is not affected by this control approach.

Local Control of Acoustical Potential Energy

Instead of using a global measure of the potential acoustic energy for global ANC, it is possible to suppress acoustic resonances by *local control of acoustic potential energy*, in particularly by minimizing the sound pressure in front of the canceling source. The cost function for this control approach is defined as

$$J_{Epot\_local} = \left|\tilde{P}(\tilde{x} = 1, j\omega)\right|^2. \tag{6.50}$$

Minimizing of (6.50) with respect to Q_s leads to

$$Q^{opt}_{Epot\_local} = -\frac{1}{\cos[2\pi\zeta] + j2\pi\xi\zeta\,\sin[2\pi\zeta]} Q_p =: H_{Epot\_localL}(\zeta) Q_p. \tag{6.51}$$

Equation (6.51) indicates that the imaginary part of the denominator depends linearly on the damping ratio ξ. Hence, viscous damping will have a measurable effect on the optimal solution.

For $\xi \neq 0$ the frequency dependent controller transfer function $H_{Epot\_local}(\zeta)$ is a complex quantity. Therefore, a real valued primary source strength Q_p will in general cause a complex secondary source strength Q_s. If the effect of viscous damping can be neglected ($\xi = 0$), (6.51) reduces to

$$Q^{opt}_{Epot\_local} = -\frac{1}{\cos[2\pi\zeta]} Q_p =: H_{Epot\_local}(\zeta, \xi = 0) Q_p, \tag{6.52}$$

compare (Curtis et al. 1987). Equation (6.52) makes clear that this control concept is also based on a 180° phase shift between primary excitation and canceling signal. However, as shown in Fig. 6.15 and Fig. 6.16 the secondary source strength peaks at the anti-resonances $\zeta_{nA} = 0.25, 0.75, 1.25, 1.75, 2.25$ of the uncontrolled system. This indicates that the application of this control strategy results in a permutation of resonance frequencies and anti-resonances.

This effect is easy to understand by evaluating the NR. For this purpose we calculate the ratio between the acoustic potential energy of the uncontrolled system (see (6.40)) and the acoustic potential energy in the case of optimal local control

$$f_{Epot\_localL} = \frac{E^V_{pot}(Q_p, Q^{opt}_{Epot\_local})}{E^{V,ini}_{pot}}, \tag{6.53}$$

and the NR obtained by controlling the local potential acoustic energy density

$$NR_{Epot\_local} = 10\log_{10}[f_{Epot\_local}]. \tag{6.54}$$

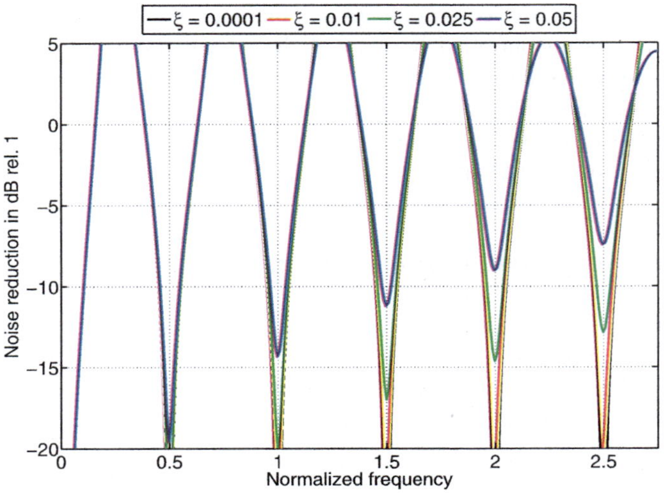

Fig. 6.8 Noise reduction by local control of potential energy density for a damping ratio of $\xi = 0.0001$ (*black*), $\xi = 0.01$ (*red*), $\xi = 0.025$ (*green*), and $\xi = 0.05$ (*blue*)

The associated results are shown in Fig. 6.8 in which the NR is plotted against the normalized frequency. These results prove that a significant NR is obtained in the acoustic resonances. However, the potential energy density of the system peaks at the former anti-resonances.

It should also be noticed that this control strategy results in a causally acting controller. As shown in (Curtis et al. 1987), the transfer function specified in (6.52) takes the form

$$\begin{aligned} H_{Epot\_local} &= -\frac{1}{\cosh[(j+\xi)2\pi\zeta]} \\ &= -\frac{1}{\cosh[(j\omega c^{-1}+\sigma)L]} \\ &=: -\frac{1}{\cosh[k'L]}, \end{aligned} \tag{6.55}$$

if a complex wave number, compare Sect. 6.3.1, is taken into account. This expression may be expanded as a *Taylor* series of complex exponential such as

$$H_{Epot\_local} \approx -2\left[e^{-k'L} - e^{-3k'L} + e^{-5k'L} - \cdots\right]. \tag{6.56}$$

The associated impulse response is an infinite sequence of delayed delta functions

$$\begin{aligned} h_{Epot\_local} \approx -2\Bigg[& e^{-\sigma L}\delta\left(t-\frac{L}{c}\right) - e^{-3\sigma L}\delta\left(t-3\frac{L}{c}\right) \\ & + e^{-5\sigma L}\delta\left(t-5\frac{L}{c}\right) - \cdots\Bigg], \end{aligned} \tag{6.57}$$

and can (ignoring the propagation loss) be interpreted as follows. A pulse of amplitude r, emitted by the primary source, arrives at the time L/c at the secondary source. To maintain zero pressure, the secondary source must emit a pulse of amplitude $-2r$. This results in a reflection of the primary pulse. The reflected pulse (with amplitude $-r$) arrives at time $2L/c$ in front of the primary source. There it is reflected. The reflected impulse with amplitude $-r$ is traveling towards the duct and will at time $3L/c$ arrive back at the secondary source. The secondary source must now emit an impulse of amplitude $2r$ to maintain zero pressure. This results in a pulse with amplitude r traveling back to the primary source. Without losses this process would continue ad infinitum.

Minimization of Total Power Input

Instead of using a measure of the global potential acoustic energy of the whole enclosure, the total power output of both primary and secondary source can be used to determine optimal secondary source strengths. Minimizing this quantity is because of (3.28) equivalent to a minimization of the acoustic energy and is known as *minimization of total power input*. The associated cost function reads

$$J_{TPI} = \mathrm{Re}\big\{P^*(\tilde{x}=0)Q_p + P^*(\tilde{x}=1)Q_s\big\}. \tag{6.58}$$

Minimizing of (6.58) with respect to Q_s leads to a relation between the primary and the optimal source strength that is independent of the viscous damping ratio

$$Q_{TPI}^{opt} = -\cos[2\pi\zeta]Q_p =: H_{TPI}(\zeta)Q_p. \tag{6.59}$$

In practice a certain amount of damping is necessary to enable the power flow measurement. Equation (6.59) clarifies that the frequency dependent controller transfer function $H_{TPI}(\zeta)$ is a real valued quantity. Therefore, a real valued primary source strength will cause a purely real valued secondary source strength. It peaks with magnitude 1 at the resonances and is zero at the anti-resonances, compare Fig. 6.15.[4]

Taking the inverse Fourier transform of $H_{TPI}(\zeta)$ leads to an impulse response that belongs to a secondary source acting non-causally

$$h_{PO}(t) = -\frac{1}{2}\left[\delta\left(t+\frac{L}{c}\right)+\delta\left(t-\frac{L}{c}\right)\right]. \tag{6.60}$$

The interpretation of (6.60) is similar to the interpretation of (6.47). The time-averaged acoustic energy of the system is identical with half the energy calculated for a wave traveling in positive direction, see (Curtis et al. 1987). It can be concluded

[4] The small shift between the curves for global control of acoustic potential energy and active control of total power input results from the fact that the kinetic acoustic energy (that is implicitly taken into account by controlling the total power input) has a significant contribution below the first acoustic mode, see (Curtis et al. 1987).

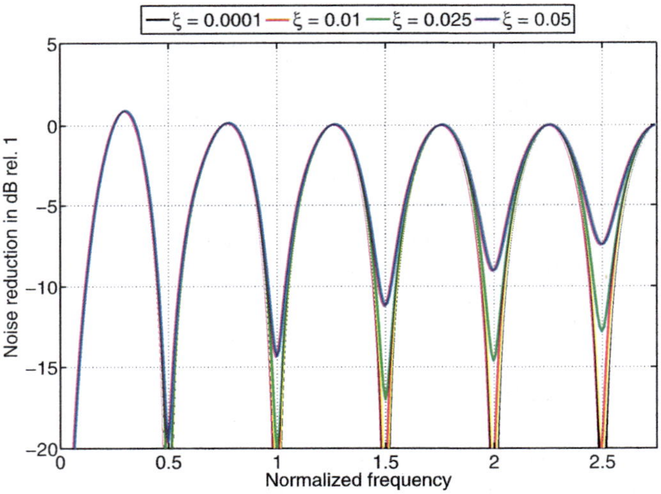

Fig. 6.9 Noise reduction by global control of acoustic energy density due to active control of total power input for a damping ratio of $\xi = 0.0001$ (*black*), $\xi = 0.01$ (*red*), $\xi = 0.025$ (*green*), and $\xi = 0.05$ (*blue*)

that an a-priory measure of the primary source strength, as in the case of periodic excitation, is also required for this control strategy.

The ratio between the acoustic potential energy of the uncontrolled system, given by (6.40), and the acoustic potential energy in the case of optimal control is

$$f_{TPI} = \frac{E_{pot}^{V}(Q_p, Q_{TPI}^{opt})}{E_{pot}^{V,ini}}. \tag{6.61}$$

The NR obtained by controlling total power input is therefore given by

$$NR_{TPI} = 10 \log_{10}[f_{TPI}]. \tag{6.62}$$

The effect of active control based on total power input is shown in Fig. 6.9 in which NR_{TPI} is plotted against the normalized frequency ζ. Effect of active control means that the noise reduction obtained by ANC is compared to the uncontrolled system having the same damping ratio.

It can be seen that an increase in viscous damping results in a decrease of NR that can be realized in the acoustic resonances $\zeta_{nR} = 0.5, 1.0, 1.5, 2.0, 2.5$. The acoustic potential energy is amplified above the first anti-resonance $\zeta_{1A} = 0.25$, because minimizing the total power input into the system is equivalent to a minimization of the (total) acoustic energy density (including the "kinetic" part) that differs at low frequencies from the acoustic potential energy.

At the other anti-resonances $\zeta_{nA} = 0.75, 1.25, 1.75, 2.25$ (the kinetic energy plays here not an important role) the potential acoustic energy is not effected by this control approach.

Local Control of Sound Pressure and Particle Velocity

It is also possible to suppress acoustic resonances by *local control of energy density*, in particularly by minimizing both sound pressure and particle velocity in front of the canceling source. The cost function for this control approach may be defined as

$$J_{E\_local} = \beta \frac{1}{\rho_\infty c^2} \left| P(\tilde{x} = 1, j\omega) \right|^2 + (1 - \beta) \frac{\rho_\infty}{2} \left| V(\tilde{x} = 1, j\omega) \right|^2, \quad (6.63)$$

where β is a real valued weighting parameter. Minimizing of (6.63) with respect to Q_s leads to

$$Q^{opt}_{E\_local} = H_{E\_local}(\zeta) Q_p = \left[H_{E\_local,R}(\zeta) - j H_{E\_local,I}(\zeta) \right] Q_p, \quad (6.64)$$

where

$$\begin{aligned} H_{PU,R}(\zeta) &= \frac{2\beta \cos[2\pi\zeta]}{H_{Den}} \\ H_{PU,I}(\zeta) &= \frac{2\pi\zeta\xi \sin[2\pi\zeta]}{H_{Den}} \\ H_{Den} &= \cos[4\pi\zeta]\left\{1 - 2\beta + \xi^2\left[1 - 4\pi^2\zeta^2 + \beta\left(8\pi^2\zeta^2\right)\right]\right\} \\ &\quad + \sin[4\pi\zeta]\left\{4\pi\zeta\xi^2[1 - \beta]\right\} - 1 + \xi^2\left[\beta - 1 - 4\pi^2\zeta^2\right]. \end{aligned} \quad (6.65)$$

Equation (6.64) indicates that a real valued primary source strength Q_p will in general cause a complex optimal source strength for the canceling source. The imaginary part of the controller transfer function $H_{E\_local}(\zeta)$ is present as long as $\xi \neq 0$. The same hold for H_{Den}.

For $\xi \neq 0$ and $\beta = 0.5$, (6.65) reduces to (6.59).[5] This shows that global suppression of acoustic resonances is possible with local action only in front of the secondary source, if the acoustic energy density in front of the canceling source is minimized. In contrast to the minimization of total power input no information about power input of the primary source is needed. For $\xi \neq 0$ and $\beta = 1$, (6.65) reduces to (6.51), because of missing information about the particle velocity.

The dependence of the control source strength on frequency is for $\xi = 0.025$ shown in Fig. 6.15 and Fig. 6.16. These figures clarify that the real part is maximized at the acoustic resonances and minimized at the anti-resonances. Its amplitude decreases with frequency. The imaginary part shows a converse behavior.

The ratio between the potential acoustic energy of the uncontrolled system (see (6.40)) and the potential acoustic energy in the case of optimal local control is given by

$$f_{E\_local} = \frac{E^V_{pot}(Q_p, Q^{opt}_{E\_local})}{E^{V,ini}_{pot}}. \quad (6.66)$$

The NR obtained by controlling the local acoustic energy density is therefore given by

[5] This also implies that the secondary source is acting non-causally.

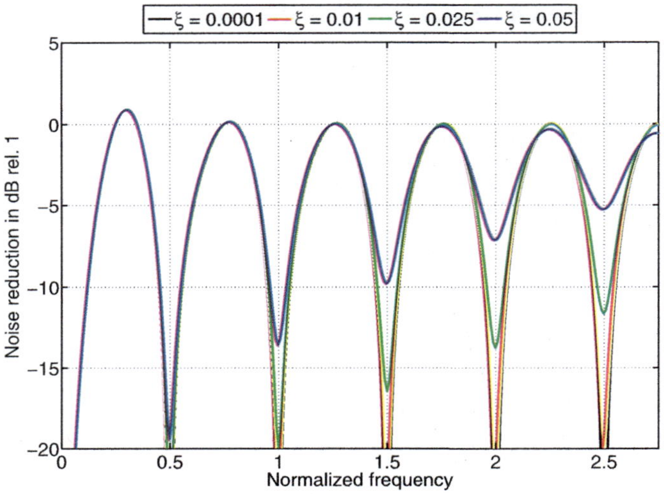

Fig. 6.10 Noise reduction by local control of sound pressure and particle velocity for a damping ratio of $\xi = 0$ (*black*), $\xi = 0.01$ (*red*), $\xi = 0.025$ (*green*), and $\xi = 0.05$ (*blue*)

$$NR_{E\_local} = 10\log_{10}[f_{E\_local}]. \tag{6.67}$$

The global effect of controlling the energy density in front of the canceling source is shown in Fig. 6.10 in which NR_{E_local} is plotted against the normalized frequency ζ. An increase in viscous damping results in a decrease of NR that can be realized in the acoustic resonances $\zeta_{nR} = 0.5, 1.0, 1.5, 2.0, 2.5$. At the anti-resonances of the uncontrolled system $\zeta_{nA} = 0.25, 0.75, 1.25, 1.75, 2.25$ are now frequencies at which the acoustic potential energy of the controlled system is slightly amplified, especially above $\zeta_{1A} = 0.25$. However, local control of both sound pressure and particle velocity (with $\beta = 0.5$) avoids significant amplification at the anti-resonances and leads to a global NR at all other frequencies.

Maximization of Power Absorption

If active control is applied to minimize the power output of the secondary source only, the power absorption of the canceling source is maximized. By applying this control strategy the power absorption of the secondary source is used to determine optimal secondary source strengths. Therefore, this control strategy is known as *maximization of power absorption*. The associated cost function is given by

$$J_{PowerAbs} = \mathrm{Re}\left\{P^*(\tilde{x} = 1)Q_s\right\}. \tag{6.68}$$

Maximizing of (6.68) with respect to Q_s leads to an optimal source strength that is consisting of a complex controller transfer function such as

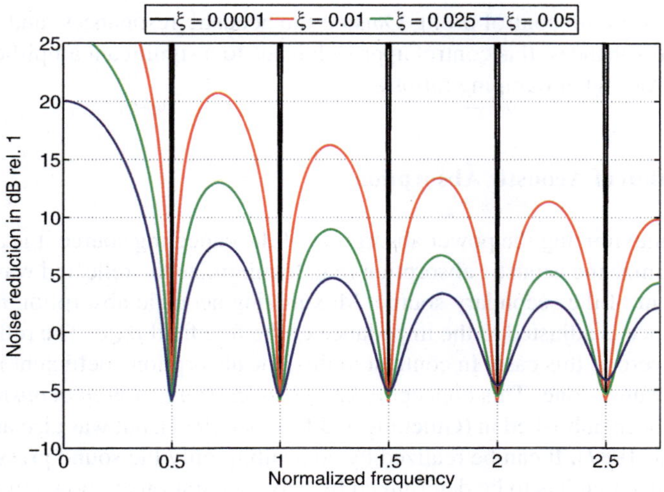

Fig. 6.11 Effect of maximizing power absorption of the secondary source on the acoustic potential energy for a damping ratio of $\xi = 0.0001$ (*black*), $\xi = 0.01$ (*red*), $\xi = 0.025$ (*green*), and $\xi = 0.05$ (*blue*)

$$\begin{aligned} Q^{opt}_{PowerAbs} &= \left(-\frac{1}{2}\cos[2\pi\zeta] + j\frac{\sin[2\pi\zeta]}{4\pi\zeta}\xi^{-1}\right)Q_p \\ &=: \left[H_{PowerAbs,R}(\zeta) + jH_{PowerAbs,I}(\zeta)\right]Q_p. \end{aligned} \tag{6.69}$$

Equation (6.69) clarifies that the imaginary part of the frequency dependent controller transfer function $H_{PowerAbs,I}(\zeta)$ is inversely proportional to the damping ratio ξ. Therefore, a certain amount of passive damping must be realized in the system, if the magnitude of the optimal source strength $Q^{opt}_{PowerAbs}$ shall be limited. The influence of the imaginary part becomes less important with increasing ζ (e.g. at higher frequencies). Please notice that $H_{PowerAbs,R}(\zeta) = 1/2 \cdot H_{TPI}(\zeta)$.

According to (6.69), a real valued primary source strength will (in general) cause a complex secondary source strength. A special situation occurs at the acoustic resonances, see Fig. 6.15 and Fig. 6.16, because $H_{PowerAbs,I}(\zeta)$ vanishes at these frequencies. Consequently, the optimal source strength of the canceling source determined by this control concept is capable of suppressing acoustic resonances. As shown in Fig. 6.16, $H_{PowerAbs,I}(\zeta)$ peaks at the anti-resonances, and is at these frequencies a source for amplification of the global acoustic potential energy.

The ratio between the potential acoustic energy of the uncontrolled system (see (6.40)) and the potential acoustic energy in the case of optimal control is given by

$$f_{PowerAbs} = \frac{E^V_{pot}(Q_p, Q^{opt}_{PowerAbs})}{E^{V,ini}_{pot}}. \tag{6.70}$$

As discussed above, a NR is only achieved in narrow bands around the acoustic resonances $\zeta_{nR} = 0.5, 1.0, 1.5, 2.0, 2.5$. (See Fig. 6.11.) An increase in viscous damping lowers the control profit. Outside the acoustic resonances, and especially at the anti-resonances, this control approach leads to a significant amplification that is lowered for higher damping ratios ξ.

Maximization of Acoustic Absorption

Instead of maximizing the power absorption of the canceling source it is also possible to maximize the acoustic absorption (e.g. to minimize the reflected wave component) in front of the secondary source. Maximizing acoustic absorption means that the impedance is adjusted to the impedance of the free field $\rho_\infty c$. The reflection coefficient is zero in this case. In contrast to this, the absorption coefficient is set to its maximum value—one. This concept is known as *maximization of acoustic absorption*. It has been published in (Guicking and Karcher 1984), but was also analyzed in (Curtis et al. 1987). It can be realized by controlling only the sound pressure of the reflected wave that has to be determined by wave separation as shown in (Guicking and Karcher 1984), (Ordufia-Bustamant and Nelson 1992) and (Kletschkowski and Sachau 2008). The associated cost function based on (6.30) reads

$$J_{AcousticAbs} = \left|P_r(\tilde{x} = 1, j\omega)\right|^2, \tag{6.71}$$

where P_r is the reflected wave component. Minimizing of (6.71) with respect to Q_s leads to an optimal source strength that consists of a complex controller transfer function such as

$$\begin{aligned} Q^{opt}_{AcousticAbs} &= (1 - 2\pi\zeta\xi)\big(-\cos[2\pi\zeta] + j\sin[2\pi\zeta]\big)Q_p \\ &=: H_{AcousticAbs}(\zeta)Q_p. \end{aligned} \tag{6.72}$$

Equation (6.72) clarifies that the real and imaginary parts of the frequency dependent controller transfer function $H_{AcousticAbs}(\zeta)$ are determined by orthogonal functions. A real valued primary source strength will—in general—cause a complex control signal. Furthermore, it can bee seen that a certain amount of viscous damping reduces the magnitude of the secondary source strength. This reduction depends linearly on frequency as shown in Fig. 6.15 and Fig. 6.16.

The real part of $Q^{opt}_{AcousticAbs}$ peaks at the resonances $\zeta_{nR} = 0.5, 1.0, 1.5, 2.0, 2.5$. The imaginary part peaks at the anti-resonances $\zeta_{nA} = 0.25, 0.75, 1.25, 1.75, 2.25$. Therefore, a real valued control signal is generated, if an acoustic mode is excited by the primary source driven by a real valued source strength. If the excitation frequency corresponds with an anti-resonance frequency of the uncontrolled system, the control signal will be an imaginary quantity. Without passive damping, the canceling source strength is reduced to

$$Q^{opt}_{AcousticAbs} = \big(-\cos[2\pi\zeta] + j\sin[2\pi\zeta]\big)Q_p =: H_{AcousticAbs}(\zeta, \xi = 0)Q_p. \tag{6.73}$$

The real part of the complex controller transfer function is in this case identical with the optimal source strength Q^{opt}_{TPI} that has been determined for active control

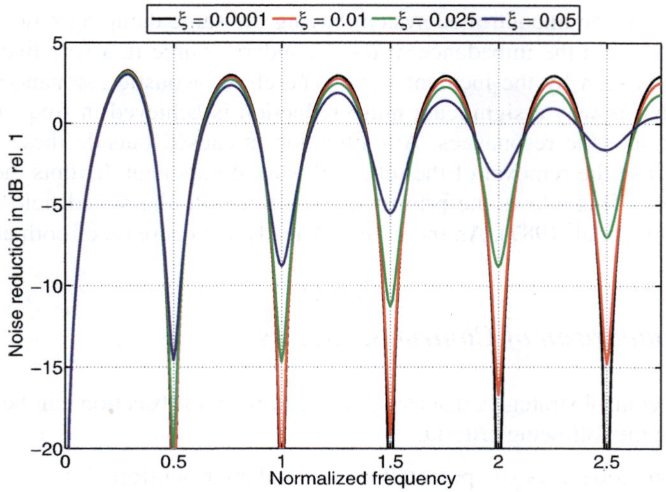

Fig. 6.12 Effect of maximizing acoustic absorption on the acoustic potential energy for a damping ratio of $\xi = 0.0001$, $\xi = 0.01$ (*red*), $\xi = 0.025$ (*green*), and $\xi = 0.05$ (*blue*)

of total power input. The differences between these two control strategies can only be evaluated in time domain (Curtis et al. 1987).

Taking the inverse Fourier transform, the impulse response corresponding to (6.73) is given by

$$h_{AcousticAbs}(t) = -\delta\left(t - \frac{L}{c}\right). \tag{6.74}$$

Equation (6.74) clarifies that the transfer function $H_{AcousticAbs}(\zeta, \xi = 0)$ describes a causally acting controller that works as follows: At $t = L/c$ a canceling impulse with amplitude r is sent out by the secondary source. The disturbance impulse with amplitude r is therefore completely canceled in front of the secondary source after passing the duct. In this case, the time-averaged acoustic energy of the system is identical with the total acoustic energy calculated for a wave traveling in positive direction, see (Curtis et al. 1987). An a-priory measure of the primary source strength is not required for this control strategy.

The ratio between the potential acoustic energy of the uncontrolled system (see (6.40)) and the potential acoustic energy in the case of optimal control reads

$$f_{AcousticAbs} = \frac{E^{V}_{pot}(Q_p, Q^{opt}_{AcousticAbs})}{E^{V,ini}_{pot}}. \tag{6.75}$$

The associated logarithmic measure is given by

$$NR_{AcousticAbs} = 10\log_{10}[f_{AcousticAbs}]. \tag{6.76}$$

The effect of active control based on maximization of acoustic absorption is shown in Fig. 6.12 in which $NR_{AcousticAbs}$ is plotted against the normalized frequency ζ. Effect of active control means that the noise reduction obtained by ANC is compared to the uncontrolled system having the same damping ratio. This control strategy alters the impedance of the secondary source in a way that free field impedance is seen by the incident wave. Therefore, acoustic resonances are suppressed. That is why a significant noise reduction is achieved in frequency bands around the acoustic resonances. Amplification is caused outside these frequency bands, because the removal of the reflected wave component disrupts the standing wave pattern. This allows the primary source to excite the sound field more efficiently (Curtis et al. 1987). An increase of ξ leads to a decrease of both effects.

6.3.3 Comparison of Control Strategies

The active control strategies discussed in the previous subsection can be evaluated considering the following criteria:

1. Effect on global acoustic potential energy and control effort,
   - At all frequencies,
   - In acoustic resonances,
   - Outside acoustic resonances,
2. Effect on local sound pressure,
   - At all frequencies,
   - In acoustic resonances,
   - Outside acoustic resonances,
3. Role of passive damping,
4. Causality of resulting control law,
5. Complexity of measurement,
6. Applicability to 3D-Problems.

(**1**) The effects of the analyzed control strategies on the global acoustic potential energy are compared in Fig. 6.13 in which the reduction of global acoustic potential energy is plotted against the integral normalized control effort. The latter has been calculated as follows[6]

$$L_{Q_s} = 10\log_{10}\left(\int_{\zeta=0}^{\zeta=2.75} \frac{Q_s^2}{Q_p^2} d\zeta\right) \quad \text{with } Q_p = 1\frac{\text{m}^3}{\text{s}}. \tag{6.77}$$

The integral noise reduction in the analyzed frequency band $0 \leq \zeta \leq 2.75$ has been calculated such as

$$NR = 10\log_{10}\left(\int_{\zeta=0}^{\zeta=2.75} f_{E_{pot}} d\zeta\right), \tag{6.78}$$

where $f_{E_{pot}}$ is the ratio between the global acoustic potential energy of the optimal controlled system and the global acoustic potential energy of the uncontrolled system as defined in the previous subsection.

---

[6]Applying (6.77) to Q_p results in $L_{Q_p} = 150.4$ dB.

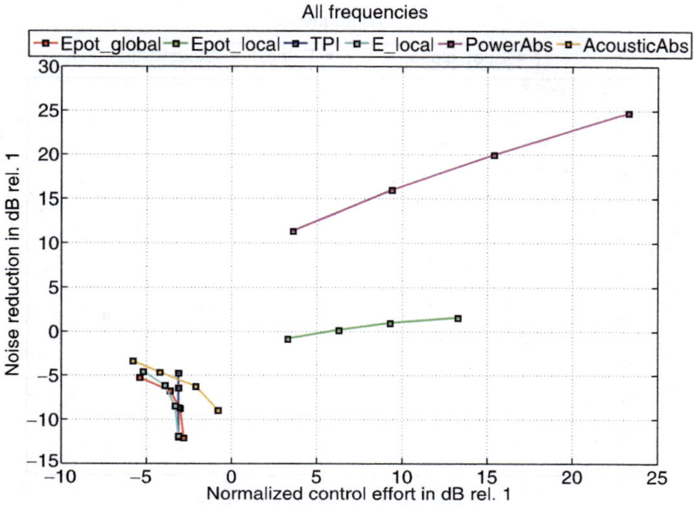

Fig. 6.13 Comparison of integral noise reduction and integral control effort realized by different control strategies (*red*: global minimization of acoustic potential energy, *green*: local minimization of potential energy, *blue*: minimization of total power input, *magenta*: maximization of power absorption, *yellow*: maximization of acoustic absorption, *cyan*: local minimization of energy density) between $\zeta = 0$ and $\zeta = 2.75$ for different values of viscous damping that has been increased from $\xi = 0.01$ (*blue curve*: below, all other curves right) over $\xi = 0.025$ and $\xi = 0.05$ to $\xi = 0.1$ (*blue curve*: top, all other curves left)

Every square in Fig. 6.13 represents a solution that correlates to a certain amount of viscous damping that has been increased from $\xi = 0.01$ (blue curve: below, all other curves right) to $\xi = 0.1$ (blue curve: top, all other curves left). These curves clarify that an increase of viscous damping results in both a decrease of the control effort, but also in a decrease of global noise reduction.

Furthermore, compare Fig. 6.13, it can be seen that only four of the analyzed control strategies—minimization of global acoustic potential energy (red curve), minimization of total power input (blue curve), local minimization of energy density (cyan curve) and maximization of acoustic absorption (yellow curve)—are capable of realizing a significant global noise reduction, if all frequencies in the analyzed frequency band are taken into account.

The remaining control strategies cannot be applied to all frequencies. Local minimization of the acoustic potential energy in front of the canceling source (green curve) has only a minor effect on the global acoustic potential energy between $\zeta = 0$ and $\zeta = 2.75$. Here, due to the shift of acoustic resonances caused by transformation of the boundary condition at $\tilde{x} = 1$ from sound hard to sound soft, the control profit at the acoustic resonances is compensated by the increase in potential energy at the former anti-resonances. Maximization of power absorption (magenta curve) at $\tilde{x} = 1$ maximizes the acoustic energy outside acoustic resonances.

Control strategies that lead to a significant integral global noise reduction are compared in Fig. 6.14. It can be seen that for every amount of viscous damping

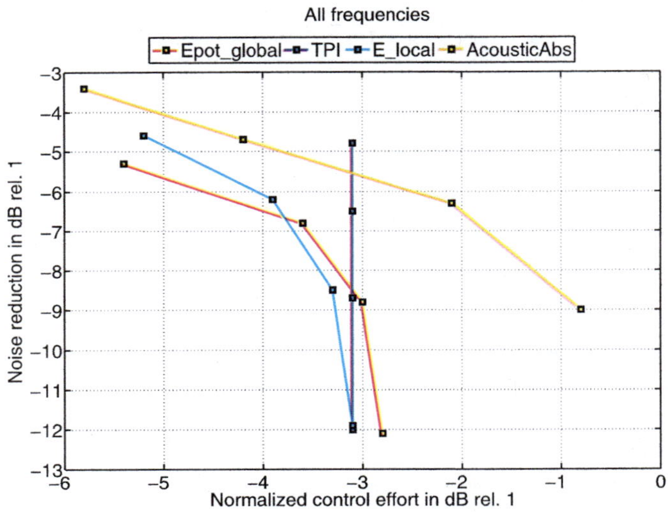

Fig. 6.14 Comparison of integral noise reduction and integral control effort realized by control strategies that enable global noise reduction

ξ minimization of acoustic potential energy leads to the highest integral noise reduction. For weakly damped enclosures ($\xi = 0.01$), the integral control effort is however lower, if the total power input or the local energy density at $\tilde{x} = 1$ is minimized. As shown in Fig. 6.14, the optimal source strength is independent on ξ for active control of total power input. At higher damping ratios $\xi = 0.05$ and $\xi = 0.1$, maximization of acoustic absorption results in the smallest integral control effort.

Optimal source strengths, calculated for a damping ratio of $\xi = 0.025$, are compared in Fig. 6.15 and Fig. 6.16. It can be seen that only two control strategies (minimization of global acoustic potential energy and minimization of total power input) cause for all frequencies a real valued optimal secondary source strength, whereas the optimal secondary source strength of all other control strategies consists (in general) of a real part and an imaginary part.

If active control is based on minimization of (a) global acoustic potential energy, (b) total power input, (c) local acoustic energy or on maximization of acoustic absorption, the magnitude of the optimal source strength is (in the analyzed frequency band) less or equal to the magnitude of the primary source strength.

This means that the control effort of these control strategies is limited. Figure 6.16 clarifies that both active minimization of local acoustic potential energy density and active maximization of absorption power must not be applied outside acoustic resonances. The imaginary part of the optimal source strengths caused by these control strategies peaks at the anti-resonances. This causes (especially for small damping ratios) amplification as shown in Fig. 6.13.

The imaginary part of the optimal secondary source strength is for all control strategies zero in the acoustic resonances, as also shown by Fig. 6.16. The magnitude of the remaining real part is except for the active maximization of power absorption

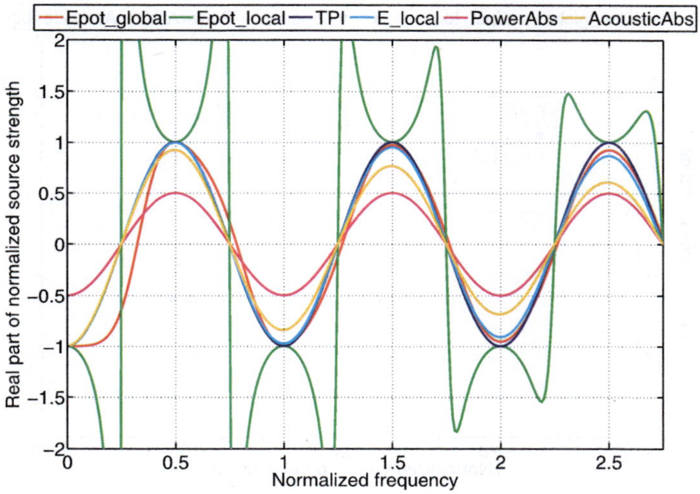

Fig. 6.15 Real part of optimal source strengths calculated for six different control strategies plotted against the normalized frequency for a damping ratio of $\xi = 0.025$

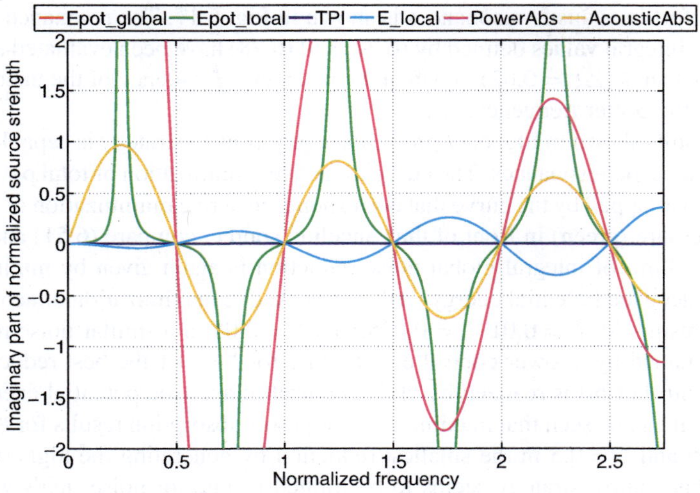

Fig. 6.16 Imaginary part of optimal source strengths calculates for six different control strategies plotted against the normalized frequency for a damping ratio of $\xi = 0.025$. *Red curve* is overlapped by *blue curve*

for all control strategies close to 1 in the first resonance $\zeta_{1R} = 0.5$. Thus, a significant noise reduction can be obtained by limited control effort, if active control is applied to an acoustic field associated to resonant frequencies of weakly damped interiors.

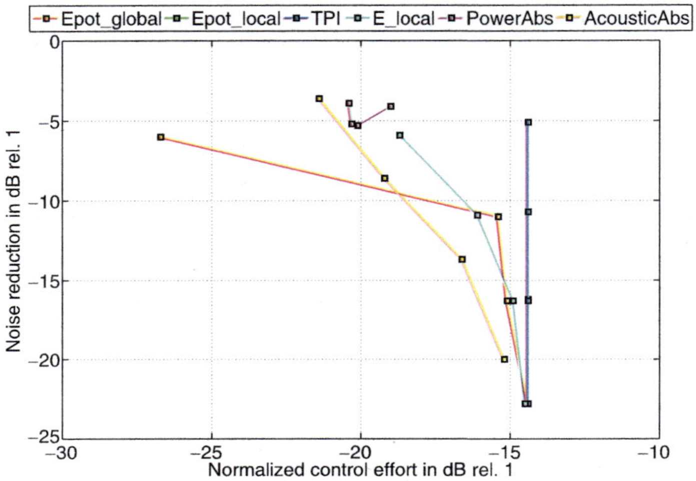

Fig. 6.17 Comparison of different control strategies for small frequency bands around acoustic resonances

The effect of different control strategies on integral noise reduction and integral control effort in acoustic resonances is shown in Fig. 6.17. The resonance zones in which the integral values defined by (6.77) and (6.78) have been evaluated are small frequency bands ($\Delta\zeta = 0.02$) in which the resonance frequency of the uncontrolled system is the center frequency.

The results shown in Fig. 6.17 prove that every control strategy is capable of suppressing acoustic resonances. The curve for active minimization of total power input (blue) is overlapped by the curve that corresponds to active minimization of the local sound pressure (green) in front of the canceling source, compare (6.51) and (6.59). The upper limit of integral global noise reduction is again given by minimization of global acoustic potential energy (red curve). Active maximization of acoustic absorption results for $\xi = 0.01$, $\xi = 0.025$, and $\xi = 0.05$ in a similar noise reduction that is obtained by a lower control effort. Only for $\xi = 0.1$ the best reduction and lowest control effort is realized by global control of acoustic potential energy. Furthermore, it can be seen that maximization of power absorption results for $\xi = 0.01$, $\xi = 0.025$ and $\xi = 0.5$ in the smallest reduction by generating the highest control effort. This control strategy seems to be limited to interior noise fields with high damping ratios. The four control strategies that reduce the integral value of global noise reduction outside acoustic resonances are compared in Fig. 6.18. The integral quantities defined by (6.77) and (6.78) have been evaluated in large frequency bands $\Delta\zeta = 0.98$ in which the anti-resonance frequency of the uncontrolled system is the center frequency. It can be seen that the highest integral attenuation at the lowest control effort is (again) obtained by controlling the global acoustic potential energy.

(2) Local effects of different control strategies are shown in Fig. 6.19–6.24. In Fig. 6.19, 6.21, and 6.23, the magnitude of the normalized sound pressure $\tilde{P}$, compare (6.34), is plotted versus the dimensionless duct length $\tilde{x}$. The local noise re-

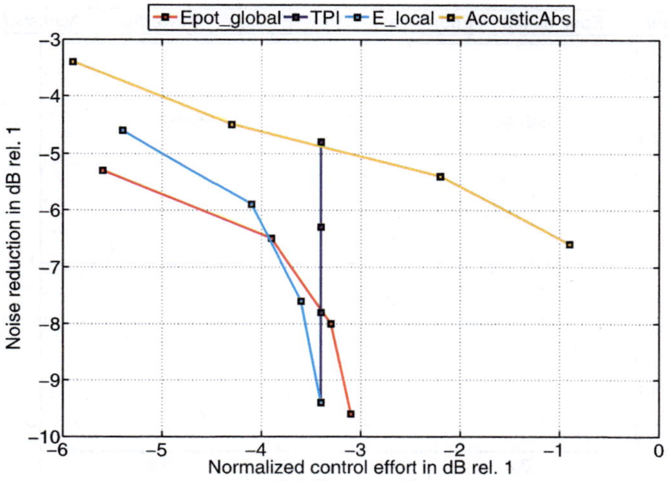

Fig. 6.18 Comparison of different control strategies for frequency bands that excludes resonance zones

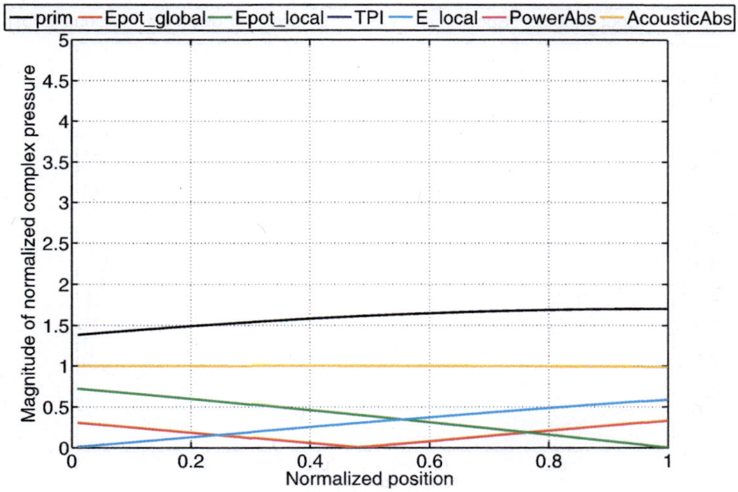

Fig. 6.19 Comparison of local sound pressure due to different control strategies (*black*: uncontrolled system, *red*: global minimization of acoustic potential energy, *green*: local minimization of potential energy, *blue*: minimization of total power input, *magenta*: maximization of power absorption, *yellow*: maximization of acoustic absorption, *cyan*: local minimization of energy density) for $\zeta = 0.1$ and a viscous damping ratio of $\xi = 0.01$. *Blue curve* is overlapped by *cyan curve*

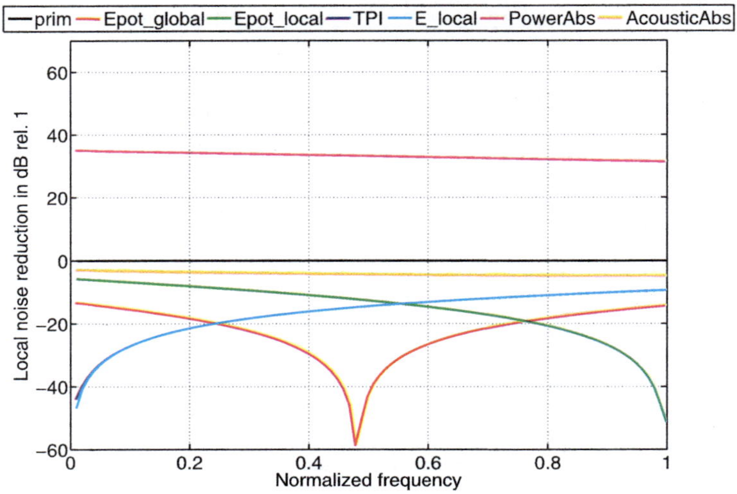

Fig. 6.20 Comparison of local noise reduction due to different control strategies for $\zeta = 0.1$ and a viscous damping ratio of $\xi = 0.01$. *Blue curve* is overlapped by *cyan curve*

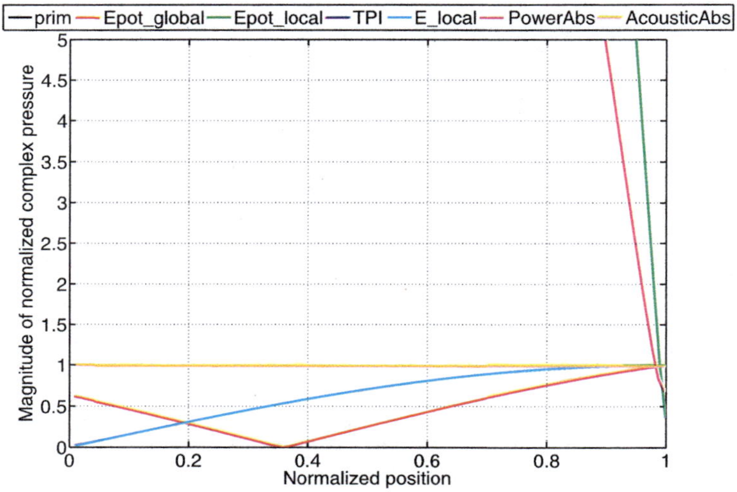

Fig. 6.21 Comparison of local sound pressure due to different control strategies (*black*: uncontrolled system, *red*: global minimization of acoustic potential energy, *green*: local minimization of potential energy, *blue*: minimization of total power input, *magenta*: maximization of power absorption, *yellow*: maximization of acoustic absorption, *cyan*: local minimization of energy density) for $\zeta = 0.25$ and a viscous damping ratio of $\xi = 0.01$. *Black curve* is overlapped by *yellow curve*

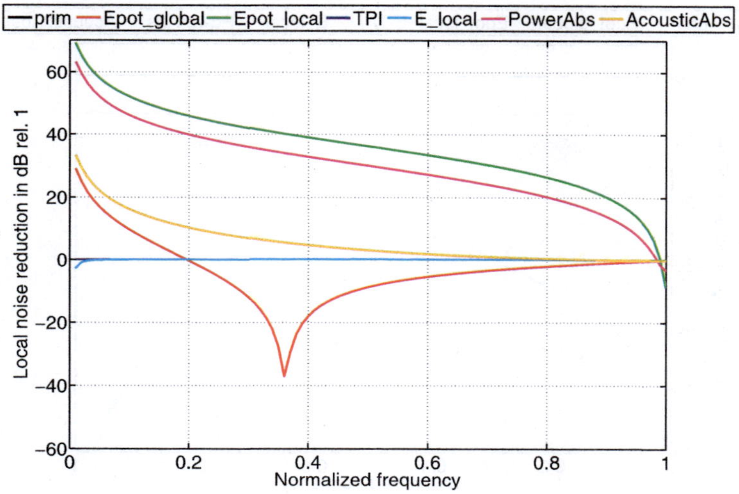

Fig. 6.22 Comparison of local noise reduction due to different control strategies for $\zeta = 0.25$ and a viscous damping ratio of $\xi = 0.01$. *Black curve* is overlapped by *blue* and *cyan curve*

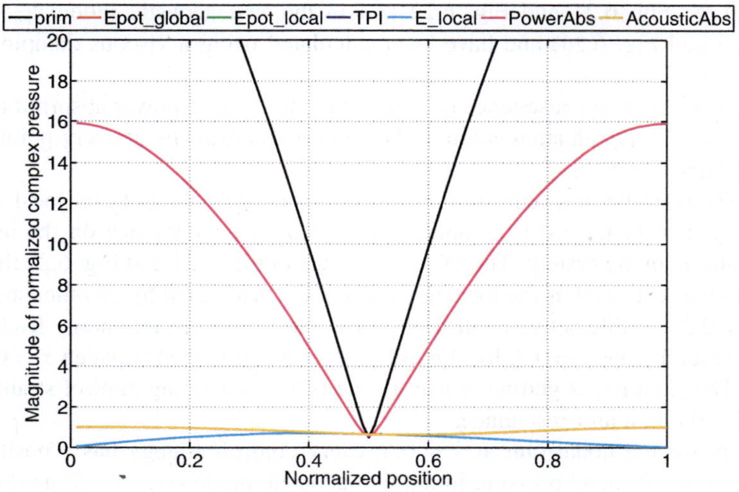

Fig. 6.23 Comparison of local sound pressure due to different control strategies (*black*: uncontrolled system, *red*: global minimization of acoustic potential energy, *green*: local minimization of potential energy, *blue*: minimization of total power input, *magenta*: maximization of power absorption, *yellow*: maximization of acoustic absorption, *cyan*: local minimization of energy density) for $\zeta = 0.5$ and a viscous damping ratio of $\xi = 0.01$. *Blue curve* is overlapped by *cyan curve*

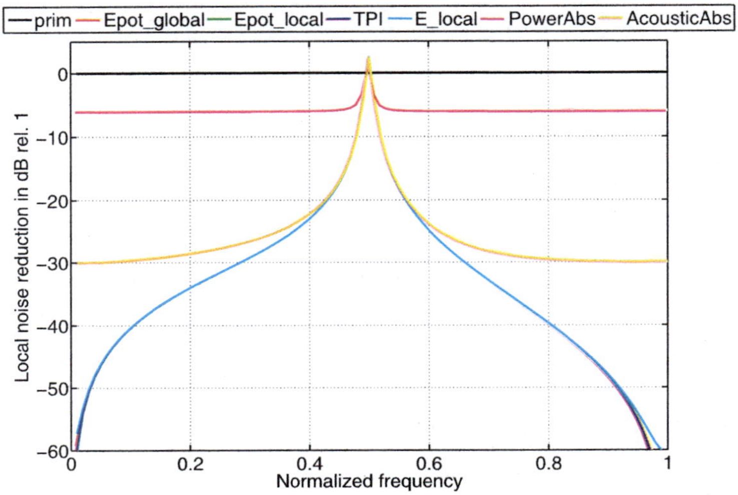

Fig. 6.24 Comparison of local noise reduction due to different control strategies for $\zeta = 0.5$ and a viscous damping ratio of $\xi = 0.01$. *Blue curve* is overlapped by *cyan curve*

duction is shown in Fig. 6.20, 6.22, and 6.24. Curves corresponding to the uncontrolled system are plotted in black. The color index for the different control strategies has been retained. The results are shown for three different frequencies (below the first anti-resonance at $\zeta = 0.1$: Fig. 6.19 and Fig. 6.20, at the first anti-resonance $\zeta_{1A} = 0.25$: Fig. 6.21 and Fig. 6.22 and in the first acoustic mode $\zeta_{1R} = 0.5$: Fig. 6.23 and Fig. 6.24) and have been calculated using a viscous damping ratio of $\xi = 0.01$.

Below the first anti-resonance only the maximization of power absorption is not capable of realizing an attenuation of the local sound pressure at every point $\tilde{x}$, see Fig. 6.19 and Fig. 6.20.

As discussed before, only minimization of total power input and local control of energy density have (in the anti-resonance case) no influence on the acoustic energy stored in the system. Therefore, as shown in Fig. 6.21 and Fig. 6.22 the local sound pressure as well as the local noise reduction is affected by all other strategies at $\zeta_{1A} = 0.25$. While active control of global acoustic potential energy leads to an amplification below $\tilde{x} = 0.2$, local noise reduction is obtained between $\tilde{x} = 0.2$ and $\tilde{x} = 1$. The situation is getting worse, if one of the remaining control strategies is applied at the first anti-resonance.

Except for the nodal pint at $\tilde{x} = 0.5$, every control strategy has a positive effect on the local sound pressure in the first acoustic mode ($\zeta_{1R} = 0.5$) as shown in Fig. 6.23 and Fig. 6.24. The local control profit is maximized, if global or local control of acoustic potential energy is applied, if local control of energy density (with $\beta = 0.5$) is used, or if the total power input into the system is minimized. However, maximization of acoustic absorption also leads to a significant local noise reduction at most positions.

(**3**) All active control strategies are influenced by the amount of passive damping. On one hand the achievable amount of noise reduction due to active control is decreased, if the amount of passive damping is increased as shown in Fig. 6.13 and Fig. 6.17. On the other hand an increase in passive damping can lead to a reduction of the required control effort as also shown in Fig. 6.17. Furthermore, to allow for power flow measurements, a certain amount of passive damping that smooths and stabilizes the dynamic response of the system is required, if power flow based control strategies such as maximization of power absorption shall be applied.

(**4**) Depending on the statistical properties of the primary disturbance, the resulting controller is allowed to act non-causally or has to act strictly causal. It has been shown that local minimization of acoustic potential energy as well as maximization of acoustic absorption result in a causally acting controller. While the first control strategy is limited to the acoustic-resonances, the latter can be applied over a wider frequency range. Global control of acoustic potential energy and minimization of total power input enable noise reduction in a wide frequency range, but non-causal parts are present in the corresponding impulse response functions of the optimal controllers. Therefore, an a-priory knowledge about the primary disturbance (as in the case of periodic excitation) is required, if these strategies shall be applied.

(**5**) The effort of a certain control strategy is not only determined by the power consumption of the canceling sources but also by the number of system components. It is obvious that global sensing with distributed sensors as required for global control of acoustic potential energy is more complex than local sensing as required for minimization of total power input.

(**6**) Finally the 3D-applicability of the control strategy has to be assured in many applications. Control strategies that are based on simple sound pressure measurement (global and local control of acoustic potential energy) can be directly adopted to 3D-problems.

As shown in (Guicking and Karcher 1984) maximization of acoustic absorption can be applied in three dimension, but a significant NR is limited to normal incident waves.

Control strategies that are based on power flow measurements (minimization of total power input, maximization of power absorption) can be applied to three dimensional enclosures, but a reliable measurement of energy flow is because of the directivity of the measured data more complex than omnidirectional sound pressure measurements. The latter technique is well understood and is capable of providing robust results by using relatively cheap sensors.

The complexity of the control task increases with the number of microphones and canceling sources. A very approximate guide to the number of components required to achieve a given reduction in energy level in an enclosed sound field is the *modal overlap* $M(\omega)$, see (Elliott 2001). This is a dimensionless parameter that quantifies the number of modes that are significantly excited in a system at any one excitation frequency.

As known from the theory of *independent modal space control* (IMSC), see (Meirovitch 1990), we will (at least) need N microphones and N canceling sources in order to control a number of N independent modes associated to N distinct resonant frequencies. To estimate the size of an active control system, it is therefore

Table 6.1 Comparison of active control strategies

Control Strategy	Minimization of global/*local* potential energy	Minimization of local acoustic energy ($\beta = 0.5$)	Minimization of total power input	Maximization of power absorption	Maximization of acoustic absorption
Effect on global acoustic potential energy	Maximum reduction in acoustic resonances, unaffected/*maximum amplification* at anti-resonances	Maximum reduction in acoustic resonances, minor amplification at anti-resonances	Maximum reduction in acoustic resonances, unaffected at anti-resonances	Minor reduction in acoustic resonances, maximum amplification outside resonances	Good reduction in acoustic resonances, minor amplification outside acoustic resonances
Effect on local sound pressure	Reduction at any point (except for nodal points) in acoustic resonances, local amplification possible/*at any uncontrolled point* in anti-resonances	Reduction at any point (except for nodal points) in acoustic resonances, minor amplification in anti-resonances	Reduction at any point (except for nodal points) in acoustic resonances, no effect in anti-resonances	Minor reduction at any point (except for nodal points) in acoustic resonances, amplification at any point outside anti-resonances	Reduction at any point (except for nodal points) in acoustic resonances, minor amplification at any point outside acoustic resonances
Role of passive damping	Decreases control profit and control effort/*fully identical*	Decreases control profit and control effort	Decreases control profit and control effort, essential for power flow measurement	Decreases control profit and control effort, essential for power flow measurement and limitation of control strength	Decreases control profit and control effort
Controller acts	Non-causal/*causal*	Non-causal (in general)	Non-causal	Non-causal	Causal
Complexity of measurement	High (due to high number of sensors)/*fair*	High, due to requirements of velocity measurement	High, due to requirements of power flow measurement	High, due to requirements of power flow measurement	Fair
Applicability to 3D-Problems	Without restrictions/*fully identical*	Possible	Possible	Possible	For normal incident waves only

necessary to determine an estimate of the number of modes that will overlap and therefore contribute to a mode that is associated with a given excitation frequency. This estimate is the modal overlap that is defined to be the average number of acoustic modes whose natural frequencies fall within the bandwidth of any one mode at a given excitation frequency.

For this reason we have to calculate the product of the *modal density* (average number of modes/Hz) and the *modal bandwidth*. The latter is according to (3.68) introduced as

$$\Delta\omega_M := 2\xi_{nR}\omega_{nR} = 4\pi\,\Delta f_{nR}. \tag{6.79}$$

Considering modes with resonant frequencies $\omega \approx \omega_{nR}$, and assuming that the damping ratio $\xi_{nR} = \xi$ is the same for each mode, we can use the approximate expression for the modal overlap of a three-dimensional enclosure that has been presented in (Elliott 2001)

$$M(\omega) = \left(\frac{L}{\pi c} + \frac{\omega S}{2\pi c^2} + \frac{\omega^2 V}{2\pi c^3}\right) \cdot 2\xi\omega. \tag{6.80}$$

It can be seen that (6.80) is based on physical parameters such as: L—the sum of linear dimensions, S—the total surface area, V—the volume of the enclosure, and c—the speed of sound. The modal overlap increases linearly with the damping ratio ξ, and, especially at higher frequencies, $M(\omega)$ increases as the cube of ω.

If we consider a maximum normalized frequency of $\zeta = 2.75$ and a damping ratio $\xi = 0.05$, the modal overlap in our benchmark system can be approximated as $M(\zeta) = 4\xi\zeta \approx 0.55$. Thus, considering one canceling source was sufficient to compare the reduction of global acoustic potential energy realized by different control strategies. The results of the discussion are summarized in Table 6.1.

6.4 Multi-channel Control of Tonal Noise

If the interior noise field is excited by harmonic disturbances and therefore dominated by tonal components, active control can effectively be applied in frequency domain, because we have an a-priory knowledge about the deterministic disturbances. In the following section all signals and transfer function are seen as the (discrete) frequency domain representation of the associated discrete-time quantities that are obtained by applying the time discrete Fourier transform (2.31), e.g. $P(e^{j\omega T}) = \mathfrak{F}_d[p(n)]$. The argument $e^{j\omega T}$ is therefore in this section dropped for convenience. We will however use the arguments n or $n+1$ to distinguish between different time steps.

6.4.1 Optimal Control of Tonal Noise

The structure of an active feed-forward control system that is applied to attenuate tonal noise is shown in Fig. 6.25. It can be seen that the $(L \times 1)$ error signal matrix

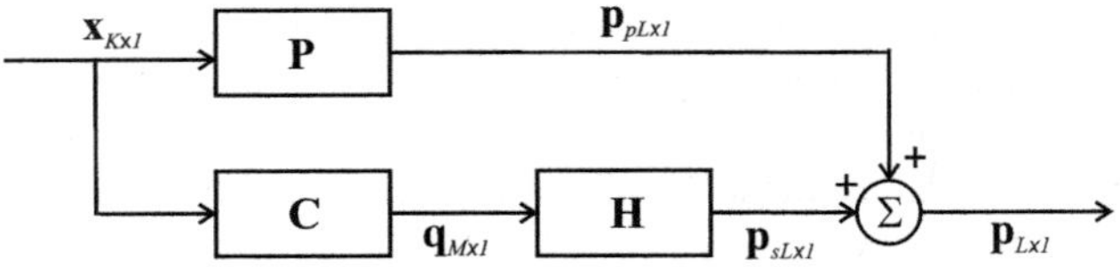

Fig. 6.25 Block diagram of a multi-channel system for active feed-forward control of tonal noise

$\mathbf{p}$ results from the superposition of two $(L \times 1)$ matrices that represent noise $\mathbf{p}_p$ and anti-noise $\mathbf{p}_s$ such as

$$\mathbf{p} = \mathbf{p}_p + \mathbf{p}_s, \tag{6.81}$$

where

$$\begin{aligned} \mathbf{p} &= [P_1\ P_2\ \ldots\ P_l\ \ldots\ P_L]^T \\ \mathbf{p}_p &= [P_{p1}\ P_{p2}\ \ldots\ P_{pl}\ \ldots\ P_{pL}]^T \\ \mathbf{p}_s &= [P_{s1}\ P_{s2}\ \ldots\ P_{sl}\ \ldots\ P_{sL}]^T. \end{aligned} \tag{6.82}$$

$\mathbf{p}_p$ results from K reference signals that are altered by the LK primary transfer paths represented by the $(L \times K)$ matrix $\mathbf{P}$. These reference signals are arranged in the $(K \times 1)$ column matrix

$$\mathbf{x} = [X_1\ X_2\ \ldots X_k\ \ldots\ X_K]^T. \tag{6.83}$$

$\mathbf{p}_s$ results from M control signals that are arranged in the $(M \times 1)$ column matrix

$$\mathbf{q} = [Q_1\ Q_2\ \ldots\ Q_m\ \ldots\ Q_M]^T = \mathbf{Cx}. \tag{6.84}$$

These control signals are altered by the LM secondary transfer paths represented by the $(L \times M)$ matrix $\mathbf{H}$ with elements H_{lm} such as

$$\mathbf{p}_s = \mathbf{Hq} \quad \text{with } P_{sl} = \sum_{m=1}^{M} H_{lm} Q_m =: \sum_{m=1}^{M} P_{slm}. \tag{6.85}$$

Considering (6.81), (6.84), and (6.85), $\mathbf{p}$ can be written such as

$$\mathbf{p} = \mathbf{p}_p + \mathbf{p}_s = \mathbf{p}_p + \mathbf{HCx} =: \mathbf{p}_p + \mathbf{Hq}. \tag{6.86}$$

If the control objective is the minimization of the weighted sum of the mean square error signals, the real valued hermitian quadratic form

$$J(n) = \mathbf{p}^H \mathbf{p}, \tag{6.87}$$

in which the parameters, compare (6.6), are given by

$$\mathbf{A} \equiv \mathbf{H}^H \mathbf{H}, \qquad \mathbf{b} = \mathbf{b}_1 = \mathbf{b}_2 \equiv \mathbf{H}^H \mathbf{p}_p, \qquad c \equiv \mathbf{p}_p^H \mathbf{p}_p, \tag{6.88}$$

has to be minimized. The control signals that have to be generated for this purpose specify the control effort or the costs of the control task. A quadratic form similar to (6.87) is therefore in the following called *cost function*. Of practical importance are the over-determined $L > M$ and the full-determined $L = M$ case.[7] The minimum of (6.87) is according to (6.13) for $L > M$ given by

$$\mathbf{q}_{opt} = -\left(\mathbf{H}^H\mathbf{H}\right)^{-1}\mathbf{H}^H\mathbf{p}_p, \qquad L > M \tag{6.89}$$

and reduces for $L = M$ to

$$\mathbf{q}_{opt} = -\mathbf{H}^{-1}\mathbf{p}_p, \qquad L = M. \tag{6.90}$$

The most common method for weighting the control effort against the mean square error is, see (Kuo and Morgan 1996) and (Elliott 2001), the introduction of a real valued weighting parameter β such as

$$J = \mathbf{p}^H\mathbf{p} + \beta\mathbf{q}^H\mathbf{q}, \tag{6.91}$$

where the cost function parameters are given by

$$\mathbf{A} \equiv \mathbf{H}^H\mathbf{H} + \beta\mathbf{I}, \qquad \mathbf{b} = \mathbf{b}_1 = \mathbf{b}_2 \equiv \mathbf{H}^H\mathbf{p}_p, \qquad c \equiv \mathbf{p}_p^H\mathbf{p}_p. \tag{6.92}$$

Using (6.92), the optimal solution is, according to (6.13), given by

$$\mathbf{q}_{opt} = -\left[\mathbf{H}^H\mathbf{H} + \beta\mathbf{I}\right]^{-1}\mathbf{H}^H\mathbf{p}_p. \tag{6.93}$$

Equation (6.93) clarifies that the weighting parameter β acts as a regularization factor, because it directly contributes to the real parts of the eigenvalues of $\mathbf{H}^H\mathbf{H}$. In the single channel case, in which (6.93) reduces to

$$Q_{opt} = -\frac{H^*}{|H|^2 + \beta}P_p \tag{6.94}$$

it is easy to see that a weighting parameter $\beta > 0$ prevents a division by zero and can therefore be used to avoid instabilities.

6.4.2 Adaptive Control of Tonal Noise

Application of the Steepest-Decent Algorithm

In the previous subsection signal processing for active control of interior noise was based on optimized but fixed controllers. In this subsection we will discuss an adaptive control scheme that can be motivated by the block-diagram shown in Fig. 6.26.

The adaptivity of the controller can be realized by implementing an iterative scheme that adjusts the control signals $\mathbf{q}$ in order to minimize (6.91). For this purpose we apply the steepest-descent algorithm, compare Sect. 6.1.3

$$\mathbf{q}(n+1) = \mathbf{q}(n) - \frac{\mu}{2}\mathbf{g}(n) \tag{6.95}$$

[7] As shown in (Elliott 2001) the matrix $\mathbf{H}^H\mathbf{H}$ is not positive definite for an under-determined system ($L < M$) and will have at least $M - L$ eigenvalues that are equal to zero. Therefore the solution of the under-determined minimization problem is not unique and will not be considered in the present work.

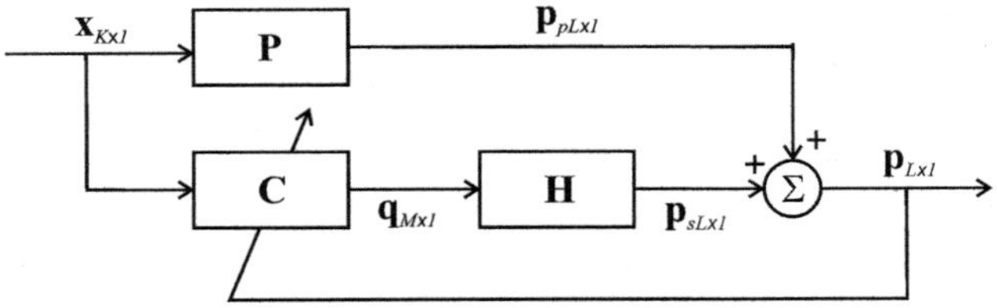

Fig. 6.26 Block diagram of a multi-channel adaptive controller with sinusoidal input to the multi-channel primary transfer path

that considering (6.16), (6.86), and (6.92) gives us

$$\begin{aligned}\mathbf{q}(n+1) &= \mathbf{q}(n) - \mu\big[\big(\mathbf{H}^H\mathbf{H} + \beta\mathbf{I}\big)\mathbf{q}(n) + \mathbf{H}^H\mathbf{p}_p(n)\big] \\ &= (1-\mu\beta)\mathbf{q}(n) - \mu\mathbf{H}^H\mathbf{p}(n).\end{aligned} \tag{6.96}$$

Equation (6.96) indicates that the error signal $\mathbf{p}(n)$ is altered by $\mathbf{H}$. Therefore, it is necessary to replace $\mathbf{H}$ by $\hat{\mathbf{H}}$ that represents a model of the multi-channel secondary path. This yields the following update equation

$$\begin{aligned}\mathbf{q}(n+1) &= (1-\mu\beta)\mathbf{q}(n) - \mu\hat{\mathbf{H}}^H\mathbf{p}(n) \\ &= \mathbf{q}(n) - \mu\big[\big(\hat{\mathbf{H}}^H\mathbf{H} + \beta\mathbf{I}\big)\mathbf{q}(n) + \hat{\mathbf{H}}^H\mathbf{p}_p(n)\big].\end{aligned} \tag{6.97}$$

Comments on Plant Modeling

The *plant models* $\hat{H}_{lm}$ that are arranged in the $(L \times M)$ matrix $\hat{\mathbf{H}}$ will in general differ from the original plant and must be identified by off-line or on-line identification procedures as described in (Kuo and Morgan 1996). An easy way to determine the lm-th plant model $\hat{H}_{lm}$ is the evaluation of the ratio between $\hat{P}_{slm}$ that is the output signal of the lm-th secondary path H_{lm}, compare (6.85), and the associated m-th harmonic control signal Q_m. This yields, compare Sect. 2.2.2 (2.44) and Fig. 2.5, the frequency response of the analyzed transfer path

$$\hat{H}_{lm} = \frac{P_{slm}}{Q_m}. \tag{6.98}$$

Another (much more complex) option that can be used to compensate (usually minor) variation in the response the secondary paths is to minimize

$$\begin{aligned}J_{H_{lm}} &= |P_{slm} - \hat{P}_{slm}|^2 = |P_s - \hat{H}_{lm}Q_m|^2 \\ &= \hat{H}^*_{lm}A\hat{H}_{lm} + \hat{H}^*_{lm}b_1 + b^*_2\hat{H}_{lm} + c,\end{aligned} \tag{6.99}$$

where

$$A = \hat{Q}^*_m Q_m, \qquad b = b_1 = b_2 = -Q^*_m P_{slm}, \qquad c = P^*_{slm}P_{slm}. \tag{6.100}$$

Equation (6.99) represents the squared difference between the (real) response of the lm-th secondary path H_{lm} and response of the lm-th secondary path model $\hat{H}_{lm}$

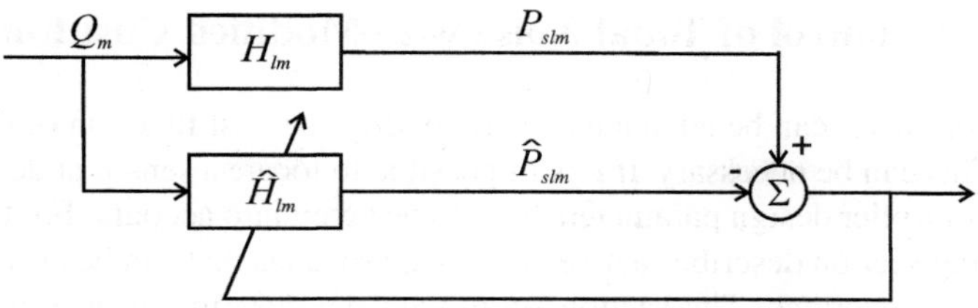

Fig. 6.27 Adaptive schema used to identify the impulse response of the lm-th secondary path

that can be minimized by means of the steepest-decent algorithm as motivated in Fig. 6.27.

Considering (6.16) as well as the identifications specified in (6.100) the update schema for the lm-th secondary path model reads

$$\hat{H}_{lm}(n+1) = \hat{H}_{lm}(n) + \mu\big[P_{slm}(n) - \hat{H}_{lm}(n)Q_m(n)\big]Q_m^*(n). \tag{6.101}$$

Comments on Stability

If the control algorithm given by (6.97) is stable, it will reach the steady-state in which $\mathbf{q}(n+1)$ is equal to $\mathbf{q}(n)$. The steady-state solution

$$\mathbf{q}_\infty = -\big(\hat{\mathbf{H}}^H\mathbf{H} + \beta\mathbf{I}\big)^{-1}\hat{\mathbf{H}}^H\mathbf{p}_p \tag{6.102}$$

differs from the theoretical optimal solution (6.93) due to the differences between the physical plant $\mathbf{H}$ and the plant model $\hat{\mathbf{H}}$.[8] The combination of (6.97) and (6.102) leads in direct analogy to the procedure described in Sect. 6.1.3 to a modified iteration schema

$$\mathbf{q}(n+1) - \mathbf{q}_\infty = \big[\mathbf{I} - \mu\big(\hat{\mathbf{H}}^H\mathbf{H} + \beta\mathbf{I}\big)\big]\big(\mathbf{q}(n) - \mathbf{q}_\infty\big) \tag{6.103}$$

that is, compare (6.26), stable, if, and only if, the in general complex eigenvalues λ_m of the matrix $\mathbf{A} = \hat{\mathbf{H}}^H\mathbf{H} + \beta\mathbf{I}$ are bounded by

$$\big|1 - \mu(\lambda_m + \beta)\big| < 1 \quad \text{for all } m \tag{6.104}$$

i.e.

$$0 < \mu < \frac{2[\mathrm{Re}(\lambda_m) + \beta]}{|\lambda_m + \beta|^2} \quad \text{for all } m. \tag{6.105}$$

Equation (6.105) makes clear that the real valued parameter β can be used to guarantee a positive real part for all eigenvalues λ_m.

---

[8] As mentioned in (Elliott 2001) convergence against the theoretical limit, defined by (6.93), can only be obtained for a fully determined system using a perfect plant model $\hat{\mathbf{H}} = \mathbf{H}$ and without any weighting $\beta = 0$.

6.5 Active Control of Tonal Noise with Modified Cost Functions

For some reasons it can be advantageous to modify the cost function of the control problem. This can be necessary, if it is impossible to locate a sensor at desired position, or if controller design parameters have to be taken into account. For this reason the upcoming section describes active control approaches that can be established on modified cost functions. The section starts with the description of a general cost function approach and is continued with subsections that are focused on the rejection of disturbances at remote sensors and parametric controller design.

Because we will for every approach identify the cost function parameters $\mathbf{A}$, $\mathbf{b}_1$, $\mathbf{b}_2$, and c, it is not necessary to quote the associated steepest decent approaches that have been elucidated in Sect. 6.1.3 for the general quadratic form.

6.5.1 Optimal Control Using a General Cost Function

A general form of a *modified cost function* that can be used to weight both the *modified mean square error* $\tilde{\mathbf{p}}$ and the *modified control effort* $\tilde{\mathbf{q}}$ can be introduced as

$$J_R = \mathrm{Re}\{J(\tilde{\mathbf{p}}, \tilde{\mathbf{q}})\}, \tag{6.106}$$

where the complex quadratic form $J(\tilde{\mathbf{p}}, \tilde{\mathbf{q}})$ is given by

$$J = \tilde{\mathbf{p}}^H \mathbf{W}_p \tilde{\mathbf{p}} + \tilde{\mathbf{q}}^H \mathbf{W}_q \tilde{\mathbf{q}} \quad \text{with } \tilde{\mathbf{p}} = \mathbf{P}\mathbf{p}_p + \mathbf{H}\mathbf{Q}\mathbf{q}, \text{ and } \tilde{\mathbf{q}} = \mathbf{Q}\mathbf{q}. \tag{6.107}$$

$\mathbf{W}_p$ and $\mathbf{W}_q$ are cost function *weighting matrices* with dimension $(L \times L)$ and $(M \times M)$, respectively. The $(L \times L)$ matrix $\mathbf{P}$ is used for an independent weight of the primary disturbance, and the $(M \times M)$ matrix $\mathbf{Q}$ is used to weight the driving signals of the secondary sources individually. Using the identifications

$$\begin{aligned} \mathbf{A} &\equiv \mathbf{Q}^H(\mathbf{H}^H\mathbf{W}_p\mathbf{H} + \mathbf{W}_q)\mathbf{Q}, & \mathbf{b}_1 &\equiv \mathbf{Q}^H\mathbf{H}^H\mathbf{W}_p\mathbf{P}\mathbf{p}_p, \\ \mathbf{b}_2 &\equiv \mathbf{Q}^H\mathbf{H}^H\mathbf{W}_p^H\mathbf{P}\mathbf{p}_p, & c &\equiv \mathbf{p}_p^H\mathbf{P}^H\mathbf{W}_p\mathbf{P}\mathbf{p}_p, \end{aligned} \tag{6.108}$$

the optimal source strength that minimizes the general cost function J_R is, compare Sect. 6.1.2 (6.9), given by

$$\begin{aligned} \mathbf{q}_{opt} &= -(\mathbf{Q}^H\mathbf{E}\mathbf{Q})^{-1}\mathbf{Q}^H\mathbf{H}^H(\mathbf{W}_p + \mathbf{W}_p^H)\mathbf{P}\mathbf{p}_p \quad \text{with} \\ \mathbf{E} &= \mathbf{H}^H(\mathbf{W}_p + \mathbf{W}_p^H)\mathbf{H} + \mathbf{W}_q + \mathbf{W}_q^H. \end{aligned} \tag{6.109}$$

6.5.2 Remote Sensor Control

Let us now consider a situation (e.g. after a design freeze) in which it is impossible to place the error sensors at the desired locations. In such a situation we have to use the admissible transducer locations (*remote* from the desired transducer locations)

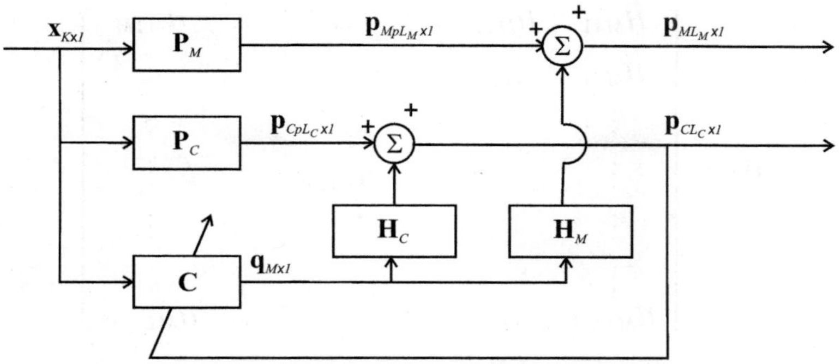

Fig. 6.28 Block diagram for adaptive multi-channel remote sensor control of tonal noise

to achieve a certain control profit at the desired positions. Such an active control approach is known as *disturbance rejection at remote error sensors*, see (Elliott 2001). However in this book we will use the more compact term *remote sensor control* (RSC). The basic structure of an adaptive RSC schema is shown in Fig. 6.28.

This figure makes clear that the input to two sets of primary paths $\mathbf{P}_C$ and $\mathbf{P}_M$ is given by a number of K reference signals represented by the column matrix $\mathbf{x}$. The output signals of the primary paths are the noise signals

$$\mathbf{p}_{pC} = [\, P_{pC1} \; P_{pC2} \ldots P_{pCl_C} \ldots P_{pCL_C} \,]^T \quad \text{with } l_C = 1, \ldots, L_C \tag{6.110}$$

and

$$\mathbf{p}_{pM} = [\, P_{pM1} \; P_{pM2} \ldots P_{pMl_M} \ldots P_{pML_M} \,]^T \quad \text{with } l_M = 1, \ldots, L_M. \tag{6.111}$$

The anti-noise is superimposed on the noise and results from the M control signals. It is represented by the column matrix

$$\mathbf{q} = [\, Q_1 \; Q_2 \ldots Q_m \ldots Q_M \,]^T \quad \text{with } m = 1, \ldots, M. \tag{6.112}$$

These signals have to pass the secondary transfer paths. As for the primary paths we consider two sets of secondary paths represented by the matrices

$$\mathbf{H}_C = \begin{bmatrix} H_{C11} & H_{C12} & \cdots & \cdots & \cdots & H_{C1M} \\ H_{C21} & \ddots & & & & \vdots \\ \vdots & & \ddots & & & \vdots \\ \vdots & & & H_{Cl_Cm} & & \vdots \\ \vdots & & & & \ddots & \vdots \\ \vdots & & & & & \vdots \\ H_{CL_C1} & \cdots & \cdots & \cdots & \cdots & H_{CL_CM} \end{bmatrix} \quad \text{with} \quad \begin{array}{l} l_C = 1, \ldots, L_C, \\ m = 1, \ldots, M, \end{array} \tag{6.113}$$

and

$$\mathbf{H}_M = \begin{bmatrix} H_{M11} & H_{M12} & \dots & \dots & \dots & H_{M1M} \\ H_{M21} & \ddots & & & & \vdots \\ \vdots & & \ddots & & & \vdots \\ \vdots & & & H_{Ml_Mm} & & \vdots \\ \vdots & & & & \ddots & \vdots \\ \vdots & & & & & \vdots \\ H_{ML_M1} & \dots & \dots & \dots & \dots & H_{ML_MM} \end{bmatrix}$$

$$\text{with} \quad \begin{array}{l} l_M = 1, \dots, L_M, \\ m = 1, \dots, M. \end{array} \tag{6.114}$$

The latter describes the transfer behavior between the L_M signals $\mathbf{p}_M$ and the M control signals. The transfer behavior between the L_C signals $\mathbf{p}_C$ and the M control signals is described by $\mathbf{H}_C$. This expresses the fact that the superposition of noise and anti-noise represented by $\mathbf{p}_C$ and $\mathbf{p}_M$ is measured at different locations. According to Fig. 6.28 these superpositions are given by

$$\mathbf{p}_C = \mathbf{p}_{pC} + \mathbf{H}_C\mathbf{q} \quad \text{and} \quad \mathbf{p}_M = \mathbf{p}_{pM} + \mathbf{H}_M\mathbf{q}. \tag{6.115}$$

The signals represented by $\mathbf{p}_C$ are used for the active control task. They are therefore called *error signals*. Such signals will in the following be indicated by the suffix C. Controlled sound field data such as $\mathbf{p}_M$ that can only be monitored, but not used in the control schema are called *monitor signals*. These signals will be indicated by the suffix M. The associated sensors are known as the *error sensors* and the *monitor sensors*. The error sensors are located at the admissible positions, whereas the monitor sensors that are related to the desired positions have to be removed from the final system.

RSC is based on minimizing the error signals, as shown in Fig. 6.28, but the objective of this control approach is the reduction of the monitor signals. For this reason it is necessary to implement a controller that is capable of shifting the control profit from the error sensors to the monitor sensors. A comprehensive review on this topic has been given by Moreau et al. (2008) and Liu and Kuo (2009). However, instead of listing the approaches described in these references, the upcoming subsection is focused on two RSC techniques that especially in tonal noise fields allow for direct computation (not only approximation[9]) of the sound pressure at the monitor positions, if modeling errors can be neglected.

These approaches have in common that they can be subdivided into a *training phase* and a *control phase*. The first is used to identify models that are based on the transfer behavior of the primary paths and the transfer behavior of the secondary

[9] An example for sound pressure approximation is the forward prediction method, see (Munn 2003) and (Moreau et al. 2008), in which the error signals at the remote locations $\mathbf{p}_M$ are approximated by the error signals obtained from physical error sensors $\mathbf{p}_C$ such as $\mathbf{p}_M = \mathbf{F}\mathbf{p}_C$, where $\mathbf{F}$ contains weighting coefficients determined by finite difference schemes.

paths. These models are used to design a controller **C** that is capable of realizing active noise reduction at the positions of the monitor sensors in the second phase. In the following we assume that the response of both primary and secondary paths is linear, time invariant and can be perfectly modeled.

Remote Sensor Control Using Virtual Microphones

This approach was originally established by Garcia-Bonito et al. (1998). Advanced approaches have been proposed in (Petersen et al. 2005), (Sachau et al. 2007a), (Kletschkowski and Sachau 2007a), and (Kletschkowski et al. 2007). Because the monitor sensor is usually called *virtual microphone* this approach is also known as *virtual microphone approach*.

Training Phase The training phase is used to determine the *transmissibility* that is the ratio between the primary noise measured at the l_M-th monitor microphone and the l_C-th error microphone such as[10]

$$T_{l_M l_C} = \frac{P_{pMl_M}}{P_{pCl_C}}. \tag{6.116}$$

Furthermore, we have to determine the transfer behavior between all monitor sensors and all control signals as well as the transfer behavior between all error sensors and all control signals to obtain the transfer matrices

$$\mathbf{H}_C = [H_{Cl_C m}] \quad \text{with } H_{Cl_C m} = \frac{P_{sCl_C m}}{Q_m} \tag{6.117}$$

and

$$\mathbf{H}_M = [H_{Ml_M m}] \quad \text{with } H_{Ml_M m} = \frac{P_{sMl_M m}}{Q_m}. \tag{6.118}$$

Please notice that according to (6.85), $P_{sCl_C m}$ is the anti-noise generated by the m-th actuator and measured at the l_C-th error sensor, whereas $P_{sMl_M m}$ is the anti-noise that is generated by the m-th actuator and measured at the l_M-th monitor sensor.

Control Phase According to Fig. 6.28 the controlled sound field is at the l-th error and monitor microphone given by

---

[10]If it is a not possible to use acoustic transducers for active control, information about the acoustical state may be obtained from non-acoustic reference signals measured by accelerometers. The transmissibility is in such a situation determined by the ratio between the primary noise measured at the l-th monitor microphone and the primary disturbance measured at the l-th accelerometer such as $T_{l_M l_C} = P_{pMl_m}/A_{pCl_C}$. For more details, especially on the more advanced approach of *radiation mode sensing*, the reader is referred to (Fahy and Gardonio 2007).

$$P_{Cl_C} = P_{pCl_C} + \sum_{m=1}^{M} H_{Cl_Cm} Q_m \quad \text{and}$$
$$P_{Ml_M} = P_{pMl_M} + \sum_{m=1}^{M} H_{Ml_Mm} Q_m. \tag{6.119}$$

Using (6.116) we can replace the primary noise measured at monitor microphone by the primary noise detected at the error microphone. This gives, considering (6.119),

$$\begin{aligned} P_{Ml_M} &= P_{pMl_M} + \sum_{m=1}^{M} H_{Ml_Mm} Q_m \\ &= T_{l_Ml_C} P_{pCl_C} + \sum_{m=1}^{M} H_{Ml_Mm} Q_m \\ &= T_{l_Ml_C} \left(P_{Cl_C} - \sum_{m=1}^{M} H_{Cl_Cm} Q_m \right) + \sum_{m=1}^{M} H_{Ml_Mm} Q_m \\ &= T_{l_Ml_C} P_{Cl_C} - T_{l_Ml_C} \sum_{m=1}^{M} H_{Cl_Cm} Q_m + \sum_{m=1}^{M} H_{Ml_Mm} Q_m. \end{aligned} \tag{6.120}$$

Equation (6.120) clarifies that during the control phase, sensing is only based on the error microphones. The real valued cost function based on the L monitor signals that are calculated by (6.120) is given by

$$J = \mathbf{p}_M^H \mathbf{p}_M. \tag{6.121}$$

The weighting matrices introduced in Sect. 6.5.1 (6.107) can be identified as

$$\mathbf{W}_p \equiv \mathbf{I}, \qquad \mathbf{W}_p \equiv \mathbf{0}, \qquad \mathbf{P} \equiv \mathbf{T}, \quad \text{and} \quad \mathbf{Q} \equiv \mathbf{I}, \tag{6.122}$$

where $\mathbf{T} = \text{diag}(T_l)$ is the $(L \times L)$ diagonal matrix of transmissibilities. The cost function parameters are therefore given by

$$\mathbf{A} = \mathbf{H}_M^H \mathbf{H}_M, \qquad \mathbf{b} = \mathbf{H}_M^H \mathbf{T} \mathbf{p}_{pC}, \qquad c = \mathbf{p}_{pM}^H \mathbf{T}^H \mathbf{T} \mathbf{p}_{pM}. \tag{6.123}$$

According to Sect. 6.5.1 (6.109), the $(M \times 1)$ column matrix of optimal control signals is given by

$$\mathbf{q}_{opt} = -\left(\mathbf{H}_M^H \mathbf{H}_M\right)^{-1} \mathbf{H}_M^H \mathbf{T} \mathbf{p}_{pC}. \tag{6.124}$$

In the single channel case ($K = L = M = 1$) in which (6.120) reduces to

$$\begin{aligned} P_M &= P_{pM} + H_M Q \\ &= T P_{pC} + H_M Q \\ &= T(P_C - H_C Q) + H_M Q \\ &= T P_C - T H_C Q + H_M Q, \end{aligned} \tag{6.125}$$

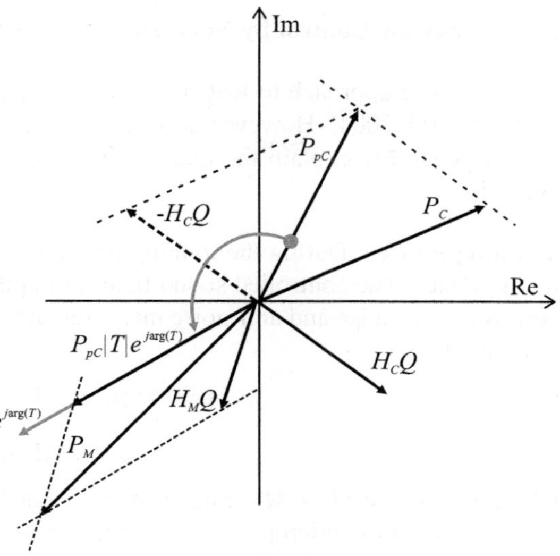

Fig. 6.29 Calculation of virtual microphone signal P_M based on the magnitude $|T|$ and the phase $\arg(T)$ of the complex transmissibility $T = P_{pM}/P_{pC}$ and the transfer numbers H_C and H_M

the optimal control signal is determined by

$$Q_{opt} = -\frac{T}{H_M} P_{pC} = -T \frac{H_C}{H_M} \frac{P_{pC}}{H_C} =: -\Gamma \frac{P_{pC}}{H_C}. \tag{6.126}$$

Equation (6.126) clarifies that P_{pC}/H_C, the optimal driving signal for noise cancellation at the control microphone, has to be altered by a in general complex number $\Gamma = TH_C/H_M$, to realize active noise canceling at the position of the monitor microphone.

A graphical representation of the single channel version of (6.120) is shown in Fig. 6.29. At first P_{pC} is determined by calculating the difference between P_C and $H_C Q$. Then P_{pC} is rotated by $\arg(T)$ and scaled (in magnitude) by $|T|$. Finally the anti-noise $H_M Q$ is added.

If the transmissibility can be approximated by $T \approx 1$, the optimal control signal Q_{opt} specified by (6.126) can be approximated by

$$Q_{opt} \approx -\frac{1}{H_M} P_{pC}. \tag{6.127}$$

This leads to the approach presented in (Garcia-Bonito et al. 1998) in which it was assumed that the primary noise is identical at both the control monitor and the error microphone. The transformation of error signal can in this case be approximated such as

$$P_M \approx P_C + (H_M - H_C)Q. \tag{6.128}$$

However, this assumption is only valid, if the distance between the two microphones is small compared to the wave length.

Remote Sensor Control by Noise Design at Error Microphones

An alternative approach to RSC based on adaptive feedback control was presented in (Pawelczyk 2004). However, according to the scope of this work, we will without loss of generality explain the main idea of this approach in view of feed-forward control.

Training Phase During the training phase active control is applied to the control microphones. The controlled sound field during this phase is determined by the superposition of noise and anti-noise measured at the control as well as at the monitor microphones

$$\begin{aligned} \mathbf{p}_{M,t} &= \mathbf{p}_{pM} + \mathbf{H}_M \mathbf{q}_t, \\ \mathbf{p}_{C,t} &= \mathbf{p}_{pC} + \mathbf{H}_C \mathbf{q}_t, \end{aligned} \tag{6.129}$$

where $\mathbf{H}_M$ is the $(L \times M)$ transfer matrix that contains the transfer behavior between all monitor microphones and all control signals. Consequently $\mathbf{H}_C$ is the $(L \times M)$ transfer matrix that determines the pressure response at the control microphones caused by the control signals. The suffix t is used to indicate the training phase.

The cost function that is to be minimized in the training phase is given by the mean square error at the monitor microphones

$$J_t = \mathbf{p}_{M,t}^H \mathbf{p}_{M,t}. \tag{6.130}$$

Considering the identifications

$$\mathbf{A} = \mathbf{H}_M^H \mathbf{H}_M, \qquad \mathbf{b} = \mathbf{H}_M^H \mathbf{p}_{pM}, \quad \text{and} \quad c = \mathbf{p}_{pM}^H \mathbf{p}_{pM}, \tag{6.131}$$

the optimal solution of (6.130) is according to (6.13) given by

$$\mathbf{q}_{opt,t} = -\left(\mathbf{H}_M^H \mathbf{H}_M\right)^{-1} \mathbf{H}_M^H \mathbf{p}_{pM}. \tag{6.132}$$

The remaining error at the error microphones in the training phase is given by

$$\begin{aligned} \mathbf{p}_{C,t} &= \mathbf{p}_{pC} + \mathbf{H}_C \mathbf{q}_{opt,t} \\ &= \mathbf{p}_{pC} - \mathbf{H}_C \left(\mathbf{H}_M^H \mathbf{H}_M\right)^{-1} \mathbf{H}_M^H \mathbf{p}_{pM}. \end{aligned} \tag{6.133}$$

Control Phase The information about the remaining error at the error microphones stored in $\mathbf{p}_{C,t}$ is used to modify the error signal at the error microphones during the control phase in which the monitor microphones are usually removed. Therefore, only the error signals obtained from the error microphones remain for the active control task.

The stored information about the remaining error signal at the error microphones in the tuning phase is now used to modify the error measured at the error microphones during the control phase such as

$$\begin{aligned} \tilde{\mathbf{p}}_{C,c} &= \mathbf{p}_{C,c} - \mathbf{p}_{C,t} \\ &= \mathbf{H}_C \mathbf{q}_c + \mathbf{H}_C \left(\mathbf{H}_M^H \mathbf{H}_M\right)^{-1} \mathbf{H}_M^H \mathbf{p}_{pM}, \end{aligned} \tag{6.134}$$

where the suffix c is used to indicate the control phase. The cost function to be minimized in the control phase is

$$J_c = \tilde{\mathbf{p}}_{C,c}^H \tilde{\mathbf{p}}_{C,c}. \tag{6.135}$$

Identification of the weighting matrices introduced in Sect. 6.5.1 (6.107)

$$\mathbf{W}_p \equiv \mathbf{I}, \qquad \mathbf{W}_p \equiv \mathbf{0}, \qquad \mathbf{P} \equiv \mathbf{H}_C \left(\mathbf{H}_M^H \mathbf{H}_M\right)^{-1} \mathbf{H}_M^H, \quad \text{and} \quad \mathbf{Q} \equiv \mathbf{I}, \tag{6.136}$$

yields the cost function parameters

$$\mathbf{A} = \mathbf{H}_C^H \mathbf{H}_C, \qquad \mathbf{b} = \mathbf{H}_C^H \mathbf{P} \mathbf{p}_{pM}, \quad \text{and} \quad c = \mathbf{p}_{pM}^H \mathbf{P}^H \mathbf{P} \mathbf{p}_{pM}. \tag{6.137}$$

The column matrix that contains the optimal canceling signals is, considering (6.13) as well as the identifications specified in (6.137), given by

$$\begin{aligned} \mathbf{q}_{opt,c} &= -\left(\mathbf{H}_C^H \mathbf{H}_C\right)^{-1} \mathbf{H}_C^H \mathbf{P} \mathbf{p}_{pM} \\ &= -\left(\mathbf{H}_C^H \mathbf{H}_C\right)^{-1} \mathbf{H}_C^H \mathbf{H}_C \left(\mathbf{H}_M^H \mathbf{H}_M\right)^{-1} \mathbf{H}_M^H \mathbf{p}_{pM}. \end{aligned} \tag{6.138}$$

By using the transmissibility matrix, compare (6.122), and considering (6.136), it is possible to write the optimal solution specified by (6.138) such as

$$\mathbf{q}_{opt,c} = -\left(\mathbf{H}_C^H \mathbf{H}_C\right)^{-1} \mathbf{H}_C^H \mathbf{P} \mathbf{T} \mathbf{p}_{pC}. \tag{6.139}$$

For a fully determined control systems in which the number of control sources equals the number of error sensors ($L = M$) the optimal solution (6.138) reduced to

$$\mathbf{q}_{opt,c} = -\mathbf{H}_M^{-1} \mathbf{p}_{pM} \tag{6.140}$$

that is the optimal solution for ANC applied to the monitor microphones. Equation (6.140) also clarifies that the title *noise design* is well-founded, because adjusting the noise at the error microphones to the value that was measured during optimal control applied to the monitor microphones in the training phase results in an optimal reduction of the disturbance at the positions of the removed monitor microphones in the control phase.

In the single channel case, where we have one error signal P_C, one monitor signal P_M, one control signal Q and two secondary paths H_C and H_M, the optimal control signal is (as in the virtual microphone approach) determined by

$$Q_{opt,c} = -\frac{1}{H_M} P_{pM} = -T \frac{H_C}{H_M} \frac{P_{pC}}{H_C} =: -\Gamma \frac{P_{pC}}{H_C}. \tag{6.141}$$

Equation (6.141) again clarifies that P_{pC}/H_C, the optimal driving signal for noise cancellation at the control microphone, has to be altered by the in general complex number $\Gamma = T H_C / H_M$, to realize ANC at the position of the monitor microphone.

A graphical interpretation of this approach is for the single channel case shown in Fig. 6.30. It clarifies that in the training phase ANC is applied to cancel noise P_{pM} at the monitor microphone. At the same time $P_{C,t}$ remains at the error microphone. Active cancellation of $P_{pC} - P_{C,t}$ at the error microphone in the control phase designs the anti-noise $H_C Q_{opt,t}$ at this sensor. At the same time the anti-noise $H_M C Q_{opt,t}$ is generated at the position of the monitor microphone. Thus, the primary noise P_{pM} is canceled at this position.

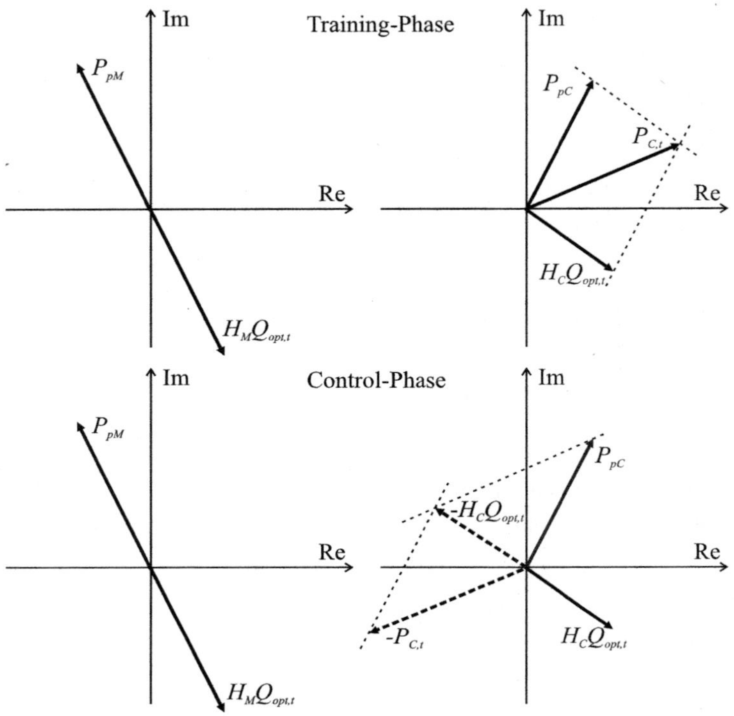

Fig. 6.30 Remote sensor control by sound design at the error microphone. Training phase (*top*): ANC is applied to cancel noise P_{pM} at monitor microphone (*top-left*). At the same time $P_{C,t}$ remains at the control microphone (*top-right*). Control phase (*below*): Active cancellation of $P_{pC} - P_{C,t}$ at the error microphone designs $H_C Q_{opt,t}$ at this sensor (*below right*), but cancels the noise at the position of the monitor microphone (*below left*)

6.5.3 Parametric Controller Design

In contrast to the previous subsection that is focused on RSC in order to shift the control profit from the error microphones to the monitor microphones, the upcoming subsection describes *parametric controller design* in which weighting matrices will be used for cost function modification. Therefore, we will in the following subsection only consider error sensors signals.

As we will see, the application of this approach enables the design of adaptive control schemes that are robust against uncertainties in the primary noise field as well as in the secondary transfer paths. The approach is also capable of making the adaptive control schema robust against plant modeling errors.

To introduce the concept of parametric controller design, we will start with the *nominal design* approach in which we have no uncertainties and perfectly modeled plants. Afterwards, we will address the *robust design* problem.

Nominal Design

The nominal design problem is based on the assumptions that we have

1. No uncertainties in the primary noise, i.e. $\mathbf{p}_p = \mathbf{p}_{p0}$,
2. No uncertainties in the transfer behavior of the secondary paths, i.e. $\mathbf{H} = \mathbf{H}_0$,
3. A perfectly modeled secondary paths such as $\hat{\mathbf{H}}_0 = \mathbf{H}_0$.

The L error signals are collected in the $(L \times 1)$ column matrix $\mathbf{p}$ that represents the superposition of noise $\mathbf{p}_{p0}$ and anti-noise $\mathbf{H}_0\mathbf{q}$ such as

$$\mathbf{p} = \mathbf{p}_{p0} + \mathbf{H}_0\mathbf{q}. \tag{6.142}$$

The $(L \times M)$ matrix $\mathbf{H}_0$ contains the transfer behavior of the LM secondary paths, and the $(M \times 1)$ column matrix $\mathbf{q}$ contains the M control signals. The cost function for nominal parametric controller design is established on the weighted squared sound pressure and on the weighted squared control effort such as

$$J_0 = \mathbf{p}^H \mathbf{W}_p \mathbf{p} + \mathbf{q}\mathbf{W}_q\mathbf{q}, \tag{6.143}$$

where $\mathbf{W}_p$ and $\mathbf{W}_q$ are positive definite hermitian weighting matrices. The parameters of this quadratic form can considering (6.142) in direct analogy to (6.108) be identified to

$$\begin{aligned} \mathbf{A}_0 &\equiv \mathbf{H}_0^H \mathbf{W}_p \mathbf{H}_0 + \mathbf{W}_q, \qquad \mathbf{b}_{01} = \mathbf{b}_{02} \equiv \mathbf{H}_0^H \mathbf{W}_p \mathbf{p}_{p0}, \\ c_0 &\equiv \mathbf{p}_{p0}^H \mathbf{W}_p \mathbf{p}_{p0}. \end{aligned} \tag{6.144}$$

Equation (6.144) clarifies that we have no transformation of the primary noise signals and no transformation of the control signals. This is expressed by

$$\mathbf{P} = \mathbf{I} \quad \text{and} \quad \mathbf{Q} = \mathbf{I}. \tag{6.145}$$

Considering (6.144) and (6.145), the optimal solution for the nominal design problem defined by (6.143) is because of (6.109) given by

$$\mathbf{q}_{opt,0} = -\left(\mathbf{H}_0^H \mathbf{W}_p \mathbf{H}_0 + \mathbf{W}_q\right)^{-1} \mathbf{H}_0^H \mathbf{W}_p \mathbf{p}_{p0}. \tag{6.146}$$

The *nominal control performance* can be defined as the logarithmic ratio between the sum of squared sound pressures in optimal controlled state and the sum of squared sound pressures in the uncontrolled (initial) state such as

$$NR_0 = 10\log_{10}\left(\frac{\mathbf{p}^H(\mathbf{q}_{opt,0})\mathbf{p}(\mathbf{q}_{opt,0})}{\mathbf{p}_{p0}^H \mathbf{p}_{p0}}\right). \tag{6.147}$$

In real world applications, the control performance specified by (6.147) is limited, because we have to avoid nonlinearities that can be caused by nonlinear distortion of canceling sources. This constraint requires a limitation of the control signals and is usually expressed in terms of the *maximum admissible control signal* $Q_{adm}^{\max}$ such as

$$Q_{adm}^{\max} - Q_{opt,0}^{\max} > 0 \quad \text{with } Q_{opt,0}^{\max} = \max(\mathbf{q}_{opt,0}). \tag{6.148}$$

Because we deal with positive definite weighting matrices $\mathbf{W}_p$ and $\mathbf{W}_q$, the cost function parameter matrix $\mathbf{A}$, specified in (6.144) is a hermitian matrix too. For this reason, the associated iterative mapping for the control signals that is considering (6.142), (6.144), and (6.145) according to (6.16) given by

$$\begin{aligned}\mathbf{q}(n+1) &= \mathbf{q}(n) - \mu\big[\big(\mathbf{H}_0^H\mathbf{W}_p\mathbf{H}_0 + \mathbf{W}_q\big)\mathbf{q}(n) + \mathbf{H}_0^H\mathbf{W}_p\mathbf{p}_{p0}(n)\big] \\ &= (\mathbf{I} - \mu\mathbf{W}_q)\mathbf{q}(n) - \mu\mathbf{H}_0^H\mathbf{W}_p\mathbf{p}(n).\end{aligned} \tag{6.149}$$

Equation (6.149) is a stable iteration, if, and only, if the step size μ is, compare Sect. 6.1.3 (6.29), for all eigenvalues λ_m of the matrix $\mathbf{A}_0 = \mathbf{H}_0^H\mathbf{W}_p\mathbf{H}_0 + \mathbf{W}_q$ bounded by

$$0 < \mu < \frac{2}{\lambda_m}. \tag{6.150}$$

Because we consider the nominal state, (6.150) defines the criteria for *nominal stability* that is for all eigenvalues ensured, if we use the greatest eigenvalue to specify the upper bound

$$0 < \mu < \frac{2}{\lambda_{\max}} \quad \text{with } \lambda_{\max} = \max(\lambda_m). \tag{6.151}$$

The *instantaneous nominal control performance* at the time step n can analogue to (6.147) be defined as the logarithmic ratio between sum of squared sound pressures at time step n and the sum of squared sound pressures in the uncontrolled (initial) state such as

$$NR_0(n) = 10\log_{10}\left(\frac{\mathbf{p}^H[\mathbf{q}(n)]\mathbf{p}[\mathbf{q}(n)]}{\mathbf{p}_{p0}^H\mathbf{p}_{p0}}\right). \tag{6.152}$$

Adaptive control based on optimal hermitian weighting matrices $\mathbf{W}_{p,opt}$ and $\mathbf{W}_{q,opt}$ must ensure that we achieve the best nominal control performance, compare (6.147) with a stable update of control signals according to (6.149). Furthermore, every control signal must be less than the maximum admissible control signal as required by the constraint (6.148).

These requirements are not independent and can be contradictory. The goal is to find the best balance between the single requirements. This leads us to the nominal optimization problem that can be formulated as follows:

$$\underset{\mathbf{W}_p,\mathbf{W}_q}{\text{minimize}}\, \mathbf{p}^H(\mathbf{q}_{opt,0})\mathbf{p}(\mathbf{q}_{opt,0}) \tag{6.153}$$

by considering the constraints

$$\begin{aligned} C_1&:\ Q_{adm}^{\max} - Q_{opt,0}^{\max} > 0, \\ C_2&:\ \lambda_{\min}(\mathbf{W}_p, \mathbf{W}_q) > 0, \\ C_3&:\ \lambda_l(\mathbf{W}_p) > 0 \quad \text{for all } l, \\ C_4&:\ \lambda_m(\mathbf{W}_q) > 0 \quad \text{for all } m. \end{aligned} \tag{6.154}$$

C_1 avoids nonlinearities and C_2 ensures that all eigenvalues are positive. This is required to guarantee nominal stability according to (6.151). C_3 and C_4 are taken into account to guarantee that the controller design parameter matrices belong to the group of positive definite matrices, necessary to obtain a convex cost function.

Robust Design

In contrast to the nominal design problem, the robust design problem considers

1. Uncertainties in the primary noise, i.e. $\mathbf{p}_p = \mathbf{p}_{p0} + \Delta\mathbf{p}_p$,
2. Uncertainties in the transfer behavior of the secondary paths, i.e. $\mathbf{H} = \mathbf{H}_0 + \Delta\mathbf{H}$,
3. Plant modeling errors $\hat{\mathbf{H}} = \mathbf{H}_0 + \Delta\hat{\mathbf{H}}$.

The L error signals are again collected in the $(L \times 1)$ column matrix $\mathbf{p}$ that represents the superposition of noise $\mathbf{p}_p$ and anti-noise $\mathbf{Hq}$ such as

$$\mathbf{p} = \mathbf{p}_p + \mathbf{Hq}. \tag{6.155}$$

The $(L \times M)$ matrix $\mathbf{H}$ contains the transfer behavior of the LM secondary paths (including the plant uncertainties $\Delta\mathbf{H}$), and the $(M \times 1)$ column matrix $\mathbf{q}$ contains the M control signals. The cost function for robust parametric controller design is established on the weighted squared sound pressure and on the weighted squared control effort such as

$$J = \mathbf{p}^H \mathbf{W}_p \mathbf{p} + \mathbf{q} \mathbf{W}_q \mathbf{q}, \tag{6.156}$$

where $\mathbf{W}_p$ and $\mathbf{W}_q$ are positive definite hermitian weighting matrices. The parameters of this quadratic form can considering (6.155) in direct analogy to (6.108) be identified to

$$\mathbf{A} \equiv \mathbf{H}^H \mathbf{W}_p \mathbf{H} + \mathbf{W}_q, \qquad \mathbf{b}_1 = \mathbf{b}_2 \equiv \mathbf{H}^H \mathbf{W}_p \mathbf{p}, \qquad c \equiv \mathbf{p}_p^H \mathbf{W}_p \mathbf{p}_p. \tag{6.157}$$

Equation (6.157) clarifies that we have no transformation of the primary noise signals and no transformation of the control signals. This is expressed by

$$\mathbf{P} = \mathbf{I} \quad \text{and} \quad \mathbf{Q} = \mathbf{I}. \tag{6.158}$$

Considering (6.157) and (6.158), the optimal solution for the robust design problem defined by (6.156) is because of (6.109) given by

$$\mathbf{q}_{opt} = -\left(\mathbf{H}^H \mathbf{W}_p \mathbf{H} + \mathbf{W}_q\right)^{-1} \mathbf{H}^H \mathbf{W}_p \mathbf{p}_p. \tag{6.159}$$

The *robust control performance* can be defined as the logarithmic ratio between the sum of squared sound pressures in optimal controlled state and the sum of squared sound pressures in the uncontrolled (initial) state such as

$$NR = 10 \log_{10} \left(\frac{\mathbf{p}^H(\mathbf{q}_{opt}) \mathbf{p}(\mathbf{q}_{opt})}{\mathbf{p}_p^H \mathbf{p}_p} \right). \tag{6.160}$$

In real world applications, the control performance specified by (6.160) is as in the nominal case limited, because we have to avoid nonlinearities that can be caused by nonlinear distortion of canceling sources. This constraint requires a limitation of the control signals and is again expressed in terms of the maximum admissible control signal $Q_{adm}^{\max}$ such as

$$Q_{adm}^{\max} - Q_{opt}^{\max} > 0 \quad \text{with } Q_{opt}^{\max} = \max(\mathbf{q}_{opt}). \tag{6.161}$$

Because we still deal with positive definite weighting matrices $\mathbf{W}_p$ and $\mathbf{W}_q$, the cost function parameter matrix $\mathbf{A}$, specified in (6.157) is also still a hermitian matrix. For this reason, the associated iterative mapping for the control signals that is considering (6.155), (6.157), and (6.158) according to (6.16) given by

$$\begin{aligned}\mathbf{q}(n+1) &= \mathbf{q}(n) - \mu\big[\big(\mathbf{H}^H\mathbf{W}_p\mathbf{H} + \mathbf{W}_q\big)\mathbf{q}(n) + \mathbf{H}^H\mathbf{W}_p\mathbf{p}_p(n)\big] \\ &= (\mathbf{I} - \mu\mathbf{W}_q)\mathbf{q}(n) - \mu\mathbf{H}^H\mathbf{W}_p\mathbf{p}(n).\end{aligned} \tag{6.162}$$

The problem with the second line in (6.162) is that we have to replace $\mathbf{H}$ by the plant model $\hat{\mathbf{H}}$. This yields considering (6.155)

$$\begin{aligned}\mathbf{q}(n+1) &= \mathbf{q}(n) - \mu\big[\big(\hat{\mathbf{H}}^H\mathbf{W}_p\mathbf{H} + \mathbf{W}_q\big)\mathbf{q}(n) + \hat{\mathbf{H}}^H\mathbf{W}_p\mathbf{p}_p(n)\big] \\ &= \mathbf{q}(n) - \mu\big[\hat{\mathbf{A}}\mathbf{q}(n) + \hat{\mathbf{H}}^H\mathbf{W}_p\mathbf{p}_p(n)\big].\end{aligned} \tag{6.163}$$

Equation (6.163) is, compare Sect. 6.1.3 (6.25)–(6.27), stable, if, and only, if

$$0 < \mu < \frac{2\,\mathrm{Re}(\hat{\lambda}_m)}{|\hat{\lambda}_m|^2} \leq \frac{2}{\mathrm{Re}(\hat{\lambda}_m)} \tag{6.164}$$

holds for all in general complex eigenvalues $\hat{\lambda}_m$ of the in general non-hermitian matrix $\hat{\mathbf{A}} = \hat{\mathbf{H}}^H\mathbf{W}_p\mathbf{H} + \mathbf{W}_q$ as introduced in (6.163). To motivate the approximation of these eigenvalues, we follow the argumentation presented in (Kochan 2009) in which it is assumed that the uncertainties $\Delta\mathbf{p}_p$ and $\Delta\mathbf{H}$ as well as the plant modeling errors $\Delta\hat{\mathbf{H}}$ are small compared to the nominal quantities. For this reason, it is possible to neglect terms of higher order. This yields

$$\hat{\mathbf{A}} \approx \mathbf{A}_0 + \mathbf{H}_0^H\mathbf{W}_p\Delta\mathbf{H} + \Delta\hat{\mathbf{H}}\mathbf{W}_p\mathbf{H}_0 =: \mathbf{A}_0 + \Delta\mathbf{A}. \tag{6.165}$$

As also described in (Kochan 2009) it is possible to approximate the eigenvalues of the non-hermitian matrix $\hat{\mathbf{A}}$ using the normalized eigenvectors $\mathbf{a}_{0m}$ of the hermitian matrix $\mathbf{A}_0$ defined in (6.144) such as

$$\hat{\lambda}_m \approx \mathbf{a}_{0m}^H\hat{\mathbf{A}}\mathbf{a}_{0m} = \lambda_m + \mathbf{a}_{0m}^H\Delta\mathbf{A}\mathbf{a}_{0m}. \tag{6.166}$$

Because we consider uncertainties in the primary noise field and the plant as well as plant modeling errors, (6.164) defines the criteria for *robust stability* that is for all eigenvalues ensured, if we use the eigenvalue with the greatest real part to specify the upper bound

$$0 < \mu < \frac{2}{\max[\mathrm{Re}(\hat{\lambda}_m)]}. \tag{6.167}$$

The *instantaneous robust control performance* at the time step n can analogue to (6.160) be defined as the logarithmic ratio between sum of squared sound pressures at time step n and the sum of squared sound pressures in the uncontrolled (initial) state such as

$$NR(n) = 10\log_{10}\left(\frac{\mathbf{p}^H[\mathbf{q}(n)]\mathbf{p}[\mathbf{q}(n)]}{\mathbf{p}_p^H\mathbf{p}_p}\right). \tag{6.168}$$

Adaptive control based on optimal hermitian weighting matrices $\mathbf{W}_{p,opt}$ and $\mathbf{W}_{q,opt}$ must ensure that we achieve the best robust control performance, compare (6.160) with a stable update of control signals according to (6.163). Furthermore, every control signal must be less than the maximum admissible control signal as required by the constraint (6.161).

These requirements are not independent and can be contradictory. The goal is to find the best balance between the single requirements. This leads us to the robust optimization problem that can be formulated as follows:

$$\underset{\mathbf{W}_p,\mathbf{W}_q}{\text{minimize}}\, \mathbf{p}^H(\mathbf{q}_{opt})\mathbf{p}(\mathbf{q}_{opt}) \tag{6.169}$$

by considering the constraints

$$\begin{aligned} C_1:&\quad Q_{adm}^{\max} - Q_{opt}^{\max} > 0,\\ C_2:&\quad \min\{\mathrm{Re}[\hat{\lambda}_m(\mathbf{W}_p,\mathbf{W}_q)]\} > 0,\\ C_3:&\quad \lambda_l(\mathbf{W}_p) > 0 \quad \text{for all } l,\\ C_4:&\quad \lambda_m(\mathbf{W}_q) > 0 \quad \text{for all } m. \end{aligned} \tag{6.170}$$

C_1 avoids nonlinearities and C_2 ensures that the real part of all eigenvalues is positive. This is necessary to guarantee robust stability according to (6.167). C_3 and C_4 are again taken into account to guarantee that the controller design parameter matrices belong to the group of positive definite matrices, necessary to obtain a convex cost function.

However, to solve the optimization problem defined by (6.169) and (6.170) we need a measure of the sum of squared sound pressures in the optimal controlled state as well as a measure of the maximum optimal canceling signal. Both quantities depend on the uncertainties in the primary noise, the uncertainties in the transfer behavior of the secondary paths and the plant modeling errors and can be estimated by the *worst-case approach* that has been presented in (Kochan et al. 2009a) and (Kochan 2009).

This approach is based on the assumption that the optimal solution for the canceling signals $\mathbf{q}_{opt}$ specified by (6.159) can be approximated by the steady state solution of the iterative mapping (6.163). The latter is in direct analogy to (6.102) given by

$$\mathbf{q}_\infty = -(\hat{\mathbf{H}}^H\mathbf{W}_p\mathbf{H} + \mathbf{W}_q)^{-1}\hat{\mathbf{H}}^H\mathbf{W}_p\mathbf{p}_p. \tag{6.171}$$

Furthermore, it is assumed that the nominal plant is perfectly modeled such as

$$\hat{\mathbf{H}} = \mathbf{H}_0, \quad \text{i.e. } \Delta\hat{\mathbf{H}} = \mathbf{0}. \tag{6.172}$$

Uncertainties are taken into account for the primary noise as well as for the secondary transfer paths. This is expressed as

$$\mathbf{p}_p = \mathbf{p}_{p0} + \varDelta\mathbf{p}_p \quad \text{and} \quad \mathbf{H} = \mathbf{H}_0 + \Delta\mathbf{H}. \tag{6.173}$$

Considering (6.171), (6.172) and (6.173), the column matrix of optimal source strengths can be approximates by

$$\mathbf{q}_{opt} \approx \mathbf{q}_\infty = -\left(\hat{\mathbf{H}}_0^H \mathbf{W}_p \mathbf{H}_0 + \hat{\mathbf{H}}_0^H \mathbf{W}_p \varDelta\mathbf{H} + \mathbf{W}_q\right)^{-1} \hat{\mathbf{H}}_0^H \mathbf{W}_p (\mathbf{p}_{p0} + \varDelta\mathbf{p}_p). \tag{6.174}$$

Applying the *Sherman-Morrison-Woodbury formula*,[11] and considering the optimal nominal controller matrix

$$\mathbf{C}_0 = -\left(\hat{\mathbf{H}}_0^H \mathbf{W}_p \mathbf{H}_0 + \mathbf{W}_q\right)^{-1} \hat{\mathbf{H}}_0^H \mathbf{W}_p \tag{6.175}$$

it is possible to express the estimate of optimal control signal matrix such as

$$\mathbf{q}_{opt} = \mathbf{C}_0 \mathbf{p}_{p0} + \mathbf{C}_0 \varDelta\mathbf{p}_{p0} + \mathbf{C}_0 (\mathbf{I} - \varDelta\mathbf{H}\mathbf{C}_0)^{-1} \varDelta\mathbf{H}\mathbf{C}_0. \tag{6.176}$$

To obtain a linear approximation we have to replace $(\mathbf{I} - \varDelta\mathbf{H}\mathbf{C}_0)^{-1}$ by $\mathbf{I} + \varDelta\mathbf{H}\mathbf{C}_0$. This is, according to (Elliott 2001), possible, if the magnitude of $\mathbf{C}_0 \varDelta\mathbf{H}\mathbf{C}_0 \varDelta\mathbf{H}\mathbf{C}_0$ is much less than that of $\varDelta\mathbf{H}\mathbf{C}_0$. This last approximation leads

$$\mathbf{q}_{opt} = \mathbf{q}_{opt,0} + \varDelta\mathbf{q}_{opt}, \tag{6.177}$$

where

$$\mathbf{q}_{opt,0} := \mathbf{C}_0 \mathbf{p}_{p0} \quad \text{and} \quad \varDelta\mathbf{q}_{opt} := \mathbf{C}_0 \varDelta\mathbf{p}_p + \mathbf{C}_0 \varDelta\mathbf{H}\mathbf{C}_0 \mathbf{p}_{p0}. \tag{6.178}$$

Equation (6.178) enables an approximation of $Q_{opt}^{\max}$ as well as the determination of an upper bound for the sum of squared sound pressures. This upper bound is determined using the triangle inequality and neglecting the quadratic terms. The resulting estimation of the worst-case residual noise at the error sensors can according to (Kochan 2009) be expressed by

$$\max\left[\mathbf{p}^H(\mathbf{q}_{opt})\mathbf{p}(\mathbf{q}_{opt})\right] \leqslant \left\{ \begin{array}{c} \|\mathbf{H}_0 \mathbf{q}_{opt,0} + \mathbf{p}_{p0}\|_2 + \cdots \\ + \max\|\Delta\mathbf{H}\mathbf{q}_{opt,0}\|_2 + \max\|\mathbf{H}_0\mathbf{C}_0\varDelta\mathbf{p}_p\|_2 + \cdots \\ + \max\|\mathbf{H}_0\mathbf{C}_0\varDelta\mathbf{H}\mathbf{q}_{opt,0}\|_2 + \max\|\varDelta\mathbf{p}_p\|_2 \end{array} \right\}^2, \tag{6.179}$$

where $\| \ \|_2$ specifies the *Euclidean norm* for column matrices and matrices such as $\|\mathbf{x}\|_2 := \sqrt{\mathbf{x}^H\mathbf{x}}$ and $\|\mathbf{A}\|_2 := \sqrt{\mathrm{tr}(\mathbf{A}^H\mathbf{A})}$. Minimizing the worst-case approximation (6.179) instead of the sum of squared sound pressure in (6.169) leads to the

[11] The Sherman-Morrison-Woodbury formula states that inverse matrix to the matrix $\mathbf{A} + \mathbf{O}\mathbf{P}^H$ is given by

$$(\mathbf{A} + \mathbf{O}\mathbf{P}^H)^{-1} = \mathbf{A}^{-1} - \mathbf{A}^{-1}\mathbf{O}(\mathbf{I} + \mathbf{P}^H\mathbf{A}^{-1}\mathbf{O})\mathbf{P}^H\mathbf{A}^{-1}.$$

To apply this formula we have to use the identifications $\mathbf{A} \equiv \hat{\mathbf{H}}_0^H \mathbf{W}_p \mathbf{H}_0$, $\mathbf{O} \equiv \hat{\mathbf{H}}_0^H \mathbf{W}_p$, and $\mathbf{P}^H \equiv \varDelta\mathbf{H}$. For more details see (Golub and Loan 1996).

worst case solution of the robust optimization problem, if in addition to the nominal plant $\mathbf{H}_0$ and the nominal disturbance $\mathbf{p}_{p0}$, the uncertainties $\Delta\mathbf{H}$ and $\Delta\mathbf{p}_p$ can be determined by measurements or numerical simulations.

A more simple but effective approach to solve the robust design problem is to use the effort weighting parameter β, compare Sect. 6.4.1 (6.91), as a design parameter. Following this approach that is in great detail described in (Elliott 2001), the parameter design matrices are given by

$$\mathbf{W}_p = \mathbf{I} \quad \text{and} \quad \mathbf{W}_q = \beta\mathbf{I}. \tag{6.180}$$

In direct analogy to Sect. 6.4.2, the effort weighting parameter β contributes to the real part of the eigenvalues of the matrix

$$\hat{\mathbf{A}} = \mathbf{H}^H\mathbf{H} + \beta\mathbf{I} + \Delta\mathbf{A} =: \bar{\mathbf{A}} + \beta\mathbf{I}. \tag{6.181}$$

Thus, the robust stability criteria is in direct analogy to (6.105) in this special case given by

$$0 < \mu < \frac{2\,\mathrm{Re}(\bar{\lambda}_m + \beta)}{|\bar{\lambda}_m + \beta|^2} \leq \frac{2}{\mathrm{Re}(\bar{\lambda}_m + \beta)}, \tag{6.182}$$

where $\bar{\lambda}_m$ are the eigenvalues of the matrix

$$\bar{\mathbf{A}} = \mathbf{H}^H\mathbf{H} + \Delta\mathbf{A}. \tag{6.183}$$

Equation (6.182) clarifies that a positive weighting parameter β can be used to guarantee positive real parts for all eigenvalues $\bar{\lambda}_m$. To ensure robust stability by the single weighting parameter approach for all eigenvalues, the upper step size bound must be specified by

$$0 < \mu < \frac{2}{\max[\mathrm{Re}(\bar{\lambda}_m + \beta)]}. \tag{6.184}$$

Comments on Remote Sensors Control with a Parametric Controller

A parametric controller can also be used to enable RSC. In contrast to the virtual microphone approach and RSC based on noise design at the sensors that are based on mathematical algorithms used to calculate the missing monitor signals, RSC with a parametric controller yields an optimization problem for the controller weighting matrices $\mathbf{W}_p$ and $\mathbf{W}_q$.

As shown in Fig. 6.28, the goal of RSC is the minimization of the squared sound pressures measured at a certain set of monitor microphones. If active control would be directly applied to the monitor sensors, the column matrix of optimal source strengths is according to (6.132) given by

$$\mathbf{q}_{opt,M} = -\left(\mathbf{H}_M^H\mathbf{H}_M\right)^{-1}\mathbf{H}_M^H\mathbf{p}_{pM}. \tag{6.185}$$

Parametric control of disturbances measured at a certain set of error sensors can be establishes on a cost function similar to (6.156) such as

$$J = \mathbf{p}_C^H \mathbf{W}_p \mathbf{p}_C + \mathbf{q}\mathbf{W}_q\mathbf{q}. \tag{6.186}$$

The associated column matrix of optimal source strengths is in direct analogy to (6.159) given by

$$\mathbf{q}_{opt,C} = -\left(\mathbf{H}_C^H \mathbf{W}_p \mathbf{H} + \mathbf{W}_q\right)^{-1} \mathbf{H}_C^H \mathbf{W}_p \mathbf{p}_{pC}. \tag{6.187}$$

To enable RSC $\mathbf{q}_{opt,C}$ must match $\mathbf{q}_{opt,M}$. This leads an optimization problem that can be formulated as follows:

$$\underset{\mathbf{W}_p,\mathbf{W}_q}{\text{minimize}}\left(\mathbf{q}_{opt,C}^H - \mathbf{q}_{opt,M}^H\right)\left(\mathbf{q}_{opt,C} - \mathbf{q}_{opt,M}\right). \tag{6.188}$$

An alternative formulation of this optimization problem is based on the squared sound pressure measured at the monitor microphones, considering that the canceling sources will in practice be driven by $\mathbf{q}_{opt,C}$ as specified by (6.187). This alternative formulation reads

$$\underset{\mathbf{W}_p,\mathbf{W}_q}{\text{minimize}}(\mathbf{p}_{pM} + \mathbf{H}_M \mathbf{q}_{opt,C})^H (\mathbf{p}_{pM} + \mathbf{H}_M \mathbf{q}_{opt,C}). \tag{6.189}$$

6.6 Multi-channel Control of Stochastic Disturbances

If the interior noise field is caused by stochastic disturbances, signal processing for active control is much more sophisticated. In order to reject delays resulting from Fourier and inverse Fourier transformation optimal control of stochastic disturbances as well as adaptive control of stochastic disturbances will be discussed in discrete-time domain in the upcoming section.

6.6.1 Optimal Control of Stochastic Disturbances

The block diagram of an active feed-forward control schema that is applied to attenuate stochastic disturbances is shown in Fig. 6.31. The control task is based on L error signals, M control signals, and K reference signals. The controller is represented by the $(MKI \times 1)$ column matrix $\mathbf{w}(n)$ that contains MKI control filter coefficients $w_{mki}(n)$.

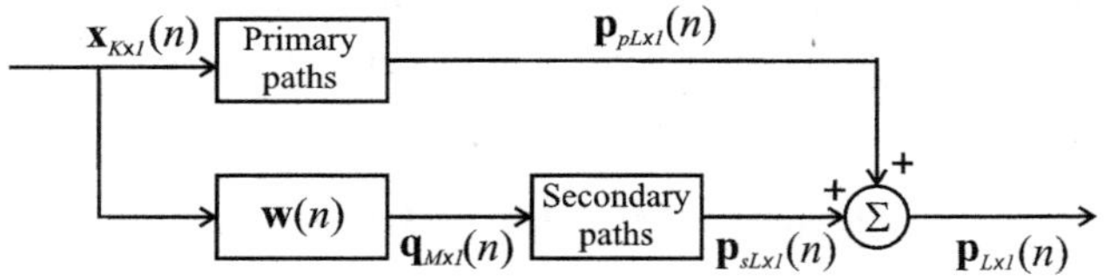

Fig. 6.31 Block diagram for multi-channel control of stochastic disturbances in time domain

The $(L \times 1)$ column matrix $\mathbf{p}(n) = [p_1(n)\ p_2(n)\ \dots\ p_l(n)\ \dots\ p_L(n)]^T$ contains the error signals that result for every discrete time step n from the superposition of noise $\mathbf{p}_p(n)$ and anti-noise $\mathbf{p}_s(n)$. The sound pressure at the l-th error microphone is therefore given by

$$p_l(n) = p_{pl}(n) + p_{sl}(n) = p_{pl}(n) + \sum_{m=1}^{M}\sum_{j=0}^{J-1} h_{lmj} q_m(n-j). \qquad (6.190)$$

The anti-noise results from the M control signals $q_m(n)$ that are altered by the LM secondary paths each modeled by an J-th order FIR filter. The driving signal of the m-th canceling loudspeaker

$$q_m(n) = \sum_{k=1}^{K}\sum_{i=0}^{I-1} w_{mki}(n) x_k(n-i) \qquad (6.191)$$

is calculated using contributions from K reference signals $x_1(n), x_2(n), \dots, x_K(n)$, each filtered by an I-th oder FIR filter with coefficients $w_{mki}(n)$. Combining (6.190) and (6.191) results in

$$p_l(n) = p_{pl}(n) + \sum_{m=1}^{M}\sum_{j=0}^{J-1}\sum_{k=1}^{K}\sum_{i=0}^{I-1} h_{lmj} w_{mki}(n) x_k(n-i-j). \qquad (6.192)$$

Equation (6.192) clarifies that the acoustic pressure measured at the l-th error microphone consists of the primary disturbance $p_{pl}(n)$ and the secondary contributions of M controllers. By introducing the *filtered reference signal*

$$x'_{lmk}(n) = \sum_{j=0}^{J-1} h_{lmj} x_k(n-j) \qquad (6.193)$$

(6.192) can be rewritten such as

$$p_l(n) = p_{pl}(n) + \sum_{m=1}^{M}\sum_{k=1}^{K}\sum_{i=0}^{I-1} w_{mki}(n) x'_{lmk}(n-i). \qquad (6.194)$$

The objective of the control task is minimizing the expectation of both the sum of the squared error signals and the sum of squared control signals. We therefore introduce the discrete-time domain equivalent to (6.91) such as

$$\begin{aligned} J(n) &= E\left[\sum_{l=1}^{L} p_l^2(n) + \beta \sum_{i=0}^{MK(I-1)} w_i^2(n)\right] \\ &= E\big[\mathbf{p}^T(n)\mathbf{p}(n) + \beta \mathbf{w}^T(n)\mathbf{w}(n)\big], \end{aligned} \qquad (6.195)$$

where the $(MKI \times 1)$ column matrix

$$\mathbf{w}(n) = \big[\mathbf{w}_0^T(n)\ \mathbf{w}_1^T(n)\ \dots\ \mathbf{w}_{I-1}^T(n)\big]^T \qquad (6.196)$$

is consisting of I $(MK \times 1)$ submatrices

$$\mathbf{w}_i(n) = \left[w_{11i}(n)\; w_{12i}(n) \ldots w_{1Ki}(n)\; w_{21i}(n) \ldots w_{MKi}(n) \right]^T. \quad (6.197)$$

If we further introduce the $(L \times MKI)$ filter matrix

$$\mathbf{X}(n) = \begin{pmatrix} \mathbf{x}_1^T(n) & \mathbf{x}_1^T(n-1) & \cdots & \mathbf{x}_1^T(n-I+1) \\ \mathbf{x}_2^T(n) & \mathbf{x}_2^T(n-1) & \cdots & \mathbf{x}_2^T(n-I+1) \\ \vdots & \vdots & & \vdots \\ \mathbf{x}_l^T(n) & \mathbf{x}_l^T(n-1) & \cdots & \mathbf{x}_l^T(n-I+1) \\ \vdots & \vdots & & \vdots \\ \mathbf{x}_L^T(n) & \mathbf{x}_L^T(n-1) & \cdots & \mathbf{x}_L^T(n-I+1) \end{pmatrix} \quad (6.198)$$

that contains the filtered reference signals, defined by (6.193) and arranged in LI column matrices $\mathbf{x}_l(n)$, each with dimension $(MK \times 1)$ such as

$$\mathbf{x}_l(n) = \left[x'_{l11}(n)\; x'_{l12}(n) \ldots x'_{l1K}(n)\; x'_{l21}(n) \ldots x'_{lMK}(n) \right]^T, \quad (6.199)$$

we can write the real valued quadratic form (6.195) analog to (6.6) such as

$$J(n) = \mathbf{w}^T(n)\mathbf{A}(n)\mathbf{w}(n) + \mathbf{w}^T(n)\mathbf{b}_1(n) + \mathbf{b}_2^T\mathbf{w}(n) + c(n), \quad (6.200)$$

where

$$\mathbf{A} \equiv E\left[\mathbf{X}^T\mathbf{X} + \beta\mathbf{I}\right], \qquad \mathbf{b}_1 = \mathbf{b}_2 \equiv E\left[\mathbf{X}^T\mathbf{p}_p\right], \qquad c \equiv E\left[\mathbf{p}_p^T\mathbf{p}_p\right]. \quad (6.201)$$

The optimal solution that minimizes (6.200) is according to Sect. 6.1.2 (6.13) given by

$$\mathbf{w}_{opt} = -\left\{E\left[\mathbf{X}^T(n)\mathbf{X}(n) + \beta\mathbf{I}\right]\right\}^{-1} E\left[\mathbf{X}^T(n)\mathbf{p}_p(n)\right]. \quad (6.202)$$

It is obvious that inverting the $(MKI \times MKI)$ matrix $E[\mathbf{X}^T(n)\mathbf{X}(n)]$ results in a high computational effort,[12] even if a block Toeplitz form, see (Moschytz and Hofbauer 2000), results from ordering the elements of $\mathbf{x}_l(n)$ as described in (6.199).

6.6.2 Adaptive Control of Stochastic Disturbances

As in the case of harmonic disturbances it can be advantageous, if the control algorithm is capable of adapting the control signals with respect to the error signals.

For this reason the upcoming subsection presents four approaches that can be used to establish adaptive feed-forward control for stochastic and of course also tonal disturbances in time domain. Common to all these approaches is a rough simplification—the underlying cost function is established on the actual values, and not on the expectation values.

[12] As shown in (Elliott 2001) a control system that uses 6 reference signals, 4 canceling sources and 128 coefficients in each filter would lead to a filter matrix with 3072×3072 elements. Assembling and inverting of such a filter matrix in every time step n is not the goal of fast signal processing.

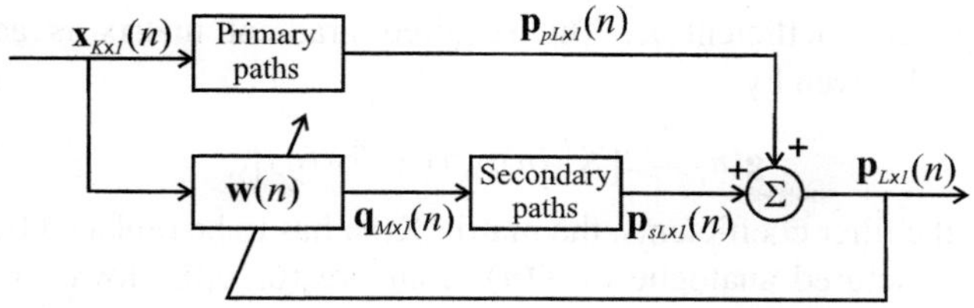

Fig. 6.32 Multi-channel adaptive feed-forward control of stochastic disturbances in time domain

Filtered Reference Approach

Application of Steepest-Decent Algorithm A block diagram of time domain adaptive control applied to attenuate stochastic input to the multi-channel primary path is shown in Fig. 6.32. The information processing is based on K reference signals, M control signals, and L error signals.

In the following it is assumed that the response of the LM secondary paths is modeled by LM FIR filters, each of length J. It will be shown that these *plant models* are used for pre-filtering of reference signals.

The M controller outputs $q_m(n)$ are generated by MK FIR control filters. Each control filter contains I filter coefficients that are updated by an iterative mapping that is derived from the steepest-decent algorithm.

As mentioned in Sect. 6.2, the control schema shown in Fig. 6.32 acts causal, if the minimum acoustic delay $\tau_{Min} = \Delta x_{Min}/c$ that is determined by the smallest distance Δx_{Min} between a reference sensor and an error sensor is greater than the time span $\tau_{s,Min} = \Delta x_{s,Min}/c$, the minimum delay of a secondary path. The latter is determined by $\Delta x_{s,Min}$, the smallest distance between a source and an error sensor. The time left for the signal processing task is $\tau_{ANC} = \tau_{Min} - \tau_{Min,s}$.

To simplify the stochastic control task, the cost function used for adaptive control is defined on actual values of error signals and control signals such as

$$\begin{aligned} J(n) &= \sum_{l=1}^{L} p_l^2(n) + \beta \sum_{i=0}^{MK(I-1)} w_i^2(n) \\ &= \mathbf{p}^T(n)\mathbf{p}(n) + \beta \mathbf{w}^T(n)\mathbf{w}(n) \\ &= \mathbf{w}^T(n)\mathbf{A}(n)\mathbf{w}(n) + \mathbf{w}^T(n)\mathbf{b}_1(n) + \mathbf{b}_2^T\mathbf{w}(n) + c(n), \end{aligned} \tag{6.203}$$

where

$$\mathbf{A} \equiv \mathbf{X}^T\mathbf{X} + \beta\mathbf{I}, \qquad \mathbf{b}_1 = \mathbf{b}_2 \equiv \mathbf{X}^T\mathbf{p}_p, \qquad c \equiv \mathbf{p}_p^T\mathbf{p}_p. \tag{6.204}$$

All matrices are structured as in the previous subsection. Adaptive control of stochastic disturbances based on (6.203) allows for effective application of the steepest-decent algorithm

$$\mathbf{w}(n+1) = \mathbf{w}(n) - \frac{\mu}{2}\mathbf{g}(n) \tag{6.205}$$

with convergence coefficient μ. The required gradient matrix is according to Sect. 6.1.2 (6.13) given by

$$\mathbf{g}(n) = 2\big[\mathbf{X}^T(n)\mathbf{p}(n) + \beta\mathbf{w}(n)\big]. \tag{6.206}$$

To update the filter coefficients, the matrix $\mathbf{X}(n)$ has to be replaced by the matrix $\hat{\mathbf{X}}(n)$ that is structured analogue to $\mathbf{X}(n)$, compare (6.198). However, the matrix $\hat{\mathbf{X}}(n)$ contains reference signals that are filtered using secondary path models $\hat{h}_{lmj}$ such as

$$\hat{x}'_{lmk}(n) = \sum_{j=0}^{J-1} \hat{h}_{lmj} x_k(n-j). \tag{6.207}$$

That is why this approach is called *filtered reference approach*. Because the sum of squared error signals is minimized considering the control effort by a scalar weighting factor that, in the absence of an error signal, tends the control filter coefficients to decay or "leak" to zero, this approach is also known as the "leaky" *filtered reference least mean square* (FxLMS) algorithm. Combining (6.205) and (6.206), yields the iterative mapping for the control filter coefficients

$$\begin{aligned}\mathbf{w}(n+1) &= (1-\mu\beta)\mathbf{w}(n) - \mu\hat{\mathbf{X}}^T(n)\mathbf{p}(n)\\ &= \mathbf{w}(n) - \mu\big[\hat{\mathbf{X}}^T(n)\mathbf{p}_p(n) + \big(\hat{\mathbf{X}}^T\mathbf{X} + \beta\mathbf{I}\big)\mathbf{w}(n)\big].\end{aligned} \tag{6.208}$$

Considering (6.190), (6.191), (6.196), (6.197), (6.198) and (6.199) the update schema for the mki-th control filter coefficient according to (6.208) reads

$$w_{mki}(n+1) = (1-\mu\beta)w_{mki}(n) - \mu\sum_{l=1}^{L} \hat{x}'_{lmk}(n-i)p_l(n). \tag{6.209}$$

The algorithmic formulation of the multi-channel FxLMS algorithm with K reference signals, M control signals, L error signals, I coefficients in each control filter, and J coefficients for each secondary path model is summarized in the left column of Table 6.2. The right column of this table contains the number of multiplication operations required per sample. It can be seen that the complexity of the FxLMS is given by a number of

$$C_{FxLMS} = KM\big[I + (I+J)L\big] + IKM \tag{6.210}$$

multiplies at each iteration, a quantity that grows precipitously as the numbers of input sensors, output actuators, and error sensors is increased.

Comments on Plant Modeling The required plant models can again be identified using the steepest-decent algorithm, compare Sect. 6.4.1 Fig. 6.27. If we want to approximate the response of the lm-th secondary transfer path h_{lm} by a FIR filter of order J such as

$$\begin{aligned}J_{hlm}(n) &= \big|p_{slm}(n) - \hat{p}_{slm}\big|^2\\ &= \big|p_{slm}(n) - \hat{h}_{lm} * q_m(n)\big|^2\end{aligned}$$

Table 6.2 The multi-channel filtered reference least mean square algorithm in time domain with K reference signals, M control signals, L error signals, I coefficients in each control filter, and J coefficients for each secondary path model

FxLMS algorithmic equations	Multiplications per iteration
for $m = 1$ to M do	
Compute controller output:	
$q_m(n) = \sum_{k=1}^{K} \sum_{i=0}^{I-1} w_{mki}(n) x_k(n-i)$	IKM
for $k = 1$ to K do	
for $l = 1$ to L do	
Pre-filtering of reference signals:	
$\hat{x}'_{lmk}(n) = \sum_{j=0}^{J-1} \hat{h}_{lmj} x_k(n-j)$	$JKLM$
end	
for $i = 0$ to $I-1$ do	
Update of control filter coefficients:	
$w_{mki}(n+1) = (1-\mu\beta) w_{mki}(n) - \mu \sum_{l=1}^{L} \hat{x}'_{lmk}(n-i) p_l(n)$	$IKM + IKLM$
end	
end	
end	

$$\approx \left| p_s(n) - \sum_{j=0}^{J-1} \hat{h}_{lmj} q_m(n-j) \right|^2, \tag{6.211}$$

the iterative mapping for the j-th coefficient of the $(J \times 1)$ column matrix $\hat{\mathbf{h}}_{lm}$ that is used to approximate the impulse response of the secondary path h_{lm} reads

$$\hat{h}_{lmj}(n+1) = \hat{h}_{lmj}(n) + \mu \left[p_{slm}(n) - \sum_{j=0}^{J-1} \hat{h}_{lmj}(n) q_m(n-j) \right] q_m(n-j). \tag{6.212}$$

Comments on Stability If the iterative algorithm given by (6.208) is stable, it will reach the steady-state in which $\mathbf{w}(n+1)$ is equal to $\mathbf{w}(n)$. The steady-state solution

$$\mathbf{w}_\infty = -\left(\hat{\mathbf{X}}^T \mathbf{X} + \beta \mathbf{I}\right)^{-1} \hat{\mathbf{X}}^T \mathbf{p}_p \tag{6.213}$$

differs from the theoretical optimal solution (6.202), because the actual values are used instead of the expectation values and due to differences between the physical plants and the plant models. The combination of (6.208) and (6.213) leads in direct analogy to procedure described in Sect. 6.1.3 to the modified iteration schema

$$\mathbf{w}(n+1) - \mathbf{w}_\infty = \left[\mathbf{I} - \mu\left(\hat{\mathbf{X}}^T \mathbf{X} + \beta \mathbf{I}\right)\right]\left(\mathbf{w}(n) - \mathbf{w}_\infty\right). \tag{6.214}$$

The expected deviation between the set of control filter coefficients at discrete time $n+1$ and the control filter coefficients in the steady state $E[\mathbf{w}(n+1) - \mathbf{w}_\infty]$

is therefore determined by the matrix $E[\hat{\mathbf{X}}^H\mathbf{X}+\beta\mathbf{I}]$. This clarifies that the iterative mapping (6.214) is, compare (6.26), stable, if, and only if, the in general complex eigenvalues λ_m of the matrix $E[\hat{\mathbf{X}}^H\mathbf{X}+\beta\mathbf{I}]$ are bounded by

$$\left|1-\mu(\lambda_m+\beta)\right|<1 \quad \text{for all } m \tag{6.215}$$

i.e.

$$0<\mu<\frac{2[\mathrm{Re}(\lambda_m)+\beta]}{|\lambda_m+\beta|^2} \quad \text{for all } m. \tag{6.216}$$

Equation (6.216) makes clear that the real valued parameter β can be used to guarantee a positive real part for all eigenvalues λ_m. If this is the case, the maximum step size is determined by the largest eigenvalue $\lambda_{\max}$.

For practical application it is recommended to use a variable convergence coefficient $\mu(n)$[13] that is defined as the quotient of the dimensionless power normalized convergence coefficient $\tilde{\mu}$ and the estimate of a filtered reference signal power $\hat{\Pi}_x(n)$ such as

$$\mu(n)=\frac{\tilde{\mu}}{\mathrm{Max}[\hat{\Pi}_x(n),\Pi_{x,Min}]} \quad \text{with } 0<\tilde{\mu}<2 \tag{6.217}$$

to guarantee stability. $\Pi_{x,Min}$ defines a lower bound that can be estimated e.g. from the self noise of the electric control circle. $\hat{\Pi}_x(n)$ can be calculated using a moving average based on the last $\tilde{I}$ data points of a filtered reference signal $\hat{x}'$

$$\hat{\Pi}_x(n)=\hat{\Pi}_x(n-1)+\frac{\hat{x}'^2(n)-\hat{x}'^2(n-\tilde{I})}{\tilde{I}}, \tag{6.218}$$

see (Kuo and Morgan 1996) or, as also suggested in (Kuo and Morgan 1996), by a recursive update schema with leakage parameter ν such as

$$\hat{\Pi}_x(n)=(1-\nu)\hat{\Pi}_x(n-1)+\nu\hat{x}'^2(n) \quad \text{with } \nu\approx 1/\tilde{I}. \tag{6.219}$$

Modified Filtered Reference Approach

To speed up the adaption process of the FxLMS algorithm, it is possible to implement the *modified filtered reference least mean square* (MFxLMS) algorithm proposed by Bjarnason (1992). This control schema, see also (Elliott 2001), ensures rapid adaption of the controller by removing the effects of the secondary path from the error signal.

To understand the advantage of this approach, it is necessary to follow the information flow shown in Fig. 6.33. This figure clarifies that the modification of the

---

[13] As shown in (Elliott 2001) and (Kuo and Morgan 1996) stability of the single channel least mean square algorithm is assured, if $0<\mu<2/(\tilde{I}\cdot E[x^2(n)])$, where $\tilde{I}$ can be chosen to equal the length of the control filter I.

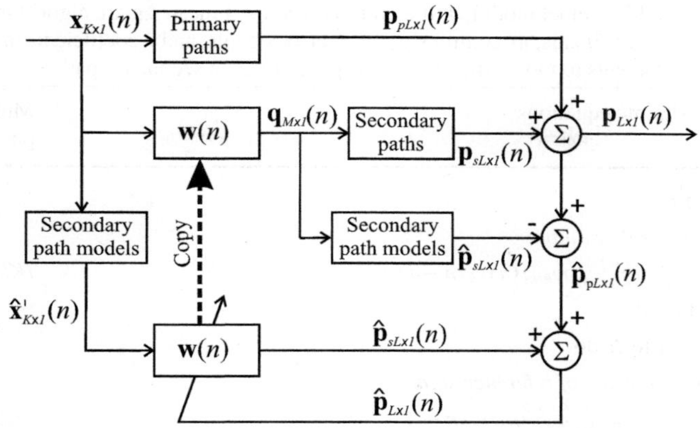

Fig. 6.33 Block diagram for multi-channel adaptive control of stochastic disturbances in time domain using the modified filtered reference approach

control structure leads to a control loop in which $\hat{\mathbf{p}}_{sL\times 1}(n)$, the output of the adaptive control filter, must not pass a secondary path (or a secondary path model) before it is returned and used to update the control filter coefficients.

For this reason, the MFxLMS algorithm effectively removes the delays in the response of the secondary paths from the adaption of the controller.

The l-th modified error signal $\hat{p}_l(n)$ is obtained by (i) subtracting an estimate of the l-th anti-noise signal from the original error signal $p_l(n)$ to calculate an approximation of the l-th primary noise signal

$$\hat{p}_{pl}(n) = p_l(n) - \sum_{m=1}^{M}\sum_{j=0}^{J-1} \hat{h}_{lmj} q_m(n-j), \tag{6.220}$$

where $q_m(n)$ is the m-th controller output as specified by (6.191), and (ii) by adding an estimation of the l-th canceling signal $\hat{p}_{sl}(n)$ to $\hat{p}_{pl}(n)$ in oder to obtain the l-th modified error signal

$$\begin{aligned}\hat{p}_l(n) &= \hat{p}_{pl}(n) + \sum_{m=1}^{M}\sum_{j=0}^{J-1}\sum_{k=1}^{K}\sum_{i=0}^{I-1} \hat{h}_{lmj} w_{mki}(n) x_k(n-i-j)\\ &= \hat{p}_{pl}(n) + \sum_{m=1}^{M}\sum_{k=1}^{K}\sum_{i=0}^{I-1} w_{mki}(n) x'_{lmk}(n-i).\end{aligned} \tag{6.221}$$

Equation (6.221) is equivalent to (6.194), if $\hat{h}_{lmj}$ is replaced by h_{lmj}. The iterative mapping for the control filter coefficient update is now performed using the modified error such as

$$w_{mki}(n+1) = (1-\mu\beta) w_{mki}(n) - \mu \sum_{l=1}^{L} \hat{x}'_{lmk}(n-i)\hat{p}_l(n). \tag{6.222}$$

Table 6.3 The multi-channel modified filtered reference least mean square algorithm in time domain with K reference signals, M control signals, L error signals, and I coefficients in each control filter using J coefficients to model the impulse response of each secondary path

MFxLMS algorithmic equations	Multiplications per iteration
for $m = 1$ to M do	
Compute controller output:	
$q_m(n) = \sum_{k=1}^{K} \sum_{i=0}^{I-1} w_{mki}(n) x_k(n-i)$	*IKM*
for $l = 1$ to L do	
for $k = 1$ to K do	
Pre-filtering of reference signals:	
$\hat{x}'_{lmk}(n) = \sum_{j=0}^{J-1} \hat{h}_{lmj} x_k(n-j)$	*JKLM*
end	
end	
end	
for $l = 1$ to L do	
Compute estimate of disturbance:	
$\hat{p}_{pl}(n) = p_l(n) - \sum_{m=1}^{M} \sum_{j=0}^{J-1} \hat{h}_{lmj} q_m(n-j)$	*JLM*
Compute estimate of error signal:	
$\hat{p}_l(n) = \hat{p}_{pl}(n) + \sum_{m=1}^{M} \sum_{k=1}^{K} \sum_{i=0}^{I-1} w_{mki}(n) x'_{lmk}(n-i)$	*IKLM*
end	
for $m = 1$ to M do	
for $k = 1$ to K do	
for $i = 0$ to $I - 1$ do	
Update of control filter coefficients:	
$w_{mki}(n+1) = (1 - \mu\beta) w_{mki}(n) - \mu \sum_{l=1}^{L} \hat{x}'_{lmk}(n-i) \hat{p}_l(n)$	*IKLM* + *IKM*
end	
end	
end	

The algorithmic formulation of the multi-channel MFxLMS algorithm with K reference signals, M control signals, L error signals, I coefficients in each control filter, and J coefficients for each secondary path model is summarized in the left column of Table 6.3. The right column of this table contains the number of multiplication operations required per sample. It can be seen that the complexity of the MFxLMS is given by

$$C_{MFxLMS} = KM\big[I + (I + 2J)L\big] + JLM + IKM. \tag{6.223}$$

Filtered Error Approach

To reduce the computational complexity, adaptive control of stochastic disturbances can also be realized using the *filtered error least mean square* (FeLMS) algorithm, discussed in (Elliott 2001). This algorithm is also known as the *adjoint* LMS algorithm, see (Wan 1996).

However, in contrast to (Elliott 2001) we will (only) derive an algorithmic formulation considering the cost function (6.203). We again use the steepest-decent algorithm to establish an iterative mapping that is used to update the control filter weights such as

$$w_{mki}(n+1) = w_{mki}(n) - \frac{\mu}{2}\frac{\partial J(n)}{\partial w_{mki}(n)}. \tag{6.224}$$

Considering (6.190), (6.191), (6.196), and (6.197) the partial derivative of $J(n)$ according to $w_{mki}(n)$ reads

$$\frac{\partial J(n)}{\partial w_{mki}(n)} = 2\sum_{l=1}^{L}\sum_{j=0}^{J-1} p_l(n)h_{lmj}x_k(n-i-j) + 2\beta w_{mki}(n). \tag{6.225}$$

If we introduce the time shift $n' := n - j$ we can rewrite (6.225) such as

$$\begin{aligned}\frac{\partial J(n)}{\partial w_{mki}(n)} &= 2\sum_{l=1}^{L}\sum_{j=0}^{J-1} p_l(n'+j)h_{lmj}x_k(n'-i) + 2\beta w_{mki}(n'+j) \\ &= 2f_m(n')x_k(n'-i) + 2\beta w_{mki}(n'+j),\end{aligned} \tag{6.226}$$

where

$$f_m(n') = \sum_{l=1}^{L}\sum_{j=0}^{J-1} p_l(n'+j)h_{lmj} \tag{6.227}$$

is the m-th *filtered error signal*. The problem with using (6.227) in a real-time implementation is that an instantaneous value of the filtered error $f_m(n)$ cannot be implemented in a causal system. The solution to this problem is to delay both the reference signals and the filtered error signals by a delay of J samples. The final form of the iterative mapping using the FeLMS is therefore given by

$$w_{mki}(n+1) = (1-\mu\beta)w_{mki}(n) - \mu f_m(n-J)x_k(n-i-J) \tag{6.228}$$

where

$$f_m(n-J) = \sum_{l=1}^{L}\sum_{j=0}^{J-1} h_{lmj}p_l(n-J+j). \tag{6.229}$$

In practice we would considering the transformation $j' = J - j$ use the filtered error signal

Table 6.4 The multi-channel filtered error least mean square algorithm in time domain with K reference signals, M control signals, L error signals, and I coefficients in each control filter using J coefficients to model the impulse response of each secondary path

FeLMS algorithmic equations	Multiplications per iteration
for $m = 1$ to M do	
Compute controller output:	
$q_m(n) = \sum_{k=1}^{K} \sum_{i=0}^{I-1} w_{mki}(n) x_k(n-i)$	*IKM*
Pre-filtering of error signals:	
$\hat{f}_m(n-J) = \sum_{l=1}^{L} \sum_{j'=0}^{J-1} \hat{h}_{lm(J-j')} p_l(n-j')$	*JLM*
for $k = 1$ to K do	
for $i = 0$ to $I - 1$ do	
Update of control filter coefficients:	
$w_{mki}(n+1) = (1-\mu\beta) w_{mki}(n) - \mu \hat{f}_m(n-J) x_k(n-i-J)$	*2IKM*
end	
end	
end	

$$\hat{f}_m(n-J) = \sum_{l=1}^{L} \sum_{j=0}^{J-1} \hat{h}_{lmj} p_l(n-J+j)$$
$$= \sum_{l=1}^{L} \sum_{j'=0}^{J-1} \hat{h}_{lm(J-j')} p_l\big(n-j'\big), \tag{6.230}$$

where $\hat{h}_{lm(J-j')}$ are the coefficients of the impulse response of the lm-th secondary path model. For notational simplicity it is assumed that this plant model is also based on J coefficients. Equation (6.230) ensures that the error signal p_l is now causally filtered using a time-reversed version of the plant impulse response model.

The algorithmic formulation of the multi-channel FeLMS in time domain is summarized in the left column of Table 6.4. The right column of Table 6.4 contains the number of multiplication operations required per sample. It can be seen that the complexity of the FeLMS algorithm is given by

$$C_{FeLMS} = (2IK + JL)M + IKM. \tag{6.231}$$

Fast Exact Filtered Reference Approach

The fast exact implementation of the FxLMS algorithm also known as *fast exact filtered reference least mean square* (FEFxLMS) algorithm that has been proposed by Douglas, see (Douglas 1997) and (Douglas 1999), avoids the additional delay problem of the FeLMS by reordering the terms in the adaption equation and leads to the convergence speed of the filtered reference approach by saving the advantage in complexity that is provided by the filtered error approach. Because of this advantage, the single-channel as well as the multi-channel FEFxLMS is presented in the following paragraphs.

Single-Channel Fast Exact Filtered Reference Approach The single-channel FEFxLMS may be applied to minimize a cost function that is for simplicity only established on the squared error signal such as

$$J(n) = p^2(n) = p_p(n) + \sum_{i=1}^{I} \sum_{j=1}^{J} h_j w_i(n) x(n-i-j). \tag{6.232}$$

The associated update equation for the control filter coefficients is, compare Table 6.2, given by

$$w_i(n+1) = w_i(n) - \mu p(n) \sum_{j=1}^{J} \hat{h}_j x(n-i-j). \tag{6.233}$$

To derive the implementation we define the filtered error signal as

$$\varepsilon_j(n) = h_j \mu p(n) =: h_j p^{\mu}(n) \tag{6.234}$$

that is used to write the iterative mapping (6.233) in the form

$$w_i(n+1) = w_i(n) - \sum_{j=1}^{J} \varepsilon_j(n) x(n-i-j). \tag{6.235}$$

The relation in (6.235) can be expressed for J successive time steps such as

$$\begin{aligned} w_i(n+1) &= w_{i-J+1}(n) - \sum_{p=0}^{J-1} \sum_{j=0}^{J} \varepsilon_j(n-J+i-p) x(n-J-i-p) \\ &= w_{i-J+1}(n) - \sum_{p=1}^{J-1} \sum_{j=p+1}^{J} \varepsilon_j(n-J+i-p) x(n-J-i-p) \ldots \\ &\quad - \sum_{p=0}^{J-1} \sum_{j=1}^{J-p} \varepsilon_j(n-p) x(n-i-j-p). \end{aligned} \tag{6.236}$$

The summation on the right-hand side of (6.236) can be expanded in a particular useful way that allows for defining the i-th *auxiliary coefficient* $\hat{w}_i(n)$ as

$$\hat{w}_i(n) = w_{i-J+1}(n) - \sum_{p=1}^{J-1} \sum_{j=p+1}^{J} \varepsilon_j(n-J+i-p) x(n-J-i-p). \tag{6.237}$$

The i-th auxiliary coefficient at discrete time $n+1$ is therefore given by

$$\hat{w}_i(n+1) = w_{i-(J-1)+1}(n)\ldots \\ -\sum_{p=1}^{J-1}\sum_{j=p+1}^{J} \varepsilon_j\big(n-(J-1)+i-p\big)x\big(n-(J-1)-i-p\big). \tag{6.238}$$

Using (6.237) and (6.238), the difference between the auxiliary filter coefficients $\hat{w}_i(n+1)$ and $\hat{w}_i(n)$ can be expressed in terms of the control filter coefficients w_i such as

$$\begin{aligned}\Delta\hat{w}_i &= w_{i-(J-1)+1}(n) - w_{i-J+1}(n)\ldots \\ &\quad -\sum_{p=1}^{J-1}\sum_{j=p+1}^{J} \varepsilon_j\big(n-(J-1)+i-p\big)x\big(n-(J-1)-i-p\big)\ldots \\ &\quad -\sum_{p=1}^{J-1}\sum_{j=p+1}^{J} -\varepsilon_j(n-J+i-p)x(n-J-i-p) \\ &= -x(n-J-i)\sum_{p=1}^{J}\varepsilon_p(n-J+p).\end{aligned} \tag{6.239}$$

By introducing the *additive filtered error* signals

$$e_j(n) = \sum_{p=1}^{j}\varepsilon_j(n-j+p) \tag{6.240}$$

it is possible to write (6.239) more compact such as

$$\Delta\hat{w}_i = -e_J(n)x(n-J-i). \tag{6.241}$$

Considering (6.241), the update equation for the i-th auxiliary coefficient reads

$$\hat{w}_i(n+1) = \hat{w}_i(n) - e_J(n)x(n-J-i). \tag{6.242}$$

It should be noticed that a recursive update schema can be used to calculate the actual values of the additive filtered error signals such as

$$e_j(n) = \begin{cases} h_1 p^{\mu}(n) & \text{if } j=1, \\ e_{j-1}(n-1) + h_j p^{\mu}(n) & \text{if } 2 \le j \le J. \end{cases} \tag{6.243}$$

Since $e_J(n)$ is, compare (6.243), obtained by filtering the product of error signal and step size $\mu p(n)$ by the time-reversed impulse response of the secondary path $[h_J, h_{J-1}, \ldots, h_1]$, (6.242) is the single channel version of the FeLMS. What is novel is the link between the FxLMS and the FeLMS that is provided by (6.237). This equation defines the relation between the auxiliary coefficients $\hat{w}_i(n)$ and the control filter coefficients $w_i(n)$. The latter are used to compute the controller output signal. Equating (6.235) and (6.236) gives

$$w_i(n) = \hat{w}_i(n) + \sum_{j=1}^{J} \varepsilon_j(n)x(n-i-j) - \sum_{p=0}^{J-1}\sum_{j=1}^{J-p} \varepsilon_j(n-p)x(n-i-j-p). \tag{6.244}$$

Equation (6.244) can, considering (6.240), be written as

$$w_i(n) = \hat{w}_i(n) - \sum_{j=1}^{J-1} e_j(n-1)x(n-i-j-1). \tag{6.245}$$

The controller output $q(n)$ is therefore for any discrete time n given by

$$\begin{aligned} q(n) &= \sum_{i=0}^{I-1} w_i(n)x(n-i) \\ &= \sum_{i=0}^{I-1} \left[\hat{w}_i(n) - \sum_{j=1}^{J-1} e_j(n-1)x(n-i-j-1) \right] x(n-i) \\ &= \sum_{i=0}^{I-1} \hat{w}_i(n)x(n-i) - \sum_{i=0}^{I-1}\sum_{j=1}^{J-1} e_j(n-1)x(n-i-j-1)x(n-i). \end{aligned} \tag{6.246}$$

If we define the reference signal based correlation term

$$r_j(n) = \sum_{i=0}^{I-1} x(n-i-j)x(n-i), \tag{6.247}$$

it is possible to compute the controller output such as

$$q(n) = \sum_{i=0}^{I-1} \hat{w}_i(n)x(n-i) - \sum_{j=1}^{J-1} e_j(n-1)r_{j+1}(n). \tag{6.248}$$

This calculation is of reasonable complexity because the correlation term defined in (6.247) can also be recursively updated by

$$r_j(n) = r_j(n-1) + x(n)x(n-j) - x(n-I)x(n-I-j). \tag{6.249}$$

Collecting (6.234), (6.242), (6.248), (6.249), and (6.243), we obtain an algorithm that exactly computes the output signal of the FxLMS algorithm. However, the update of the auxiliary coefficients is based on the FeLMS approach.

Multi-channel Fast Exact Filtered Reference Approach We now consider the multi-channel extension of the FEFxLMS approach that will be used to minimize a cost function that is based on L squared error signals such as

$$J(n) = \sum_{l=1}^{L} p_l^2(n). \tag{6.250}$$

At first we specify the product between the step size μ and the l-th error signal

$$p_l^{\mu}(n) = \mu p_l(n). \tag{6.251}$$

If we take into account K references, L error signals, and M control signals, the update equation for the mki-th control filter coefficient is, considering Table 6.2 and (6.251), given by

$$\begin{aligned} w_{mki}(n+1) &= w_{mki}(n) - \mu \sum_{l=1}^{L} \hat{x}'_{lmk}(n-i)p_l(n) \\ &= w_{mki}(n) - \mu \sum_{l=1}^{L} \sum_{j=1}^{J} p_l^{\mu}(n) h_{lmj} x_k(n-i-j). \end{aligned} \tag{6.252}$$

If we, similar to (6.234), introduce the filtered error signal for the multi-channel FEFxLMS approach[14]

$$\varepsilon_{jm}(n) = \sum_{l=1}^{L} p_l^{\mu}(n) h_{lmj}, \tag{6.253}$$

we can write (6.252) such as

$$w_{mki}(n+1) = w_{mki}(n) - \sum_{j=1}^{J} \varepsilon_{jm}(n) x_k(n-i-j). \tag{6.254}$$

The additive filtered error of the multi-channel FEFxLMS algorithm is analog to (6.240) and considering a recursive update of this variable defined by

$$\begin{aligned} e_{jm}(n) &= \sum_{p=1}^{j} \varepsilon_{mj}(n-j+p) \\ &= \begin{cases} \sum_{l=1}^{L} h_{lm1} p_l^{\mu}(n) & \text{if } j = 1, \\ e_{(j-1)m}(n-1) + \sum_{l=1}^{L} h_{lmj} p_l^{\mu}(n) & \text{if } 2 \le j \le J. \end{cases} \end{aligned} \tag{6.255}$$

In direct analogy to the previous paragraph, the update equation for the *IMK* auxiliary coefficients reads

$$\hat{w}_{mki}(n+1) = \hat{w}_{mki}(n) - e_{Jm}(n) x_k(n-J-i). \tag{6.256}$$

To compute the controller output, we specify the multi-channel equivalent to (6.248) such as

---

[14] The index j corresponds to the j-th coefficient of the impulse response of the lm-th secondary path, whereas m corresponds to the m-th controller output signal.

$$q_m(n) = \sum_{k=1}^{K} \sum_{i=0}^{I-1} \hat{w}_{mki}(n) x_k(n-i) - \sum_{j=1}^{J-1} e_{jm}(n-1) r_{j+1}(n). \tag{6.257}$$

The correlation term $r_j(n)$ in this case is defined as

$$\begin{aligned} r_j(n) &= \sum_{k=1}^{K} \sum_{i=0}^{I-1} x_k(n-1) x_k(n-i-j) \\ &= r_j(n-1) + \sum_{k=1}^{K} \big[x_k(n) x_k(n-j) - x_k(n-I) x_k(n-I-j) \big]. \end{aligned} \tag{6.258}$$

The algorithmic formulation of the multi-channel FEFxLMS in time domain in which similar to Table 6.2 the leakage factor β was taken into account, is summarized in the left column of Table 6.5.

The right column of Table 6.5 contains the number of multiplication operations required per sample. It can be seen that the complexity of FEFxLMS algorithm is given by the following number of multiplications at each iteration

$$C_{FEFxLMS} = 3IKM + JLM + L + (J-1)(2K+M). \tag{6.259}$$

A Short Comparison of Adaptive Schemas

The complexity of the presented time domain algorithms is compared in Fig. 6.34 in which the complexity is plotted against the number of reference signals (top left), the number of controller output signals (top right), the number of error signals (below left) and the number of filter coefficients (below right).

To generate these plots a number of $K = 6$ reference signals, $M = 8$ control signals, $L = 8$ error signals, and a number of $I = J = 128$ filter coefficients, were taken into account as a basic set. In each plot one of these variables was varied, whereas the remaining three variables were kept constant.

It can be seen that the complexity of the FeLMS algorithm and the FEFxLMS algorithm is lower as the complexity of the FxLMS algorithm. The MFxLMS algorithm requires the highest computational effort. The difference in complexity increases, if the number of signals or the number of control filter coefficients increases. The FEFxLMS algorithms gives the smallest complexity.

Figure 6.34 also clarifies that in case of limited computational power, the application of the FEFxLMS algorithm can significantly reduce the computational complexity of the control task without causing additional problems in view of causality. For a detailed comparison of algorithms see (Bouchard and Quednau 2000).

To give an example for the convergence behavior, the single-channel implementations of the adaptive algorithms summarized in Tables 6.2, 6.3, 6.4, 6.5 have been used to simulate active canceling of band limited white noise (500 Hz $< f <$ 650 Hz). The therefore needed transfer paths have been taken from (Matlab 2010).

Table 6.5 The multi-channel fast exact filtered reference least mean square algorithm in time domain with K reference signals, M control signals, L error signals, and I coefficients in each control filter using J coefficients to model the impulse response of each secondary path

FEFxLMS algorithmic equations	Multiplications per iteration
for $j = 2$ to J do	
Update reference signals:	
$r_j(n) = r_j(n-1) + \sum_{k=1}^{K} [x_k(n)x_k(n-j) - x_k(n-I)x_k(n-I-j)]$	$2(J-1)K$
end	
for $l = 1$ to L do	
Multiply error and step size:	
$p_l^{\mu}(n) = \mu p_l(n)$	L
end	
for $m = 1$ to M do	
Compute controller output:	
$q_m(n) = \sum_{k=1}^{K} \sum_{i=0}^{I-1} \hat{w}_{mki}(n)x_k(n-i) - \sum_{j=1}^{J-1} e_{jm}(n-1)r_{j+1}(n)$	$IKM + (J-1)M$
Pre-filtering of first error signal:	
$e_{1m}(n) = \sum_{l=1}^{L} \hat{h}_{lm1} p_l^{\mu}(n)$	LM
for $j = 1$ to $J-1$ do	
Update all remaining filtered error signals:	
$e_{(j+1)m}(n) = e_{jm}(n-1) + \sum_{l=1}^{L} \hat{h}_{lm(j+1)} p_l^{\mu}(n)$	$(J-1)LM$
end	
for $k = 1$ to K do	
for $i = 0$ to $I-1$ do	
Update of control filter coefficients:	
$\hat{w}_{mki}(n+1) = (1-\mu\beta)\hat{w}_{mki}(n) - e_{Jm}(n)x_k(n-J-i)$	$2IKM$
end	
end	
end	

The impulse responses of both primary and secondary path are shown in Fig. 6.35. A number of 400 coefficients was used for the control filter as well as for the plant models.

The convergence behavior that has been determined in discrete-time simulations using a sampling frequency of 8 kHz and the same step size for all algorithms is shown in Fig. 6.36. It can be seen that all control schemes are capable of providing a fast and significant reduction of the initial disturbance. The convergence behavior of all implementations is in this simple example nearly identical. However, the comparison of the FxLMS algorithm and the MFxLMS algorithm shown in Fig. 6.36 (top right) gives an indication that the MFxLMS algorithm could be the best suited approach in view of convergence speed.

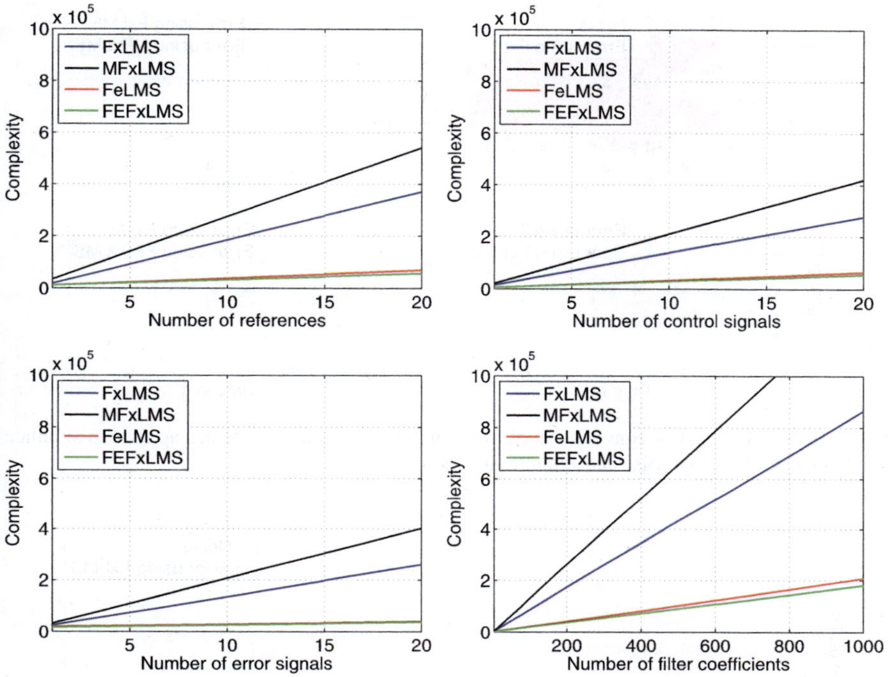

Fig. 6.34 Comparison of complexity of time domain adaptive algorithms

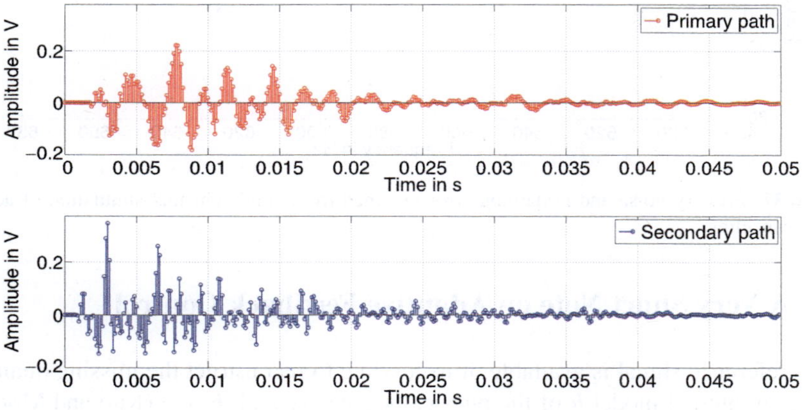

Fig. 6.35 Impulse response of primary and secondary path used for single-channel simulation of active noise canceling

The ASD of the primary noise and the ASD of the remaining error signal are shown in Fig. 6.37. It can be seen that broadband active noise control can lead to a significant and uniform reduction of disturbing noise.

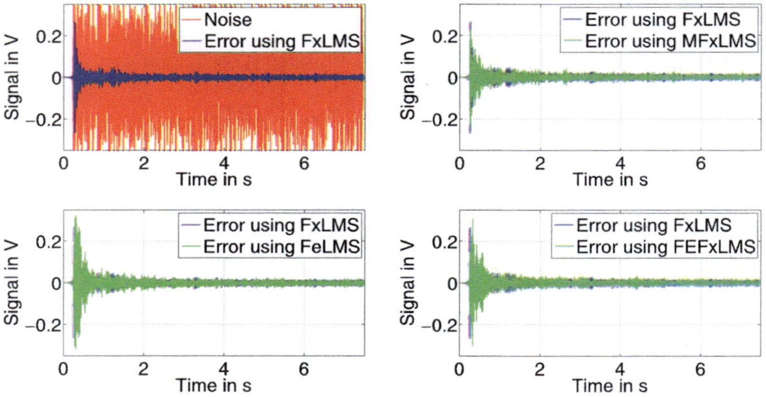

Fig. 6.36 Comparison of convergence of different adaptive control schemes applied to simulate single-channel adaptive noise canceling in time domain

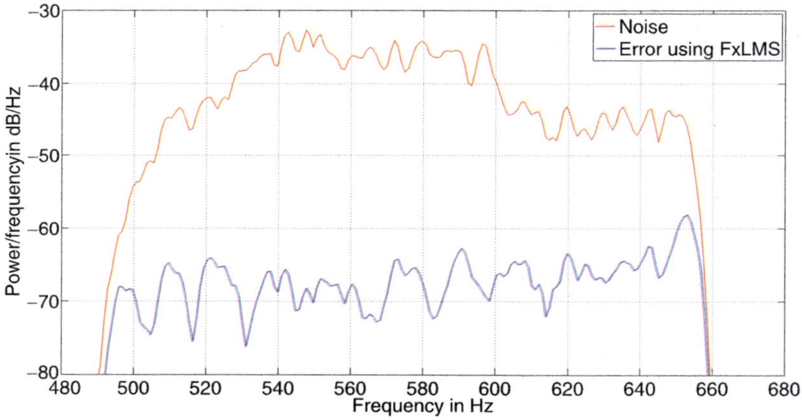

Fig. 6.37 Primary noise and remaining error obtained from single-channel simulation of active noise canceling

6.7 A Very Short Note on Adaptive Feedback Control

If no reference signal is available, it is possible to reconstruct the missing quantity using an internal model $\hat{h}$ of the physical secondary path h, see (Kuo and Morgan 1996; Pawelczyk et al. 1997) and (Elliott 2001). This approach is therefore known as *internal model control* (IMC). The missing reference signal x is calculated by subtracting the known canceling signal (controller output q filtered with a model of the secondary path $\hat{h}$) from the error signal p. The resulting estimate of the primary disturbance $\hat{p}_p$ can then be used as reference. Such an IMC controller can be implemented in a feed-forward structure with an adaptive FIR filter as shown in Fig. 6.38.

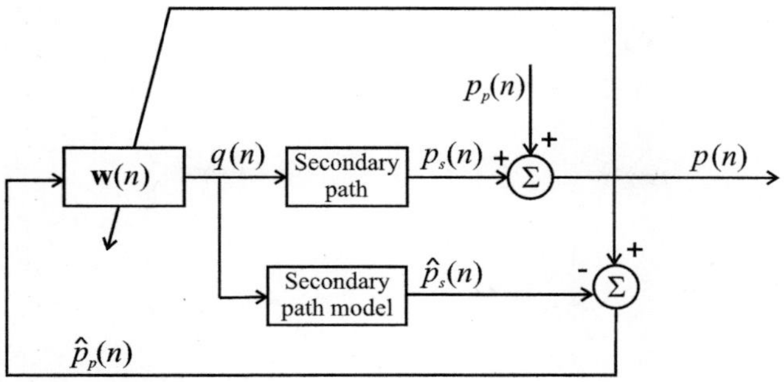

Fig. 6.38 Adaptive feedback control with internal plant model realized in feed-forward structure

The general problem caused by this approach is that no reference signal is available before the disturbance is picked up by the error sensor. The adaptive controller therefore acts with a certain delay that is determined by the phase response of the secondary path. This is acceptable for periodic (especially for tonal) noise, but critical for active reduction of broadband disturbances with non-collocated sensor-actuator pairs.

Part III
ANC-System Design: Theory

Chapter 7
ANC-System Design Tools

Abstract During ANC system design we will have to answer questions such as: "How is the acoustic pressure distributed in the interior?", "What are the positions of the primary noise sources?" or "What will be the control profit, if active noise cancellation is applied? As we know, it will in many projects be impossible to perform time and cost intensive measurement campaigns. This is especially true in an early design stage in which we have neither a functional model nor a prototype. For this reason, the distribution of acoustic field variables has to be predicted by a numerical simulation that is based on an appropriate mathematical model. In view of the second question it must be emphasized that because of standing waves localization of sound sources in weakly damped enclosed sound fields is—especially in the low frequency range—a difficult mission. Noise source identification therefore requires advanced experimental and also advanced numerical approaches. In order to answer the third question we will have to estimate the amount of noise reduction that can be realized by a feed-forward ANC-system. Therefore, it will be necessary to analyze the correlation between the reference signals and the disturbance as well as to analyze the effect of transducer locations on the control profit. It is obvious that we need some specific instruments to process work packages that are associated with these questions. For this reason the upcoming chapter is dedicated to some tools that can be used in the ANC-system design process.

7.1 Numerical Analysis of Sound Fields

7.1.1 A Short Overview on Numerical Methods

Several approaches for the numerical analysis of enclosed sound fields have been established. The most important are the *finite difference method* (FDM), the *boundary element method* (BEM), and the *finite element method* (FEM). A mathematical description of these numerical concepts is to be found in (Zwillinger 1989).

The FDM works as follows. The acoustic field is divided by a line grid and the field values such as the sound pressure are assigned to the grid intersection points. If the acoustic field is described by a rectangular Cartesian coordinate system, the resulting grid is necessarily square. The FDM has two drawbacks. It is of course also possible to establish a FDM using a cylindrical coordinate system or using

T. Kletschkowski, *Adaptive Feed-Forward Control of Low Frequency Interior Noise*, Intelligent Systems, Control and Automation: Science and Engineering 56,
DOI 10.1007/978-94-007-2537-9_7, © Springer Science+Business Media B.V. 2012

a spherical coordinate system, but it is not possible to handle arbitrary geometric forms with boundaries that do not conform with the grid line pattern. The reason is that the derivatives of a PDE are represented by finite-difference approximations, those pertaining to any given point involving values of surrounding grid points. It should also be noticed that the FDM is directly applied to the PDE. It is therefore very sensitive to local errors of field representation. Because of these drawbacks, FDA is not discussed in the present wok.

The BEM is usually applied to evaluate the acoustic field generated by a sound radiating surface in an infinitely extended volume. It is based on an approximate solution of the *Kirchhoff-Helmholtz integral equation* (KHIE). This equation states that for any harmonic sound pressure field that satisfies both the homogeneous Helmholtz equation (3.22) and the *Sommerfeld radiation condition*,[1] the sound pressure at any point $\vec{x}$ in the unbounded infinitely extended volume is determined by the combination of the pressure distribution and the distribution of the normal component of the acoustic velocity on the radiating surface, compare (Fahy and Gardonio 2007). A numerical solution of the KHIE is usually required, because fundamental solutions are limited to a small family of radiator surface geometries. It is based on dividing the radiating surface into discrete elements and a fundamental solution that describes the sound radiation from a point monopole source in free-space.[2] The aim is to obtain an estimate of the surface pressure distribution that corresponds to the prescribed distribution of the normal component of the acoustic velocity. A detailed introduction to the BEM may be found in (Berbbia 1978) and (Gaul et al. 1997). A comprehensive overview is given in (von Estorff et al. 2000). Nowadays, industrial applications are widespread. An example for a state of the art commercial BEM software is LMS Sysnoise (SYSNOISE 2010). However, because we are not interested in exterior noise problems, the BEM is not discussed.

We will proceed with the FEM, see (Bathe 1996), (Zienkiewicz and Taylor 2000), and (Fahy and Gardonio 2007). Applying this approach, the fluid space is theoretically divided into contiguous elements of linear dimension considerably smaller than the smallest wave length of interest. The FEM can therefore handle arbitrary geometries. The aim is to obtain an estimate of the interior sound pressure that corresponds to prescribed boundary conditions. Examples for commercial FEM software are Ansys (ANSYS 2010), MSC Nastran (NASTRAN 2010), and Comsol (COMSOL 2010).

[1] Sommerfeld's radiation condition requires that the radial distance r from a harmonic sound source approaches infinity such as $\lim_{r\to\infty}\{r[P(\vec{x}, j\omega) - \rho_\infty c V_r(\vec{x}, j\omega)]\} = 0$, where $V_r(\vec{x}, j\omega)$ is the radial component of the acoustic velocity field. "This condition requires that as $r \to \infty$, then only outgoing waves can exist", see (Nelson and Elliott 1992).

[2] This fundamental solution is known as the free-space form of *Green's function*, see (Fahy and Gardonio 2007).

7.1.2 The Finite Element Method

As outlined in the previous subsection, several approaches have been established that can be used for the numerical analysis of enclosed sound fields. However, because of its moderate computational complexity and its capability to handle arbitrary geometries, the FEM is an appropriate choice, if standing wave fields have to be analyzed. This is especially true, in the low frequency range in which the modal overlap defined by (6.80) is smaller than 3, compare (Nelson and Elliott 1992).

An acoustic finite element scheme is usually based on wave equations that are formulated for a scalar valued quantity such as the wave equation for the acoustic pressure (3.9), see (Chargin and Gartmeier 1990), or the wave equation for the velocity potential (3.15), see (Bathe 1996) and (Fahy and Gardonio 2007). A formulation based on the acoustic velocity is also possible. If the latter is used, the vector wave equation for the acoustic velocity (3.18) has to be discretized.

Finite element schemes based on the velocity potential have the advantage that the symmetry of the resulting system matrices is preserved, if fluid-structure coupling has to be taken into account. A disadvantage is given by the fact that an artificial damping matrix without physical meaning results, if this approach is applied. Therefore, and because of the intention to outline a straight-forward approach that is based on an acoustic variable, the FEM discussed here is based on the *principle of virtual pressure* (PVP) that represents a *weak formulation* of the acoustic field problem.

The Principle of Virtual Pressure

As outlined above, we will use the wave equation for the acoustic pressure (3.9) to derive the *principle of virtual pressure*. In order to consider viscous damping effects in the fluid we will add a "volumetric drag" term to (3.9). This yields

$$\operatorname{div}\big[\operatorname{grad} p(\vec{x},t)\big] - \frac{r_f}{\rho_\infty c^2} \cdot \frac{\partial p(\vec{x},t)}{\partial t} - \frac{1}{c^2} \cdot \frac{\partial^2 p(\vec{x},t)}{\partial t^2} = 0, \tag{7.1}$$

where r_f describes the viscous properties of the fluid. Furthermore, we introduce an arbitrary variational field $\delta p(\vec{x},t)$ that is superimposed on the pressure distribution $p(\vec{x},t)$ such as $p(\vec{x},t) + \delta p(\vec{x},t)$ and postulate that this variational field fulfills the homogeneous Dirichlet BC, i.e.

$$\delta p(\vec{x}_r,t) = 0 \quad \text{on } R_p. \tag{7.2}$$

If (7.1) is multiplied by $\delta p(\vec{x},t)$ and, considering

$$\operatorname{div}(\delta p \cdot \operatorname{grad} p) = \operatorname{grad} \delta p \cdot \operatorname{grad} p + \delta p \cdot \operatorname{div}(\operatorname{grad} p), \tag{7.3}$$

integrated over the whole fluid volume, we obtain

$$\int_V \left[\frac{\operatorname{grad} \delta p \cdot \operatorname{grad} p}{\rho_\infty} + \frac{r_f \delta p}{\rho_\infty c^2} \frac{\partial p}{\partial t} + \frac{\delta p}{\rho_\infty c^2} \frac{\partial^2 p}{\partial t^2} \right] dV$$
$$= \int_V \frac{\operatorname{div}(\delta p \cdot \operatorname{grad} p)}{\rho_\infty} dV. \tag{7.4}$$

Applying Gauss's theorem and using (3.2), the right hand side of (7.4) can be written such as

$$\begin{aligned} \int_V \frac{1}{\rho_\infty} \operatorname{div}(\delta p \cdot \operatorname{grad} p) dV &= -\int_R \delta p \left(\vec{n} \cdot \frac{\partial \vec{v}}{\partial t} \right) dR \\ &= -\int_R \delta p \frac{\partial v}{\partial t} dR \\ &= -\int_R \delta p \cdot a dR, \end{aligned} \tag{7.5}$$

where a is the normal component of the *acoustic acceleration* on the boundary. The negative sign of the "loading" term indicates that the surface normal vector $\vec{n}$ is taken to point out of the fluid. Because the boundary of the interior is described by Dirichlet BCs defined on the pressure boundary R_p, Neumann BCs defined on the velocity boundary R_v, and Robin BCs defined on the impedance (or mixed) boundary R_Z, compare Sect. 3.2, we can split the right hand side of (7.5) into three integrals

$$\int_R \delta p \cdot a dR = \int_{R_p} \delta p \cdot a dR_p + \int_{R_v} \delta p \cdot a dR_v + \int_{R_Z} \delta p \cdot \left(Z_{R_Z}^{-1} \frac{\partial p}{\partial t} \right) dR_Z, \tag{7.6}$$

where in the last summand the normal component of the acoustic acceleration has been replaced by the time derivative of the acoustic pressure using (3.2) and (3.12). The first summand on the right hand side of (7.6) vanishes because of (7.2). For this reason the PVP reads

$$\int_V \left[\frac{\operatorname{grad} \delta p \cdot \operatorname{grad} p}{\rho_\infty} + \frac{r_f}{\rho_\infty c^2} \frac{\partial p}{\partial t} + \frac{\delta p}{\rho_\infty c^2} \frac{\partial^2 p}{\partial t^2} \right] dV + \cdots$$
$$+ \int_{R_Z} \delta p \cdot \left(Z_{R_Z}^{-1} \frac{\partial p}{\partial t} \right) dR_Z = -\int_{R_v} \delta p \cdot a dR_v. \tag{7.7}$$

Equation (7.7) is the equivalent variational ("weak") form of the wave equation (7.1). If (7.7) is for a given pressure distribution $p(\vec{x}, t)$ fulfilled for any arbitrary variational field $\delta p(\vec{x}, t)$, the pressure distribution describes the dynamic equilibrium and satisfies (7.1) as well as the boundary conditions. If the PVP is for a given pressure distribution only fulfilled for specific variational fields, the pressure distribution satisfies only the integral mean of (7.1) and the integral mean of the boundary conditions. The virtual pressure distribution is in this case a weighting function.

Finite Element Approximation

To solve the (7.7), we approximate the pressure distribution as well as the virtual pressure in the e-th finite element by interpolation functions also known as *shape functions*. If we use the same functions for the pressure and the virtual pressure distribution (Galerkin's method), this approximation reads

$$p_e(\vec{x},t) = \mathbf{h}_e^T \mathbf{p}_e, \qquad \delta p_e = \mathbf{h}_e^T \delta \mathbf{p}_e, \tag{7.8}$$

where

$$\mathbf{h}_e = \left[h_{e1}(\vec{x})\; h_{e2}(\vec{x})\; \ldots\; h_{eN}(\vec{x})\right]^T \tag{7.9}$$

is the $(N \times 1)$ column matrix of N interpolation functions of the e-th finite element, and

$$\mathbf{p}_e = \left[p_{e1}(t)\; p_{e2}(t)\; \ldots\; p_{eN}(t)\right]^T \tag{7.10}$$

is the $(N \times 1)$ column matrix that contains the N nodal degrees of freedom of the e-th finite element. The $(N \times 1)$ column matrix for the virtual pressure at the N nodes of the e-the finite element is analogue to (7.10) given by

$$\delta \mathbf{p}_e = \left[\delta p_{e1}(t)\; \delta p_{e2}(t)\; \ldots\; \delta p_{eN}(t)\right]^T. \tag{7.11}$$

To proceed, we analyze the PVP for the e-th element. For this purpose $p_e(\vec{x},t)$ and $\delta p_e(\vec{x},t)$ are replaced considering (7.8)–(7.11). This yields a set of N ODEs

$$\mathbf{M}_e \ddot{\mathbf{p}}_e(t) + \mathbf{D}_{r_f e} \dot{\mathbf{p}}_e(t) + \mathbf{D}_{Z_e} \dot{\mathbf{p}}_e(t) + \mathbf{C}_e \mathbf{p}_e(t) = \mathbf{f}_e(t) \tag{7.12}$$

with

$$\begin{aligned}
\mathbf{M}_e &= \frac{1}{\rho_\infty c^2} \int_{V_e} \mathbf{h}_e \mathbf{h}_e^T \, dV_e \\
\mathbf{D}_{r_f e} &= \frac{r_{f_e}}{\rho_\infty c^2} \int_{V_e} \mathbf{h}_e \mathbf{h}_e^T \, dV_e = r_{f_e} \mathbf{M}_e \\
\mathbf{D}_{Z_e} &= Z_{R_{Z_e}}^{-1} \int_{V_e} \mathbf{h}_e \mathbf{h}_e^T \, dV_e = \frac{\rho_\infty c^2}{Z_{R_{Z_e}}} \mathbf{M}_e \\
\mathbf{C}_e &= \frac{1}{\rho_\infty} \int_{V_e} [\mathrm{grad}(\mathbf{h}_e)][\mathrm{grad}(\mathbf{h}_e^T)] \, dV_e \\
\mathbf{f}_e(t) &= -\int_{R_{v_e}} \mathbf{h}_{R_{v_e}} a_e(t) \, dR_{v_e}.
\end{aligned} \tag{7.13}$$

$\mathbf{M}_e$ is the *generalized element mass matrix*, $\mathbf{D}_{r_f e}$ and $\mathbf{D}_{Z_e}$ are mass proportional *generalized element damping matrices* that represent viscous damping effects in the fluid and the effect of absorbing boundaries, $\mathbf{C}_e$ is the *generalized element stiffness matrix*, and $\mathbf{f}_e(t)$ is the *element loading matrix*.

The link between the *local degrees of freedom* of the e-th element that are arranged in the $(N \times 1)$ column matrix $\mathbf{p}_e$ and the G *global degrees of freedom* that are arranged in the $(G \times 1)$ column matrix $\mathbf{p}$ is provided by the $(G \times N)$ "connection" matrix $\mathbf{A}_e$ such as

$$\mathbf{p}_e(t) \mapsto \mathbf{p}(t) = \mathbf{A}_e \mathbf{p}_e(t). \tag{7.14}$$

The same transformation is applied to the column matrix of the virtual pressure

$$\delta \mathbf{p}_e(t) \mapsto \delta \mathbf{p}(t) = \mathbf{A}_e \delta \mathbf{p}_e(t). \tag{7.15}$$

Summation over all E elements of the discretized sound field leads considering (7.14) and (7.15) to a set of G coupled ordinary ODEs for the global degrees of freedom. This set has to be solved considering the BCs. It is given by

$$\mathbf{M}\ddot{\mathbf{p}}(t) + \mathbf{D}_{r_f}\dot{\mathbf{p}}(t) + \mathbf{D}_Z\dot{\mathbf{p}}(t) + \mathbf{C}\mathbf{p}(t) = \mathbf{f}(t), \tag{7.16}$$

where the *system mass matrix* $\mathbf{M}$, the *system damping matrices* $\mathbf{D}_{r_f}$ and $\mathbf{D}_Z$, the *system stiffness matrix* $\mathbf{C}$, and the *system load matrix* $\mathbf{f}(t)$ are defined as

$$\begin{aligned}
\mathbf{M} &= \sum_{e=1}^{E} \mathbf{A}_e^T \mathbf{M}_e \mathbf{A}_e \\
\mathbf{D}_{r_f} &= \sum_{e=1}^{E} \mathbf{A}_e^T \mathbf{D}_{r_f e} \mathbf{A}_e \\
\mathbf{D}_Z &= \sum_{e=1}^{E} \mathbf{A}_e^T \mathbf{D}_{Z_e} \mathbf{A}_e \\
\mathbf{C} &= \sum_{e=1}^{E} \mathbf{A}_e^T \mathbf{C}_e \mathbf{A}_e \\
\mathbf{f}(t) &= \sum_{e=1}^{E} \mathbf{A}_e^T \mathbf{f}_e(t).
\end{aligned} \tag{7.17}$$

In case of harmonic excitation with $\mathbf{f}(t) = \mathrm{Re}[\hat{\mathbf{f}}(j\omega)e^{j\omega t}]$, the real valued solution $\mathbf{p}(t) = \mathrm{Re}[\hat{\mathbf{p}}(j\omega)e^{j\omega t}]$ is determined by the complex magnitudes of the global degrees of freedom that are arranged in the $(G \times 1)$ column matrix

$$\hat{\mathbf{p}} = \left[P_1(j\omega)\ P_2(j\omega)\ \ldots\ P_N(j\omega)\right]^T. \tag{7.18}$$

$\hat{\mathbf{p}}$ can be determined, from a set of G coupled algebraic equations which has to be solved considering the BCs. This set is given by

$$\mathbf{K}\hat{\mathbf{p}} := \left[-\omega^2\mathbf{M} + j\omega(\mathbf{D}_{r_f} + \mathbf{D}_Z) + \mathbf{C}\right]\hat{\mathbf{p}} = \hat{\mathbf{f}}, \tag{7.19}$$

where $\mathbf{K}$ is the frequency dependent *generalized complex stiffness matrix*.

If the effects of viscous damping and absorbing boundaries have not been taken into account because of missing data on the physical parameters, (7.19) reduces to

$$\left[-\omega^2\mathbf{M} + \mathbf{C}\right]\hat{\mathbf{p}} = \hat{\mathbf{f}}. \tag{7.20}$$

However, to study the influence of viscous damping on the results it is still possible to solve (7.20) considering a complex wave number $k = (1 - j\xi)\omega c^{-1}$, compare (3.75). This approach results in a modified form of (7.20) that is given by

$$\left[-\omega^2\mathbf{M} + j\omega\mathbf{D} + \mathbf{C}\right]\hat{\mathbf{p}} = \hat{\mathbf{f}}, \tag{7.21}$$

where

$$\mathbf{D} = \xi(2 - j\xi)\omega\mathbf{M} =: d_f\mathbf{M} \tag{7.22}$$

is a complex and frequency dependent mass proportional damping matrix. The real part of the damping parameter d_f is given by $2\xi\omega$. If the excitation frequency ω is close to an acoustic resonance such as $\omega \approx \omega_{nR}$, the real part of d_f equals the modal bandwidth, compare (6.79).

A comparison of (7.13) and (7.22) indicates that a uniform damping parameter $r_{f_e} = r_f$ (for all elements) is given by $\mathrm{Re}[d_f] = 2\xi\omega$. Thus, the introduction of a complex wave number is for small (modal) damping ratios $\xi \ll 1$ equivalent to the consideration of a "volumetric drag" term.

For more details on special shape functions, finite element types, the implementation of the FEM and the solution of ODEs and/or algebraic equations resulting from a finite element approach, see (Bathe 1996), (Zienkiewicz and Taylor 2000) and (Zwillinger 1989).

7.2 Inverse Noise Source Identification

7.2.1 A Note on Source Identification Methods

Noise pollution of enclosed interiors such as aircraft cabins results in a limitation of human comfort. The good news is, that a source free enclosure of volume V can actively be shielded against sound flux from outside by arranging appropriate compensation sources along the boundary R of V, see (Guicking 2002). This statement can be illustrated by the one-dimensional benchmark problem of Sect. 6.3.1. The appropriate compensation source that actively shields the whole one-dimensional wave-guide would be a single secondary ring source that is arranged in coincidence with the primary ring source and driven in exact antiphase to the primary source. This simple example clarifies that the sound sources at the interior boundaries have to be identified, if active noise reduction techniques shall be applied with (global) success.

Several localization methods have been established ranging from simple sound pressure or sound intensity measurements to more sophisticated methods such as *beamforming*, see (Christensen and Hald 2004), or *acoustic holography* that is presented in (Maynard et al. 1985) and (Hald 1989). As stated in an actual review, see (Hundeck 2008), a drawback in applying these techniques is that free field conditions are required. An alternative approach is given by inverse methods such as the *inverse boundary element method* (IBEM), see (Holland and Nelson 2004) that can be used for reconstructing acoustic quantities (in the enclosure as well as on the boundary) based on acoustic pressure measurements. An up to date overview on inverse reconstruction methods can be found in (Wu 2008).

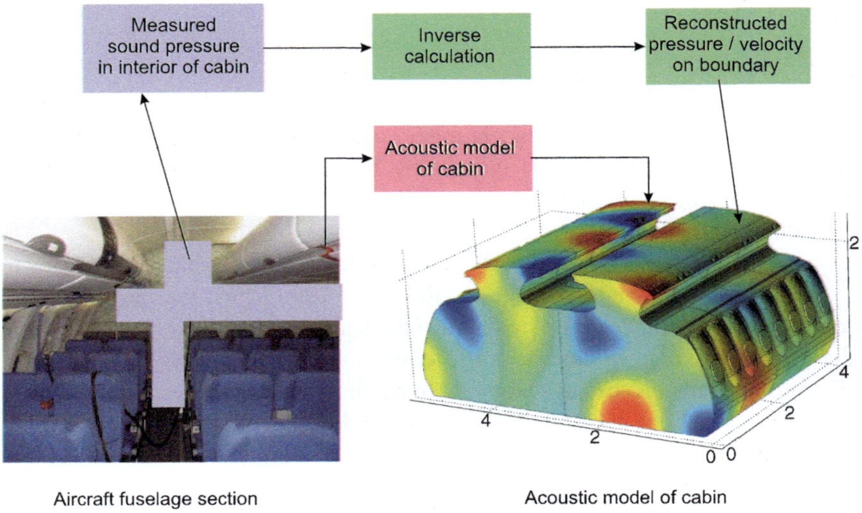

Fig. 7.1 Idea of noise source identification by the inverse finite element method. Measurement data taken from a certain subspace of the interior in combination with an acoustic finite element model are used to reconstruct the distribution of sound pressure and particle velocity at the interior boundary (adapted from Weber et al. 2009a)

However, this review does not consider the *inverse finite element method*.[3] Its application to noise source identification in enclosed sound fields was proposed in (Drenckhan and Sachau 2004). Further publications were focused on the influence of smoothing and regularization to the IFEM, compare (Drenckhan and Sachau 2005) and also (Sachau and Drenckhan 2006), on error criteria (Sachau et al. 2007b) as well as on the application of this method to aircraft cabins as presented in (Sachau et al. 2005b), and recently in (Weber et al. 2008a, 2008b, 2008c, 2009a, 2009b, 2009c), (Weber 2009), and in (Sachau et al. 2009). Numerical investigations based on the IFEM that were focused on the reconstruction of the pressure distribution in damped interiors (but not on source localization at the boundary) were also presented in (Anderssohn et al. 2006).

The basic idea of the IFEM is illustrated by Fig. 7.1. It motivates an application in which the IFEM reconstructs both the distribution of sound pressure and acoustic velocity on the *boundary*—B of an aircraft cabin. For this purpose measured sound pressure data are taken from a certain subspace of the interior that is referred as the *measurement region*—M. The remaining volume in which no measurement is performed is called *transition region*—T. The measurement data taken from M are used as input for the inverse calculation that reconstructs the sound field in T as

[3]From the mathematical point of view the names *inverse* BEM and *inverse* FEM may be misleading, because these methods are only applied to derive a set of equations that are used to solve the inverse boundary value problem. However, from an engineering point of view it is important to know where this set comes from: a discretized boundary or a discretized domain.

well as on B. Examining sound pressure and particle velocity on these boundaries, conclusions can be drawn to the location of acoustic hot spots.

The main advantages of the IFEM compared to the IBEM approach are that the measurement data taken from the subdomain M can be directly interpolated using the nodal degrees of freedom of the associated finite element model, and that completely no measure of the impedance of the outer boundary B is needed.

Because only the boundary is discretized, direct conversion of data measured in the interior into nodal degrees of freedom is not possible, if the IBEM is applied. For this reason an IBEM analysis requires a forward calculation in order to determine acoustic transfer functions between every boundary element and relevant points in the sound field. Therefore, it is necessary to incorporate an assumption of the impedance at those parts of B that are not modeled as an acoustic source.

The resulting transfer functions are arranged in a transfer function matrix that is inverted and multiplied with a column matrix that contains measured data of the interior sound pressures. It is obvious that the solution strongly depends on the assumptions for the boundary impedances.

The IFEM requires no information on the boundary impedances. These information are implicitly taken into account by both magnitude and phase of the sound pressure data taken from M. The upcoming section contains a review of the IFEM. Its application to noise source identification in a cross-section of a long-range airliner is presented in part IV of this book.

7.2.2 *The Inverse Finite Element Method*

Formulation of the Inverse Problem

To derive the IFEM we assume a source free volume V that is split into the measurement region M and the transition region T. V may be restricted by a velocity boundary $B \equiv R_V$ on which the normal component of the acoustic velocity is prescribed as shown in Fig. 7.2. The objective is to reconstruct the acoustic variables in T and on B using values of the acoustic pressure measured in M.

If V is discretized by the FEM, compare Sect. 7.1.2, the problem is represented by (7.19). Because of the spatial domain decomposition it is possible to write (7.19) such as

$$\mathbf{K}\hat{\mathbf{p}} := \begin{bmatrix} \mathbf{K}_1 & \mathbf{K}_2 & \mathbf{K}_3 \\ \mathbf{K}_4 & \mathbf{K}_5 & \mathbf{K}_6 \\ \mathbf{K}_7 & \mathbf{K}_8 & \mathbf{K}_9 \end{bmatrix} \begin{bmatrix} \hat{\mathbf{p}}_{MK} \\ \hat{\mathbf{p}}_{TU} \\ \hat{\mathbf{p}}_{BU} \end{bmatrix} = \begin{bmatrix} \mathbf{0} \\ \mathbf{0} \\ \hat{\mathbf{f}}_{BU} \end{bmatrix} =: \hat{\mathbf{f}}, \tag{7.23}$$

where $\mathbf{K}_i$ with $i = 1, \ldots, 9$ are the submatrices of complex stiffness matrix $\mathbf{K}$. The first index of the three sound pressure sub-matrices $\hat{\mathbf{p}}_{ij}$ and $\hat{\mathbf{f}}_{ij}$ denotes the subdomain of the decomposed calculation domain. The second index denotes whether the variable is known (K) or unknown (U). Because we assume that all sources are located on B, it is obvious that the right hand side sub-matrices associated to $\hat{\mathbf{p}}_{MK}$

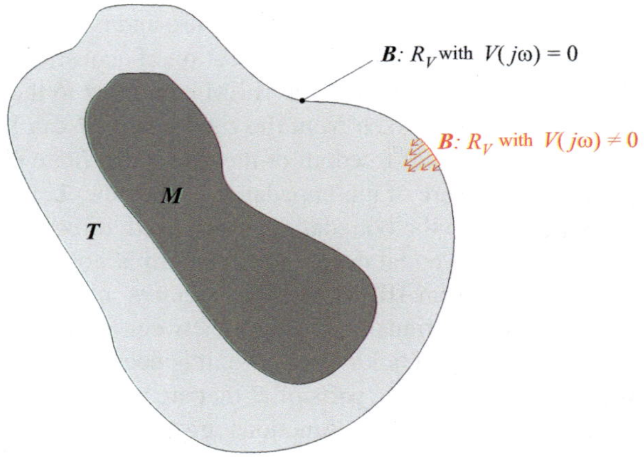

Fig. 7.2 Spatial domain decomposition for inverse finite element method. Calculation domain is split into an outer boundary (B), a transition region (T), and into a measurement domain (M)

and $\hat{\mathbf{p}}_{TU}$ are given by $\hat{\mathbf{f}}_{MK} = \mathbf{0}$ and $\hat{\mathbf{f}}_{TK} = \mathbf{0}$. Therefore, it is possible to split (7.23) into two sets of algebraic equations that can be solved successively. These sets are given by

$$\mathbf{K}_{red}\hat{\mathbf{p}}_{red} := \begin{bmatrix} \mathbf{K}_2 & \mathbf{K}_3 \\ \mathbf{K}_5 & \mathbf{K}_6 \end{bmatrix} \begin{bmatrix} \hat{\mathbf{p}}_{TU} \\ \hat{\mathbf{p}}_{BU} \end{bmatrix} = \begin{bmatrix} -\mathbf{K}_1\hat{\mathbf{p}}_{MK} \\ -\mathbf{K}_4\hat{\mathbf{p}}_{MK} \end{bmatrix} =: \hat{\mathbf{f}}_{red} \tag{7.24}$$

and

$$[\mathbf{K}_8\ \mathbf{K}_9\ -\mathbf{I}] \begin{bmatrix} \hat{\mathbf{p}}_{TU} \\ \hat{\mathbf{p}}_{BU} \\ \hat{\mathbf{f}}_{BU} \end{bmatrix} = [-\mathbf{K}_7\hat{\mathbf{p}}_{MK}], \tag{7.25}$$

where $\mathbf{K}_{red}$ is the (new portioned) reduced stiffness matrix, $\hat{\mathbf{p}}_{red}$ is the new portioned column matrix of unknowns, and $\hat{\mathbf{f}}$ is the new portioned column matrix of known sound pressure values. Evaluation of (7.24) yields the unknown pressure distribution in T and on B.

Based on these data it is possible to determine the unknown velocities on B by solving (7.25). Evaluation of (7.24) means that we have to minimize the cost function

$$J_{red}(\hat{\mathbf{p}}_{red}) := \mathbf{r}_{red}^H \mathbf{r}_{red} \quad \text{with } \mathbf{r}_{red} := \mathbf{K}_{red}\hat{\mathbf{p}}_{red} - \hat{\mathbf{f}}_{red}. \tag{7.26}$$

As stated in (Drenckhan and Sachau 2004), the condition number of $\mathbf{K}_{red}$ in the least square problem (7.26) can be very high, especially, if (7.24) is under-determined and, if (in addition) realistic measurement errors are taken into account. Therefore, regularization is required to stabilize the solution $\hat{\mathbf{p}}_{red}$.

However, before we proceed with a modified least square problem and comments on regularization, we pay attention to the one-dimensional case considering the benchmark problem described in Sect. 6.3.1. If losses are neglected ($\xi = 0$), the solution for the acoustic pressure that is specified by (6.34) and (6.36) reduces to

$$P(\tilde{x}, j\omega) = \frac{\rho_\infty c}{S}\left[\frac{\cos[2\pi\zeta(1-\tilde{x})]}{j\sin(2\pi\zeta)}Q_p(j\omega) + \frac{\cos(2\pi\zeta\tilde{x})}{j\sin(2\pi\zeta)}Q_s(j\omega)\right]$$
$$=: Z_p(\tilde{x}, j\omega)Q_p(j\omega) + Z_s(\tilde{x}, j\omega)Q_s(j\omega). \quad (7.27)$$

To determine the two source strengths $Q_p(j\omega)$ and $Q_s(j\omega)$, we need two independent measures of the interior sound pressure at the positions $\tilde{x}_1$ and $\tilde{x}_2$ such as

$$\mathbf{p} := \begin{bmatrix} P_1(\tilde{x}_1, j\omega) \\ P_2(\tilde{x}_1, j\omega) \end{bmatrix} = \begin{bmatrix} Z_{1p}(\tilde{x}_1, j\omega) & Z_{1s}(\tilde{x}_1, j\omega) \\ Z_{2p}(\tilde{x}_2, j\omega) & Z_{2s}(\tilde{x}_2, j\omega) \end{bmatrix}\begin{bmatrix} Q_p(j\omega) \\ Q_s(j\omega) \end{bmatrix}$$
$$=: \mathbf{Zq}, \quad (7.28)$$

where Z_{1p} and Z_{2p} are the *transfer impedances* that describe the pressure response at $\tilde{x}_1$ and $\tilde{x}_2$ caused by the primary source, and Z_{1s} and Z_{2s} are the transfer impedances that describe the pressure response at $\tilde{x}_1$ and $\tilde{x}_2$ cased by the secondary source. The solution of the fully determined inverse problem specified by (7.28) reads

$$\mathbf{q} = \mathbf{Z}^{-1}\mathbf{p} =: \mathbf{Np}, \quad (7.29)$$

where $\mathbf{N}$ is the *complex compliance matrix*. The primary and secondary source strengths are determined by the elements of $\mathbf{N}$ as well as by the complex sound pressure data for the positions $\tilde{x}_1$ and $\tilde{x}_2$ such as

$$Q_p = \frac{jS}{\rho_\infty c}\sin(2\pi\zeta)\left[-\frac{\cos(2\pi\zeta\tilde{x}_2)}{D}P_1 + \frac{\cos(2\pi\zeta\tilde{x}_1)}{D}P_2\right]$$
$$Q_s = \frac{jS}{\rho_\infty c}\sin(2\pi\zeta)\left[\frac{\cos[2\pi\zeta(\tilde{x}_2-1)]}{D}P_1 - \frac{\cos[2\pi\zeta(\tilde{x}_1-1)]}{D}P_2\right], \quad (7.30)$$

with

$$D = \cos(2\pi\zeta\tilde{x}_1)\cos\big[2\pi\zeta(\tilde{x}_2-1)\big] - \cos(2\pi\zeta\tilde{x}_2)\cos\big[2\pi\zeta(\tilde{x}_1-1)\big]. \quad (7.31)$$

This solution fails, if the

- Sound field in the one-dimensional wave-guide is excited in an acoustic resonance with $\zeta_{nR} = n/2$ with $n = 0, 1, \ldots,$
- Under-determined case ($\tilde{x}_1 = \tilde{x}_2$) is analyzed,
- Combination of ζ, $\tilde{x}_1$ and $\tilde{x}_2$ yields $D = 0$.

Furthermore, (7.29) clarifies that measurement errors that can—in analogy to Sect. 6.5.3—be expressed as $\mathbf{p} = \mathbf{p}_0 + \Delta\mathbf{p}$ cause a (proportional) deviation from the exact solution that is given by $\Delta\mathbf{q} = \mathbf{N}\Delta\mathbf{p}$. Because of the non-linear relation between the transfer impedances and the normalized positions, compare (7.27), a position mismatch causes a non-linear deviation from the correct solution.

Modified Formulation of the Inverse Problem

Solution of (7.26) is based on the reduced stiffness matrix $\mathbf{K}_{red}$. For this reason some information about the discretized system that is stored in the system stiffness

$\mathbf{K}$ would be ignored by this approach. However, rejecting of such information can lead to solutions that minimize (7.26) but differ from the actual sound field data. For this reason it is advantageous to solve a modified least square problem as proposed in (Weber 2009). This modification is derived from the minimization of a cost function such as

$$J(\hat{\mathbf{p}}_{red}, \mathbf{l}) := \hat{\mathbf{p}}^H \mathbf{K} \hat{\mathbf{p}} + \mathbf{l}^H (\mathbf{K}_{red} \hat{\mathbf{p}}_{red} - \hat{\mathbf{f}}_{red}) \tag{7.32}$$

in which the first summand corresponds to the (time-averaged) acoustic potential energy[4] of the discretized interior, whereas the second summand represents (7.24) that has been taken into account as a constraint using the method of *Lagrange multipliers*, compare e.g. (Vapnyarskii 2001). These multipliers are arranged in the column matrix $\mathbf{l} = [\lambda_1\ \lambda_2\ \ldots\ \lambda_M]^T$, where M is the number of equations in (7.24).

The system stiffness matrix $\mathbf{K} = \mathbf{C} - \omega^2 \mathbf{M}$ is a real valued and symmetric quantity, if losses are neglected. For such a matrix the cost function specified by (7.32) reads

$$J(\hat{\mathbf{p}}_{red}, \mathbf{l}) = \hat{\mathbf{p}}_{MK}^H \mathbf{K}_1 \hat{\mathbf{p}}_{MK} + 2\hat{\mathbf{p}}_{MK}^H \mathbf{K}_{23} \hat{\mathbf{p}}_{red} + \hat{\mathbf{p}}_{red}^H \mathbf{K}_{59} \hat{\mathbf{p}}_{red} + \mathbf{l}^H (\mathbf{K}_{red} \hat{\mathbf{p}}_{red} - \hat{\mathbf{f}}_{red}). \tag{7.33}$$

The stiffness sub-matrix $\mathbf{K}_1$ has been introduced in (7.23). The other stiffness sub-matrices are defined as

$$\mathbf{K}_{23} = [\,\mathbf{K}_2\ \mathbf{K}_3\,], \qquad \mathbf{K}_{47} = \begin{bmatrix} \mathbf{K}_4 \\ \mathbf{K}_7 \end{bmatrix}, \qquad \mathbf{K}_{59} = \begin{bmatrix} \mathbf{K}_5 & \mathbf{K}_6 \\ \mathbf{K}_8 & \mathbf{K}_9 \end{bmatrix}. \tag{7.34}$$

The modified version of the inverse problem is obtained by setting the partial derivatives of (7.33) with respect to $\hat{\mathbf{p}}_{red}$ and $\mathbf{l}$ to zero. This yields

$$\mathbf{K}_{mod} \hat{\mathbf{p}}_{mod} := \begin{bmatrix} 2\mathbf{K}_{59}^H & \mathbf{K}_{red}^H \\ \mathbf{K}_{red} & \mathbf{0} \end{bmatrix} \begin{bmatrix} \hat{\mathbf{p}}_{red} \\ \mathbf{l} \end{bmatrix} = \begin{bmatrix} -2\mathbf{K}_{23}^H \hat{\mathbf{p}}_{MK} \\ \hat{\mathbf{f}}_{red} \end{bmatrix} =: \hat{\mathbf{f}}_{mod}. \tag{7.35}$$

Evaluation of (7.35) requires the minimization of the associated modified cost function that is given by

$$J_{mod}(\hat{\mathbf{p}}_{mod}) := \mathbf{r}_{mod}^H \mathbf{r}_{mod} \quad \text{with } \mathbf{r}_{mod} := \mathbf{K}_{mod} \hat{\mathbf{p}}_{mod} - \hat{\mathbf{f}}_{mod}. \tag{7.36}$$

Comments on the Solution of Inverse Problems

As we have seen, the solution of the inverse problem requires minimization of cost functions, i.e. the solution of least square problems such as

$$\min_{\mathbf{x}} \|\mathbf{r}\|_2 \quad \text{with } \mathbf{r} := \mathbf{A}\mathbf{x} - \mathbf{b} \text{ and } \|\mathbf{r}\|_2 := \sqrt{\mathbf{r}^H \mathbf{r}}. \tag{7.37}$$

[4]The dimension of the acoustic potential energy is Force × Length, compare (3.27), whereas the dimension of the first summand in (7.32) is Force × Length × Time^{-2}. Thus, multiplying this summand with ω^{-2} gives an energy term.

However, if the least square problem is *ill posed* in the sense that small changes in the data can cause arbitrary large changes in the solution, and this is reflected in ill conditioning of the matrix **A**, regularization methods must be applied. Beside classical approaches such as *singular value decomposition* (SVD) (Golub and Kahan 1965) or *Tikhonov regularization* (TR) (Tikhonov 1963) certain iterative methods for solving least square problems have minimization properties, see (Hansen and O'Leary 1993), because the lower frequency components of the solution converge faster than the higher-frequency components.

The family of *conjugate gradient* methods minimizes the least square problem (7.37) by keeping the solution norm $\|\mathbf{x}_{reg}\|_2$ as small as possible. One formulation that is particularly suited to ill-posed problems is the LSQR implementation of Paige and Saunders, see (Paige and Saunders 1982). However, because the LSQR algorithm requires more storage and work per iteration, the *conjugated gradient least square* (CGLS) algorithm, see (Björck 1996), has been used in the IFEM approach.

The CGLS algorithm is defined for positive definite systems and has to be applied to the normal equations that can for complex data considering

$$\bar{\mathbf{A}} = \begin{bmatrix} \mathrm{Re}[\mathbf{A}] & \mathbf{0} \\ \mathbf{0} & \mathrm{Im}[\mathbf{A}] \end{bmatrix}, \qquad \bar{\mathbf{x}} = \begin{bmatrix} \mathrm{Re}[\mathbf{x}] \\ \mathrm{Im}[\mathbf{x}] \end{bmatrix}, \qquad \bar{\mathbf{b}} = \begin{bmatrix} \mathrm{Re}[\mathbf{b}] \\ \mathrm{Im}[\mathbf{b}] \end{bmatrix} \tag{7.38}$$

be specified as

$$\bar{\mathbf{A}}^T\bar{\mathbf{A}}\bar{\mathbf{x}} = \bar{\mathbf{A}}^T\bar{\mathbf{b}}. \tag{7.39}$$

If this iterative algorithm is applied to solve the least square problem (7.39), its k-th iteration step can be summarized as follows:

$$\begin{aligned}
\alpha_k &= \frac{\|\bar{\mathbf{A}}^T\mathbf{r}^{(k-1)}\|_2^2}{\|\bar{\mathbf{A}}\mathbf{d}^{(k-1)}\|_2^2} \\
\mathbf{x}^{(k)} &= \mathbf{x}^{(k-1)} + \alpha_k\mathbf{d}^{(k-1)} \\
\mathbf{r}^{(k)} &= \mathbf{r}^{(k-1)} - \alpha_k\bar{\mathbf{A}}\mathbf{d}^{(k-1)} \\
\beta_k &= \frac{\|\bar{\mathbf{A}}^T\mathbf{r}^{(k)}\|_2^2}{\|\bar{\mathbf{A}}^T\mathbf{r}^{(k-1)}\|_2^2} \\
\mathbf{d}^{(k)} &= \bar{\mathbf{A}}^T\mathbf{r}^{(k)} + \beta_k\mathbf{d}^{(k-1)}.
\end{aligned} \tag{7.40}$$

$\mathbf{r}^{(k)} = \bar{\mathbf{b}} - \bar{\mathbf{A}}\mathbf{x}^{(k)}$ is the k-th residual matrix and $\mathbf{d}^{(k)}$ is the k-th auxiliary iteration matrix. The CGLS algorithm is initialized with starting matrices such as $\mathbf{x}^{(0)}$, $\mathbf{r}^{(0)} = \bar{\mathbf{b}} - \bar{\mathbf{A}}\mathbf{x}^{(0)}$ and $\mathbf{d}^{(0)} = \bar{\mathbf{A}}^T\mathbf{r}^{(0)}$. α_k and β_k are auxiliary variables, which are used to calculated the solution matrix, the auxiliary iteration matrix and the residual matrix. The number of iterations plays the role of the regularization parameter.

One of the main difficulties in properly solving a discrete ill-posed problem is the determination of a suitable regularization parameter. A small one (usually) gives a good approximation to the system equations and has therefore a low residual. Such a solution is very sensitive to the influence of data errors and is said to be under-regularized. A large regularization parameter suppresses data errors, but increases

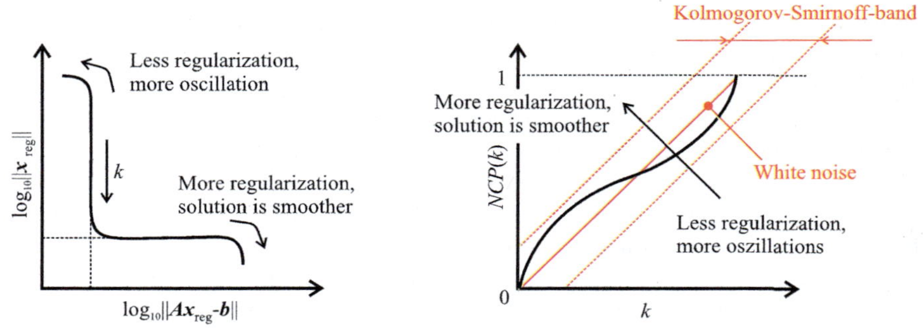

Fig. 7.3 *Left*: Generic shape of the L-curve. *Right*: Generic shape of NCP (adapted from Weber 2009)

the approximation error. This is also true for the CGLS algorithm, if $1/k$ is understood as regularization parameter.

A method commonly used to determine an optimal regularization parameter is based on the *L-curve* approach, see (Hansen and O'Leary 1993). This curve is obtained by plotting the logarithm of the norm of the regularized solution versus the logarithm of the corresponding residual norm for all regularization parameters k. A generic shape of the L-curve is shown in Fig. 7.3 (left). This curve clarifies that the optimal regularization parameter is found in the corner region, because this parameter minimizes both the solution norm and the residual norm. However, as stated in (Hansen and O'Leary 1993), the characteristic L-shape with (i) the "flat" part where the regularized solution is dominated by regularization errors and (ii) the "vertical" part where the regularized solution is dominated by the errors of the right hand side is only obtained, if

1. The discrete *Picard*[5] condition is satisfied,
2. The errors in the right hand side are essentially white noise,
3. We have a reasonable signal-to-noise ratio.

It is therefore possible that the L-curve approach fails as reported in (Hanke 1996) and (Vogel 1996). In such a case it is possible to use the *normalized cumulative periodogram* (NCP), see (Hansen et al. 2006), to determine a proper regularization parameter. The NCP method selects the solution whose residual vector, treated as a time series and using (2.36) transformed into frequency domain, is closest to white noise. Thus, it is ensured that neither high frequencies caused by under-regularization, nor low frequencies caused by over-regularization, dominate the solution. According to (Hansen et al. 2006) the NCP is defined as the normalized discrete ASD that is calculated from the components of the residuum matrix

$$\mathbf{r} := \mathbf{A}\mathbf{x} - \mathbf{b} = [R_1\ R_2\ \dots\ R_M]^T \tag{7.41}$$

[5]This condition requires that all *Fourier-coefficients* $\mathbf{u}_i^H \mathbf{b}$ (calculated as the inner product between the i-th left singular vector of the matrix $\mathbf{A}$ and the right hand side $\mathbf{b}$) on average decay faster than the singular values σ_i of $\mathbf{A}$.

such as

$$NCP(k) = \frac{\sum_{i=2}^{k+1} |\mathfrak{F}_d[R_i]|^2}{\sum_{i=2}^{n} |\mathfrak{F}_d[R_i]|^2} \quad \text{with } k = 1, 2, \ldots, M-1. \tag{7.42}$$

Because the summation starts with $i = 2$, the DC part is rejected. Hence, the NCP of white noise corresponds to the red diagonal line that is shown in Fig. 7.3 (right). The NCP is dominated by high frequency components, if the solution is under-regularized. These oscillations are suppressed, if the regularization parameter is increased. As stated above, its optimum value is found, if the NCP is as close as possible to the NCP of white noise. The difference between the NCP of the regularized solution and the NCP of white noise can be quantified by the *Kolmogorov-Smirnoff-test*, see (Fuller 1976). According to this test, a proper regularization parameter is found, if the NCP is enclosed by the *Kolmogorov-Smirnoff* bands as also shown in Fig. 7.3 (right). This test can therefore be used to establish a stop criterion.

Finally it should be noticed that the IFEM presented here also offers the opportunity for RSC and health monitoring. RSC is possible, if the IFEM is used to determine the sound pressure remote from a physical sensor. Monitoring the change of boundary values estimated by an IFEM approach enables the detection of changes in transfer paths or source strengths that can be indicators for damage processes in elastic boundary structures.

7.3 Initial Performance Estimation

As mentioned in the motivation to this chapter we want to have an idea on the control profit that can be achieved by an active control system. For this purpose we need tools that can be used in time or frequency domain to determine an *initial performance estimation* (IPE). IPE means that we are able to predict the noise reduction and/or the costs realized by a certain control approach or a specific set of actuator and sensor locations *before* the final system is developed.

Figure 7.4 clarifies that an adaptive ANC-system can be seen as a multiple-input and multiple-output system with underlying control structure. Because the more general case of RSC, compare Sect. 6.5.2, has been considered, the system output is given by M monitor signals and C error signals. The system input is given by K reference signals.

The optimal nominal control performance, see Sect. 6.5.3 (6.147), of the ANC-system is beside cost function weighting matrices, see Sect. 6.5.1, determined by the disturbance signals measured at the monitor and error sensors as well as by the transfer behavior of the secondary transfer paths. If active control is applied to reduce broadband noise, the optimal control filter coefficients are determined by statistic variables such as the cross correlation between (filtered) reference signals and disturbance signals, compare Sect. 6.6.1. Equation (6.202). How these connection can be used for IPE, will be shown in the upcoming section.

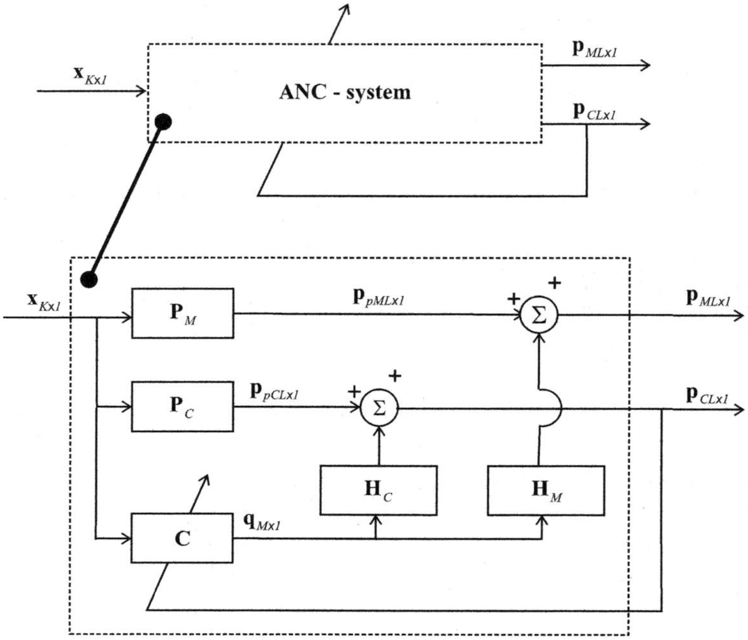

Fig. 7.4 Block diagram of an ANC-system that is seen as an adaptive multiple-input and multiple-output system considering K reference signals, M monitor signals, and C error signals

7.3.1 Coherence Analysis

One particular tool that can effectively be used for IPE in frequency domain based on the Fourier transforms of continuous or sampled data is the *coherence analysis* (CA). This tool will be explained for the case of direct control, compare Fig. 7.5, of L error signals using K reference signals and M control signal.

To consider the statistic properties of the involved signals, the estimation of the optimal nominal control performance is based on the expectation of the hermitian quadratic form (6.87) in which the sound pressure matrix $\mathbf{p}$ is defined by (6.86). According to (6.82) it contains the Fourier transforms of continuous-time or discrete-time data. However, the associated arguments $j\omega$ and $e^{j\omega T}$ have been omitted for an easy notation. The cost function that is to be analyzed is given by

$$\begin{aligned} J_0 &= E\left(\mathbf{p}^H\mathbf{p}\right) = \mathrm{tr}\left[E\left(\mathbf{p}\mathbf{p}^H\right)\right] \\ &= \mathrm{tr}\left[E\left(\mathbf{p}_p\mathbf{p}_p^H + \mathbf{p}_p\mathbf{x}^H\mathbf{C}^H\mathbf{H}^H + \mathbf{H}\mathbf{C}\mathbf{x}\mathbf{p}_p^H + \mathbf{H}\mathbf{C}\mathbf{x}\mathbf{x}^H\mathbf{C}^H\mathbf{H}^H\right)\right]. \end{aligned} \tag{7.43}$$

As described in (Elliott 2001), the expectation only needs to be taken over the stochastic parts in (7.43), which involve the disturbance $\mathbf{p}_p$ and the reference $\mathbf{x}$. If we introduce the $(K \times K)$ ASD matrix of the reference signals as

$$\mathbf{S}_{xx} = E\left(\mathbf{x}\mathbf{x}^H\right), \tag{7.44}$$

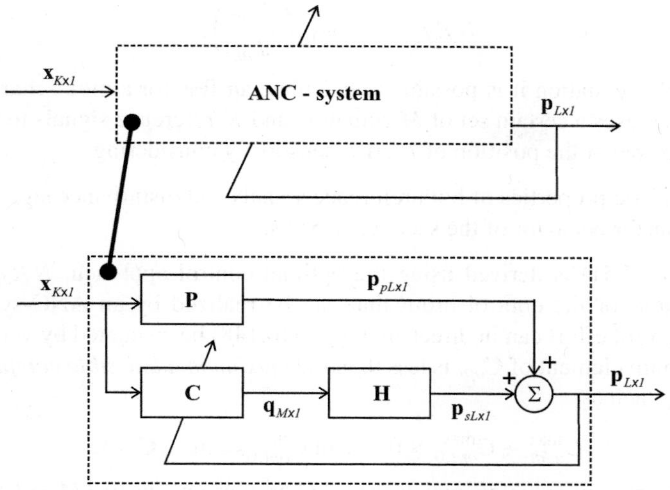

Fig. 7.5 Block diagram of an ANC-system for direct control that is seen as an adaptive multiple-input and multiple-output system considering K reference signals and L error signals

the $(L \times K)$ CSD matrix that contains the CSDs between the disturbance signals and the reference signals as

$$\mathbf{S}_{xp} = E\left(\mathbf{x}\mathbf{p}_p^H\right), \tag{7.45}$$

and the $(L \times L)$ ASD matrix of the disturbance signals as

$$\mathbf{S}_{pp} = E\left(\mathbf{p}_p\mathbf{p}_p^H\right), \tag{7.46}$$

it is possible to write (7.43) such as

$$J_0 = \mathrm{tr}\left(\mathbf{S}_{pp} + \mathbf{S}_{xp}\mathbf{C}^H\mathbf{H}^H + \mathbf{H}\mathbf{C}\mathbf{S}_{xp}^H + \mathbf{H}\mathbf{C}\mathbf{S}_{xx}\mathbf{C}^H\mathbf{H}^H\right). \tag{7.47}$$

The total derivative of (7.47) is given by

$$d\,J_0 = 2\,\mathrm{Re}\left[\mathrm{tr}\left(\mathbf{S}_{xp}^H\mathbf{H}d\mathbf{C} + \mathbf{S}_{xx}\mathbf{C}^H\mathbf{H}^H\mathbf{H}d\mathbf{C}\right)\right]. \tag{7.48}$$

Equation (7.48) allows for calculating the optimal controller response by setting $d\,J_0$ to zero. This yields considering $\mathbf{S}_{xx} = \mathbf{S}_{xx}^H$

$$\mathbf{C}_{opt} = -\left(\mathbf{H}^H\mathbf{H}\right)^{-1}\mathbf{H}^H\mathbf{S}_{xp}\mathbf{S}_{xx}^{-1}. \tag{7.49}$$

An expression for the minimum possible value of the cost function is obtained, by substituting (7.49) into (7.47). This yields

$$J_{\min} = J_{ini} - \mathrm{tr}\left[\mathbf{S}_{xp}\mathbf{S}_{xx}^{-1}\mathbf{S}_{xp}^H\mathbf{H}\left(\mathbf{H}^H\mathbf{H}\right)^{-1}\mathbf{H}^H\right], \tag{7.50}$$

where $J_{ini} = \mathrm{tr}(\mathbf{S}_{pp})$ is the associated initial cost value. The estimate of the optimal nominal controller performance for each frequency line is according to (6.147) defined as

$$NR_{J_0} = 10\log_{10}\left(\frac{J_{\min}}{J_{ini}}\right). \tag{7.51}$$

Using this equation it is possible to calculate an IPE for a multi-channel ANC-systems that uses a certain set of M actuators and K reference signals to reduce the noise measured at the position of L error sensors by considering

1. The statistic properties of both reference signals and disturbance signals,
2. The transfer behavior of the secondary paths.

Because (7.51) is derived using the optimal control approach, NR_{J_0} gives an upper bound for the control profit that can be realized by an ANC-system. The optimal control effort can in direct analogy to (6.148) be evaluated by verifying that the maximum element of $\mathbf{C}_{opt}$ is less than the *maximum admissible complex control gain* $C_{adm}^{\max}$ such as

$$C_{adm}^{\max} - C_{opt,0}^{\max} > 0 \quad \text{with } C_{opt,0}^{\max} = \max(\mathbf{C}_{opt}). \tag{7.52}$$

If the number of actuators equals the number of error sensors ($M = L$), the noise reduction specified by (7.51) is independent of the properties of the secondary transfer paths, because of $\mathbf{I} = \mathbf{H}(\mathbf{H}^H\mathbf{H})^{-1}\mathbf{H}^H$.

For practical applications it is important to choose valid positions for the reference sensors, before fixing the set of error microphones and canceling loudspeakers. Therefore, the special case of a single actuator $M = 1$ and a single error sensor $L = 1$ has to be evaluated for multiple references $K > 1$. The IPE can than be calculated using the *multiple coherence function* γ_{xd}^2 between K reference signals and one disturbance signal such as

$$NR_{J_0} = 10\log_{10}\left(1 - \gamma_{xp}^2\right) \quad \text{with } \gamma_{xp}^2 = \frac{\mathbf{S}_{xp}\mathbf{S}_{xx}^{-1}\mathbf{S}_{xp}^H}{S_{pp}}. \tag{7.53}$$

For this reason IPE based on (7.53) is named CA. Please note that $\mathbf{H} \equiv H$ is a scalar in this case, and that $\mathbf{H}(\mathbf{H}^H\mathbf{H})^{-1}\mathbf{H}^H$ is therefore equal to unity. Furthermore, $\mathbf{S}_{pp} \equiv S_{pp}$ as well as $\mathbf{S}_{xp}\mathbf{S}_{xx}^{-1}\mathbf{S}_{xp}^H$ are also scalars in this case, and so the trace operation in (7.50) can be omitted.

In the single-channel case in which only one reference signal has to be taken into account, the estimated noise reduction for each frequency line is given by

$$NR_{J_0} = 10\log_{10}\left(1 - \gamma^2\right) \quad \text{with } \gamma^2 = \frac{|S_{xd}|^2}{S_{xx}S_{dd}}, \tag{7.54}$$

where γ is the *coherence function* between one reference signal and one disturbance signal. Because this function is implemented in the most modern spectrum analyzers, an ad hoc estimation of the maximum value of active noise reduction is (in most situations) possible. Equation (7.54) clarifies that a noise reduction of up to -17 dB can be expected, if the $\gamma^2 = 0.99$, whereas a lower value of $\gamma^2 = 0.9$ yields an IPE of only -7.2 dB.

Finally, it should be noticed that the positions of the K reference sensors, M actuators and L error sensors can explicitly considered, if CA is applied to determine an IPE. For this purpose we have to replace the $(L \times 1)$ error signal matrix $\mathbf{p}$ as well as the $(M \times 1)$ column matrix $\mathbf{q}$ that contains the M control signals by modified matrices such as

$$\mathbf{p} \mapsto \tilde{\mathbf{p}} = \mathbf{L}(\mathbf{p}_p + \mathbf{H}\tilde{\mathbf{q}}), \tag{7.55}$$

and

$$\mathbf{q} \mapsto \tilde{\mathbf{q}} = \mathbf{M}\mathbf{q}, \tag{7.56}$$

where $\mathbf{L} = \operatorname{diag}(L_1, L_2, \ldots, L_L)$ is an error weighting matrix that enables the use of a certain error signal $(L_i = 1)$ or the exclusion $(L_i = 0)$ of a certain error signal from the analysis and $\mathbf{M} = \operatorname{diag}(M_1, M_2, \ldots, M_M)$ is a control signal weighting matrix that is introduced to analyze the effect of including $(M_i = 1)$ or excluding $(M_i = 0)$ a certain control source.

7.3.2 Transducer Placement Analysis

Albeit the fact that a proper coherence between the reference signals and the error signals is required to achieve a distinct noise reduction, it can be sufficient to consider only the placement of actuators and error sensors to determine an IPE. This is especially true, if active control is applied to harmonic disturbances. Such an investigation can be called *transducer placement analysis* (TPA). The IPE determined by a TPA specifies the noise reduction potential of a specific transducer setting. Because this method will be outlined in frequency domain, we again can omit the associated arguments $j\omega$ and $e^{j\omega T}$ for convenience.

As in Sect. 7.3.1, the main idea is to introduce an error weighting matrix $\mathbf{L} = \operatorname{diag}(L_1, L_2, \ldots, L_L)$ that enables the control system to use a sensor $(L_i = 1)$ or to exclude $(L_i = 0)$ a certain sensor signal from the control loop. If direct control is applied, we have only one set of error microphones. If however TPA is performed to determine an IPE of a RSC approach, the weighting matrix $\mathbf{L}$ is usually applied to include/exclude error microphones (not the later on removed monitor microphones).

In direct analogy a control signal weighting matrix $\mathbf{M} = \operatorname{diag}(M_1, M_2, \ldots, M_M)$ is introduced to analyze the effect of including $(M_i = 1)$ or excluding $(M_i = 0)$ a certain control source. If the weighting matrices of the general cost function (6.107) are identified such as

$$\mathbf{W}_p \equiv \mathbf{L}, \qquad \mathbf{W}_q \equiv \beta \mathbf{M}^{-2}, \qquad \mathbf{P} \equiv \mathbf{I}, \qquad \mathbf{Q} = \mathbf{M}, \tag{7.57}$$

where β is a positive and real valued weighting parameter as also used in (6.91), we obtain the following cost function

$$J_0 = \left[(\mathbf{p}_p + \mathbf{H}\mathbf{M}\mathbf{q})\right]^H \mathbf{L}(\mathbf{p}_p + \mathbf{H}\mathbf{M}\mathbf{q}) + \beta \mathbf{q}^H \mathbf{q}. \tag{7.58}$$

The column matrix that contains the optimal canceling signals is, considering (6.13) as well as the identification of the cost function parameters

$$\mathbf{A} \equiv \mathbf{M}\mathbf{H}^H \mathbf{L}\mathbf{H}\mathbf{M} + \beta \mathbf{I}, \qquad \mathbf{b} = \mathbf{b}_1 = \mathbf{b}_2 \equiv \mathbf{M}\mathbf{H}^H \mathbf{L}\mathbf{p}_p, \qquad c \equiv \mathbf{p}_p^H \mathbf{L}\mathbf{p}_p, \tag{7.59}$$

given by

$$\mathbf{q}_{opt}(\mathbf{L}, \mathbf{M}, \beta) = -\left[\mathbf{M}\mathbf{H}^H\mathbf{L}\mathbf{H}\mathbf{M} + \beta\mathbf{I}\right]^{-1}\mathbf{M}\mathbf{H}^H\mathbf{L}\mathbf{p}_p. \tag{7.60}$$

Equation (7.60) clarifies that $\mathbf{q}_{opt}$ depends on the transducer locations that are specified by the diagonal matrices $\mathbf{L}$ and $\mathbf{M}$. These matrices have elements that are zero (deactivated case) or one (activated case). The estimate of the optimal nominal controller performance is according to (6.147) for each frequency line defined as

$$NR_{J_0} = 10\log_{10}\left(\frac{J_0[\mathbf{q}_{opt}(\mathbf{L}, \mathbf{M})]}{\mathbf{p}_p^H\mathbf{p}_p}\right). \tag{7.61}$$

Using (7.61) it is possible to calculate an IPE for a multi-channel ANC-systems that uses a certain set of M actuators to reduce the noise measured at the position of L error sensors assuming a perfect coherence between disturbance and reference signals.

Because (7.61) is derived using the optimal control approach, the estimated noise reduction gives an upper bound for the control profit that can be realized by an ANC-system. The optimal control effort can (again) in direct analogy to (6.148) be evaluated by verifying that the maximum element of $\mathbf{q}_{opt}$ is less than the *maximum admissible control signal* $Q_{adm}^{\max}$ such as

$$Q_{adm}^{\max} - Q_{opt,0}^{\max} > 0 \quad \text{with } Q_{opt,0}^{\max} = \max(\mathbf{q}_{opt}). \tag{7.62}$$

7.3.3 Correlation Analysis

The time domain counterpart to CA is the *correlation analysis* (COA) . This tool will be introduced for direct control, compare Fig. 7.5, of discrete-time signals. In order to determine an IPE by COA, considering the statistic properties of the involved signals, we will according to Sect. 6.6.1 (6.195) use the expectation of the squared sound pressure to define a cost function such as

$$\begin{aligned} J_0(n) &= E\left[\mathbf{p}^T(n)\mathbf{p}(n)\right] \\ &= \mathbf{w}^T(n)\mathbf{A}(n)\mathbf{w}(n) + 2\mathbf{w}^T(n)\mathbf{b}(n) + c(n), \end{aligned} \tag{7.63}$$

where

$$\begin{aligned} \mathbf{A}(n) &\equiv E\left[\mathbf{X}^T(n)\mathbf{X}(n)\right] =: \mathbf{R}_{xx}(n) \\ \mathbf{b}(n) &\equiv E\left[\mathbf{X}^T(n)\mathbf{p}_p(n)\right] =: \mathbf{r}_{xp}(n) \\ c(n) &\equiv E\left[\mathbf{p}_p^T(n)\mathbf{p}_p(n)\right] =: r_{pp}(n). \end{aligned} \tag{7.64}$$

Equation (7.64) contains the discrete-time domain counterparts of the spectral densities defined by (7.44)–(7.46). $\mathbf{R}_{xx}(n)$ is the *auto correlation matrix* (ACM) that is based on the $(L \times MKI)$ matrix $\mathbf{X}(n)$ that contains the filtered reference signals, compare (6.198). $\mathbf{r}_{xp}(n)$ is the $(L \times 1)$ *cross correlation matrix* (CCM) that contains the cross correlation between the disturbance signals and the filtered reference signals. The scalar $r_{pp}(n)$ specifies the expectation of the sum of squared disturbances. The minimum cost function value for (7.63) is considering (6.14) given by

$$J_{\min} = r_{pp}(n) - \mathbf{r}_{xp}^T(n)\mathbf{R}_{xx}^{-1}(n)\mathbf{r}_{xp}(n), \tag{7.65}$$

whereas the initial value is determined by

$$J_{ini} = r_{pp}(n). \tag{7.66}$$

In direct analogy to Sect. 7.3.1, the estimate of the optimal nominal controller performance is obtained from

$$\begin{aligned}\overline{NR}_{J_0} &= 10\log_{10}\left(\frac{J_{\min}}{J_{ini}}\right)\\ &= 10\log_{10}\left(1 - \frac{\mathbf{r}_{xp}^T(n)\mathbf{R}_{xx}^{-1}(n)\mathbf{r}_{xp}(n)}{r_{pp}(n)}\right)\\ &=: 10\log_{10}(1 - \upsilon_{xp}^2),\end{aligned} \tag{7.67}$$

in which the *multiple correlation function* υ_{xp}^2 is the time-domain analogon to the multiple coherence function that has been introduced in (7.53). The symbol $\overline{NR}_{J_0}$ indicates that the IPE is calculated for the frequency band of interest and (except for harmonic excitation) not for a single frequency line. Using (7.67) it is also possible to calculate an IPE for a multi-channel ANC-systems that uses a certain set of M actuators and K reference signals to reduce the noise measured at the position of L error sensors by considering

1. The statistic properties of both reference signals and disturbance signals,
2. The transfer behavior of the secondary paths.

Because (7.67) is again derived using the optimal control approach, $\overline{NR}_{J_0}$ gives an upper bound for the control profit that can be realized by an ANC-system. However, in contrast to CA, COA requires also in the single-channel case assembling and evaluation of matrices such as the ACM. This leads a higher computational effort.

The optimal control effort can in analogy to (6.148) be evaluated by verifying that the maximum value of optimal secondary source strengths $q_{opt,0}^{\max}$ is less than the *maximum admissible control signal* $q_{adm}^{\max}$ such as

$$q_{adm}^{\max} - q_{opt,0}^{\max} > 0 \quad \text{with } q_{opt,0}^{\max} = \max(\mathbf{q}_{opt}). \tag{7.68}$$

The optimal set of control filter coefficients $\mathbf{w}_{opt}$ that is required to evaluate (7.68) by calculating the secondary source strength as specified by (6.191) is, according to (6.13) and considering (7.64), given by the *Wiener solution*

$$\mathbf{w}_{opt} = -\mathbf{R}_{xx}^{-1}(n)\mathbf{r}_{xp}(n). \tag{7.69}$$

7.3.4 Impulse Response Analysis

Another tool that can be used to determine an IPE in time domain is the *impulse response analysis* (IRA). To explain this tool, we again consider the case of direct control, compare Fig. 7.5, and in addition the single-channel case in which we have one reference signal, one error signal and one control signal. Because of the superposition of noise and anti-noise, the error signal at discrete time n is given by

$$p(n) = p_p(n) + p_s(n). \tag{7.70}$$

The noise is determined by the convolution between the impulse response of the primary path and the reference signal such as

$$p_p(n) = \sum_{p=0}^{P-1} h_{pp} x(n-p) =: \mathbf{h}_p^T \mathbf{x}, \tag{7.71}$$

where

$$\mathbf{x}(n) = \left[x(n)\; x(n-1)\; \dots\; x(n-P+1)\right]^T, \tag{7.72}$$

is the $(P \times 1)$ column matrix of (buffered) reference signals and

$$\mathbf{h}_p = [h_{p0}\; h_{p1}\; \dots\; h_{pP-1}]^T \tag{7.73}$$

is a $(P \times 1)$ column matrix that contains the P coefficients of the primary path impulse response. The anti-noise is according to (6.192) determined by

$$p_s(n) = \sum_{j=0}^{J-1} \sum_{i=0}^{I-1} h_j x(n-i-j) =: (\mathbf{H}\mathbf{w})^T \mathbf{x}, \tag{7.74}$$

where

$$\mathbf{H} = \begin{pmatrix} h_0 & 0 & \cdots & 0 \\ h_1 & h_0 & \cdots & 0 \\ \vdots & \vdots & \ddots & \vdots \\ h_{I-1} & h_{I-2} & \cdots & h_0 \\ h_I & h_{I-1} & \cdots & h_1 \\ \vdots & \vdots & \ddots & \vdots \\ 0 & 0 & \cdots & h_{J-1} \end{pmatrix}, \tag{7.75}$$

is a $(I + J - 1 \times I)$ matrix that contains the coefficients of the secondary path impulse response and

$$\mathbf{w} = [w_0\; w_1\; \dots\; w_{I-1}]^T \tag{7.76}$$

is a $(I \times 1)$ column matrix that contains the I optimal control filter coefficients. Please notice that these coefficients are not specified by (7.69). Their optimal values have to be determined by minimizing $p^2(n)$, if IRA is applied. Considering (7.70)–(7.76), the controlled sound pressure can be written as

$$p(n) = (\mathbf{h}_p + \mathbf{H}\mathbf{w})^T \mathbf{x}. \tag{7.77}$$

Thus, the norm of

$$\mathbf{h}_c = \mathbf{h}_p + \mathbf{H}\mathbf{w} \tag{7.78}$$

has to be minimized with respect to $\mathbf{w}$, if the squared sound pressure shall be minimized. For this purpose it is assumed that $P \leq J$. If the dimension of the $(P \times 1)$ column matrix $\mathbf{h}_p$ is less than the number of rows in $\mathbf{H}$, the missing coefficients have to be added to $\mathbf{h}_p$ by appending zeros until $P = I + J - 1$. Based on (7.78) the required cost function can be specified as

$$J_{IRA} = \mathbf{h}_c^T \mathbf{h}_c. \tag{7.79}$$

Considering the identifications

$$\mathbf{A} = \mathbf{H}^T\mathbf{H}, \qquad \mathbf{b} = \mathbf{H}^T\mathbf{h}_p, \qquad c = \mathbf{h}_p^T\mathbf{h}_p, \tag{7.80}$$

as well as (6.13), minimization of (7.79) yields

$$\mathbf{w}_{opt} = -\left(\mathbf{H}^T\mathbf{H}\right)^{-1}\left(\mathbf{H}^T\mathbf{h}_p\right). \tag{7.81}$$

In order to determine an IPE that is based on this optimal solution, we introduce the ratio between the squared sound pressure of the controlled system and the squared sound pressure of the uncontrolled system such as

$$f_p = \frac{p^2}{p_p^2} = \frac{[(\mathbf{h}_p + \mathbf{H}\mathbf{w}_{opt})^T\mathbf{x}][(\mathbf{h}_p + \mathbf{H}\mathbf{w}_{opt})^T\mathbf{x}]}{(\mathbf{h}_p^T\mathbf{x})(\mathbf{h}_p^T\mathbf{x})}. \tag{7.82}$$

After some algebraic manipulations this formula yields

$$\mathbf{x}^T\left[\mathbf{h}_p(\mathbf{H}\mathbf{w}_{opt})^T + (\mathbf{H}\mathbf{w}_{opt})\mathbf{h}_p^T + (\mathbf{H}\mathbf{w}_{opt})(\mathbf{H}\mathbf{w}_{opt})^T + (1 - f_p)\mathbf{h}_p\mathbf{h}_p^T\right]\mathbf{x} = 0. \tag{7.83}$$

Equation (7.83) can, considering the rule $\mathrm{tr}(\mathbf{a}\mathbf{a}^T) = \mathbf{a}^T\mathbf{a}$, be used to express the ratio f_p such as

$$f_p = \frac{\mathbf{h}_p^T\mathbf{h}_p + \mathbf{h}_p^T(\mathbf{H}\mathbf{w}_{opt}) + (\mathbf{H}\mathbf{w}_{opt})^T\mathbf{h}_p + (\mathbf{H}\mathbf{w}_{opt})^T(\mathbf{H}\mathbf{w}_{opt})}{\mathbf{h}_p^T\mathbf{h}_p}. \tag{7.84}$$

By taking the logarithm of (7.84) it is possible to obtain an IPE for the nominal state that is given by

$$\overline{NR}_{f_p} = 10\log_{10}(f_p). \tag{7.85}$$

If it is possible to identify $\mathbf{h}_p$, the impulse response of the primary path, IRA is a proper approach to determine the control filter length I that is required to obtain a specified noise reduction in the frequency band of interest. If the number of coefficients used to represent the impulse response of the primary transfer path equals the number of coefficients used to model the secondary transfer ($P = J$), the noise reduction calculated by (7.85) will lead to an infinite value, if $I = P$ filter weights are used for adaptive filtering, see (Moschytz and Hofbauer 2000).

If the number of filter coefficients is less than the number of coefficients in the primary path impulse response ($I < P$), the optimal solution is a least square solution with a non-vanishing residual error. In such a situation, evaluation of (7.85) allows for a trade of between the computational load that is determined by the control filter length, compare Fig. 6.34, and the required control profit.

In the multi-channel case, in which L microphones and M loudspeakers are used to perform active control based on one ($K = 1$) reference signal, the matrices that have to be taken into account to evaluate (7.85) are given by the ($L \times M$) block matrix

$$\mathbf{H} = \begin{pmatrix} \mathbf{H}_{11} & \mathbf{H}_{12} & \cdots & \mathbf{H}_{1m} & \cdots & \mathbf{H}_{1M} \\ \vdots & \vdots & \ddots & \vdots & \ddots & \vdots \\ \mathbf{H}_{l1} & \mathbf{H}_{l2} & \cdots & \mathbf{H}_{lm} & \cdots & \mathbf{H}_{lM} \\ \vdots & \vdots & \ddots & \vdots & \ddots & \vdots \\ \mathbf{H}_{L1} & \mathbf{H}_{L2} & \cdots & \mathbf{H}_{Lm} & \cdots & \mathbf{H}_{LM} \end{pmatrix} \tag{7.86}$$

that contains the impulse responses of the LM secondary paths, thereby each submatrix $\mathbf{H}_{lm}$ is structured as shown in (7.75), the column matrix

$$\mathbf{w}_{opt} = [\mathbf{w}_{opt,1} \ \cdots \ \mathbf{w}_{opt,m} \ \cdots \ \mathbf{w}_{opt,M}]^T \tag{7.87}$$

that contains the IM optimal control filter coefficients, thereby each sub-matrix $\mathbf{w}_{opt,m}$ is structured as shown in (7.76), and the column matrix

$$\mathbf{h}_p = [\mathbf{h}_{p1} \ \mathbf{h}_{p2} \ \ldots \ \mathbf{h}_{pP}]^T \tag{7.88}$$

that contains the LP coefficients of the primary path impulse responses, thereby each sub-matrix $\mathbf{h}_{pp}$ is structured as shown in (7.73).

7.3.5 A Short Comparison of Initial Performance Estimators

The tools discussed in this section can be used to calculate an estimate of the prospective control profit that can be realized with an ANC-system *before* a functional model is developed and integrated into the basic system.

Similarities and differences between the different tools are summarized in Table 7.1. This table clarifies that the type of required input quantities, the computational effort and the calculation result (or output quantity) differs from tool to tool.

While Fourier spectra of reference and disturbance signals are required to calculate an IPE via CA in frequency domain, the time history of these quantities must be recorded to determine an IPE by COA in (discrete-) time domain. It is obvious that the computational effort of a CA (that requires multiplication of complex quantities) is less than the computational effort of a COA, because the latter requires the evaluation of convolution integrals in order to determine cross and auto correlation functions between the involved signals.

Table 7.1 Comparison of tools for initial performance estimation

Method	CA	COA	TPA	IRA
Applied in	Frequency domain	Time domain	Frequency domain	Time domain
Available in standard spectrum analyzers	Yes	No	No	No
Computational effort	Low	High	Low	High
Required inputs	Reference and disturbance	Reference and disturbance	Disturbance and transfer functions of secondary paths	Impulse responses of primary and secondary paths and optimal control filter coefficients
Output	Estimate of noise reduction for every frequency line	Estimate of noise reduction in frequency band of interest	Estimate of noise reduction for every frequency line	Estimate of noise reduction in frequency band of interest

It should also be noticed that the coherence function γ is available in nearly all standard spectrum analyzers. Hence, single-channel CA that simply requires the analysis of two sensor data can in most situations easily be performed in on-line measurements with negligible computational effort.

In contrast to this, a COA is subdivided into a measurement part in which the time history of the investigated signals is recorded and a post-processing part in which the IPE is determined.

It is obvious that the frequency domain tools (CA and TPA) yield performance estimations $NR(j\omega)$ that are specified for each frequency line. In contrast to this, the application of a time domain tool (COA or IRA) yields an IPE $\overline{NR}$ that specifies the prospective noise reduction in a certain frequency band.

The transfer behavior the secondary paths can be considered in time as well as in frequency domain as shown in Sect. 7.3.2 and Sect. 7.3.4. This is necessary, if we want to evaluate the noise reduction potential of a specific setting. The advantage of IPE in frequency domain is that the transfer behavior is described by complex transfer functions that can (easily) be measured and evaluated with standard spectrum analyzers. However, the effect of delays that are present in the transfer paths can only be studied in time domain.

7.4 Two Short Remarks on the Effect of Sampling Frequency

7.4.1 Attenuation of Aliasing

The sampling frequency f_s can be specified by the *Nyquist-Shannon* criteria such as $f_s \geq 2f_{\max}$, where $f_{\max}$ is the highest frequency in the disturbance spectrum. However, we may have to consider the broadband nature of the primary disturbance as well as the finite fall-off rate of the analogue filter[6] used to avoid aliasing and imaging effects.

The attenuation A that can be realized at $f_s/2$ using an anti-aliasing filter with a fall-of rate of R dB/octave and a cut-off frequency f_c can according to (Elliott 2001) be approximated by

$$A \approx R \log_2\left(\frac{f_s/2}{f_c}\right) \approx 93 \text{ dB}. \tag{7.89}$$

The usefulness of (7.89) can be demonstrated by numerical examples. Considering a fall-of rate of $R = 45$ dB/octave as for the *Butterworth* filter Kemo VBF21 and a cut-off frequency of $f_c = 500$ Hz for low-pass filtering, evaluation of (7.89) yields an attenuation of 45 dB, if sampling is performed with 2 kHz. Time discretization with a sampling frequency of 1.5 kHz yields an attenuation that is less than 27 dB at $f_s/2$. If sampling is performed using a sampling frequency of only 1.1 kHz (such a sampling frequency would also meet the *Nyquist-Shannon* criteria, if the highest frequency of interest is 500 Hz), the attenuation at $f_s/2$ is less than 7 dB, and aliasing effects may affect the control task.

Please notice that (7.89) can also be used to specify a sampling frequency, if a certain attenuation at $f_s/2$ is accepted for an anti-aliasing filter with a given fall-off rate. If this formula is (for reasons of convenience) applied to estimate the attenuation of reconstruction filters (that are used to suppress imaging effects at $3f_s/2$), it will yield an underestimation of the effective suppression at $3f_s/2$. Thus, the worst case is considered, if (7.89) is applied to estimate the attenuation of a reconstruction filter.

7.4.2 Comments on Filter Delay

The sampling rate also influences the group delay of the secondary transfer path. Following the argumentation in (Elliott 2001) that is based on the assumptions that

- The analogue anti-aliasing filters and the analog reconstruction filters used in the secondary path have the same corner frequency f_c,
- An one-sample delay is caused by the digital controller (if the signal is passed though without any manipulation),

[6]Practice oriented details about design, selection and application of analog and digital filters are to be found in (Frequency Devices Inc. 2009).

- A half-sample delay is caused by the zero-order hold in the DAC,

the total group delay caused by anti-aliasing filter, A/D converter, D/A converter and reconstruction filter, can be approximated by

$$\tau_w = 1.5 \cdot T_s + \frac{N_p}{8 f_c}, \tag{7.90}$$

where N_p is the number of poles in both the analog anti-aliasing filter and the analog reconstruction filter and T_s is the sampling time. Two four-pole *Butterworth* filter ($N_p = 8$) with a corner frequency of $f_c = 500$ Hz, one used to suppress aliasing effects and one used to suppress imaging effects would (according to (7.90)) cause a group delay of approximately $\tau_w \approx 0.0027$ s. This delay has to be subtracted from the time span that is required by a single impulse to travel from the position of the reference microphone to the position of the error microphone in order to calculate the time left for causal signal processing. It corresponds to a distance of about 95 cm, if 343 m/s are considered for the speed of sound. This distance has been taken into account by placing the reference sensor in order to ensure causality.

Chapter 8
ANC-System Design Methodology

Abstract Because an ANC-system consists of a basic system (the air filled enclosure) with integrated sensors such as microphones and actuators such as loudspeakers that are driven by a controller, it is obvious that the development will result in a system with a high level of complexity. For this reason the development cannot take place separately in the domains involved. In fact, a successful development process requires a guideline for cross-domain communication and cooperation between mechanical engineering, electrical engineering and information technology to obtain a common conception of the future ANC-system that is optimized overall. For these reasons the upcoming chapter presents a design methodology for ANC-systems. This methodology is derived from the more general design methodology for mechatronic systems (VDI 2206) that has been converted and concretized in order to provide a guideline for ANC-system design. It is based on five overlapping design steps. Every design step requires the handling of specific design tasks. Successful processing of these design tasks enables successful completion of design steps. This is expressed by an increase of system maturity. The latter is assessed using a number of nine different maturity levels. This chapter is focused on theoretical aspects. The application of the proposed design methodology to specific interior noise problems using the ANC-system design tools that have been discussed in the previous chapter is shown in the next part of this book.

8.1 A Note on the Design Methodology for Mechatronic Systems

As mentioned in the motivation to this chapter, we will use the design methodology for mechatronic systems, proposed in (VDI 2206 2004), to derive a design methodology for ANC-systems. Therefore, it is necessary to introduce terms and models of this more general and more flexible development approach, which is based on three elements:

- The general problem-solving cycle on the micro-level,
- The V-model on the macro-level,
- Predefined process modules for handling recurrent working steps.

T. Kletschkowski, *Adaptive Feed-Forward Control of Low Frequency Interior Noise*, Intelligent Systems, Control and Automation: Science and Engineering 56,
DOI 10.1007/978-94-007-2537-9_8, © Springer Science+Business Media B.V. 2012

The *general problem-solving cycle on the micro-level* originally comes from *system engineering*,[1] see (Sage 1992; Danzer and Huber 1994) and (NASA 2007), but has been adopted and used in other disciplines such as business management and software engineering. According to (VDI 2206 2004), it compromises five steps:

1. Situation analysis or adoption of goal,
2. Analysis and synthesis,
3. Analysis and assessment,
4. Decision,
5. Planning of further procedure or learning.

The first step is the starting point of an elementary handling cycle in which an externally prescribed goal is adopted before the (actual) situation is analyzed[2] or an initially unclear situation is analyzed before an internal goal formulation.[3] The aim of the second step is to search for solutions to the given problem on the background of situation analysis and objective as well as to work out alternative solutions. The solution variants worked out in the second step are subsequently subjected to a detailed evaluation (on the basis of the requirements) in the third step. The evaluation process can take place by calculation, simulation or experimentation, etc. As a result we obtain recommendations for one or more alternatives for a solution that are used to prepare the fourth step—the decision. With the decision it must be established, if a satisfactory result is found or, if a return to the first step is required. The last step includes the more or less smoothly changeover into further problem-solving cycles but also a critical examination of the conducted work in the finished micro-cycle in order to improve future process sequences.

The *V-model* describes the generic procedure for the design of mechatronic systems. Designing is understood as a process that "starting from the requirements, leads to a concretization of a technical system. This concretization is expressed in components of mechatronics and the interaction of these components," see (VDI 2206 2004).

As shown in Fig. 8.1, the V-model describes a procedure that—based on the *requirements*— leads to a (mechatronic) *product*. The requirements can be interpreted as input quantities. This input defines the starting point of a design process that is formed by an actual development order in which the defined object was specified and described by requirements. The requirements also define the measure against which the product is to be assessed.

The latter is the result of a continuous macro-cycle or the increasing concretization of the future product. The degree of product maturity can be assessed by maturity or readiness levels that will discussed later on. As also mentioned in (VDI

[1]Systems engineering consists of two disciplines. The first is the technical knowledge domain in which the systems engineer operates. The second is the system engineering management. According to (Sage 1992) it can be defined as "an appropriate combination of theories and tools, carried out through use of a suitable methodology and set of system management procedures, in a useful setting appropriate for the resolution of real-world problems that are often large scale and scope."

[2]This procedure is governed by the desired state.

[3]This procedure is governed by the actual situation.

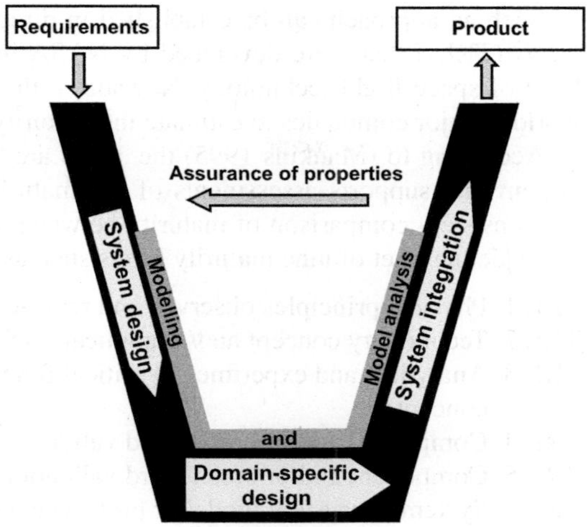

Fig. 8.1 Simplified representation of macro-cycle in the design process of a mechatronic system (adapted from VDI 2206 2004)

2206 2004), a complex mechatronic system is generally not produced within one macro-cycle. In fact it will be necessary to run through a number of macro-cycles to increase the maturity level, e.g. from the level of a laboratory specimen to the level of a *prototype* that is a physical or virtual model used to evaluate the technical feasibility of a particular system.

One macro-cycle that is described by the V-model includes four *process modules* for recurrent working steps that can bee seen as *design tasks*. These steps are:

- *System design* in which a cross-domain solution concept describing the main physical and logical operating characteristics is established. The overall functions of the future product are broken down into sub-functions for this purpose to assign suitable operating principles or solution elements. The sub-function performance is tested in the context of the system.
- *Domain-specific design* in which on the basis of the solution concept further concretization takes place separately in the involved domains.
- *System integration* in which the overall system is formed by the results from the individual domains to allow for interaction test on the system level.
- *Assurance of properties* in which the design process is continually checked on the basis of the specified solution concept to ensure that the actual system properties meet the desired system properties.

Furthermore, the V-process is usually flanked by *modeling* and *model analysis* to support forming and investigation of the system properties using simulation tools.

8.2 System Maturity Levels

In order to assess the maturity level of a particular ANC-system, we need an approach to capture evidence, and assess and communicate the system maturity in a consistent manner to stakeholders.

Such an approach can be established in direct analogy to *technology readiness levels* (TRLs) that were developed by NASA in the 1980s to assess the maturity level of space flight technology. Nowadays, this approach is used in many of the world's major companies to estimate the maturity level of an evolving technology.

According to (Mankins 1995) the TRLs are "a systematic metric/measurement system that supports assessments of the maturity of a particular technology and the consistent comparison of maturity between different types of technology." The TRLs define a set of nine maturity levels such as

TRL 1 Physical principles observed and reported.
TRL 2 Technology concept and/or application formulated.
TRL 3 Analytical and experimental critical function and/or characteristic proof-of-concept.
TRL 4 Component and/or breadboard validation in laboratory environment.
TRL 5 Component and/or breadboard validation in relevant environment.
TRL 6 System/subsystem model or prototype demonstration in a relevant environment (ground or space).
TRL 7 System prototype demonstration in a space environment (operational environment).
TRL 8 Actual system completed and "flight qualified" through test and demonstration (ground or space).
TRL 9 Actual system "flight proven" through successful mission operations.

A *relevant environment* is a testing environment that simulates the key aspects of the operational environment, whereas an *operational environment* addresses all the operational requirements and specifications required of the final system. Such an environment should be as close to mission operation conditions as circumstances allow.

If we want to assess maturity levels, we have to distinguish between the maturity level of a *technology* that can be understood as the usage and organization of methods and systems, and the *maturity level* of a certain ANC-system that is an integrated composite product (sensors, actuators, controller, etc.) providing the capability to satisfy the need of active noise reduction.

It should be noticed that the TRL can be high, even if the maturity level of a particular system is low. The ANC-technology for example is flight qualified and flight proven, see (Halvorsen and Emborg 1989) and (Emborg 1998), and has therefore achieved TRL 9, from the viewpoint of the author. Furthermore, some ANC-systems have an equivalent maturity level (e.g., noise cancellation headphones for the cockpit-crew). The maturity level of many other ANC-systems however is lower at the moment. A typical example is the ANC-system for a special working area in a new military aircraft, compare (Breitbach et al. 2009) and (Breitbach and Sachau 2009).

For this reason we will use nine *system maturity levels* (SML) to assess the maturity level of an ANC-system. These SMLs are introduced in direct analogy to the TRLs such as:

SML 1 Physical behavior of the operational environment in which the future ANC-system will operate are observed and reported.

SML 2 ANC-system concept is formulated.
SML 3 Successful proof-of-ANC-system-concept.
SML 4 Successful validation of components and overall ANC-system functionality in simple laboratory environment.
SML 5 Successful validation of components and overall ANC-system functionality in improved laboratory environment.
SML 6 Successful validation of components and overall ANC-system functionality in relevant environment.
SML 7 Successful validation of robust components and robust overall ANC-system functionality in operational environment.
SML 8 ANC-system qualified through test and demonstration in operational environment (e.g. an ANC system is flight qualified).
SML 9 ANC-system established through successful service in operational environment (e.g. an ANC system is flight proven).

Within this book an *improved laboratory environment* is seen as a testing environment that simulates some (but not all) key aspects of the operational environment. In contrast to this, a *simple laboratory environment* is understood as a testing environment that clearly differs from the operational environment, but that is appropriate to demonstrate a proof-of-concept.

8.3 A Design Methodology for ANC-Systems

8.3.1 ANC-System Design Tasks

Based on the (general) design methodology for mechatronics that was outlined in Sect. 8.1 it is possible to propose a design methodology for ANC-systems by specifying four ANC-system design tasks.

The concretization of the general VDI-methodology is illustrated by Fig. 8.2 in which one macro-cycle or one ANC-system *design step* is interpreted as a V-process. The process input is given by the requirements. The process output is the ANC-system (or a prototype of the future ANC-system).

Furthermore, it can be seen that such a macro-cycle requires the handling of four specific design tasks. *Analysis of the primary noise field* can be seen as an ANC-specific form of the system design task. It is usually performed at first to understand the interior noise problem that can be specified by the

- Statistical properties (such as arithmetic mean and variance) of disturbance,
- Frequency range, spatial distribution and magnitude of disturbance,
- Power flow into/out of the enclosure.

Furthermore, ANC-system design requires the calculation of an estimate of the prospective amount of active noise reduction using one ore more of the IPE tools that were presented in Sect. 7.3. *Analysis of the anti-noise field* also belongs to the

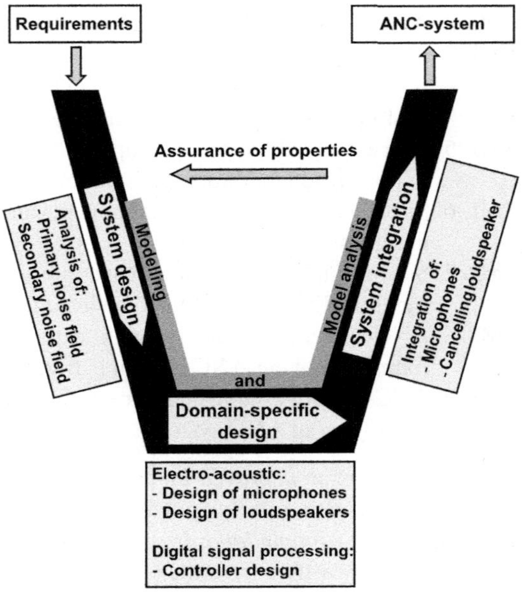

Fig. 8.2 ANC-system design process described by the V-model

system design task. It is performed to determine an IPE considering a certain setting of sensors and actuators.

Domain-specific system design includes design and/or testing of electro-acoustic components such as sensors, actuators and controller hardware as well as design and/or test of the controller software. Examples for domain specific tests of ANC-system components are the determination of sensitivity, frequency response, total harmonic distortion, etc., of actuators (according to DIN EN 60268-5 2004) and sensors (according to DIN EN 60268-4 2004) as well as the test of the controller software using an environment for numerical simulation such as Matlab/Simulink.

To assure that the specified ANC-system meets the requirements (assurance of properties) *functional testing* is required *after* system integration. It is obvious that the effort of the system integration task depends on the SML. If the SML is 3, system integration means that the connection between the single system components enables the proof-of-concept. Functional testing at higher SMLs also requires that the connected ANC-system components are integrated into a laboratory, relevant or operational environment.

8.3.2 ANC-System Design Steps

As mentioned in Sect. 8.1, a complex mechatronic system is generally not produced within one macro-cycle. This is especially true for ANC-systems. For this reason it will be necessary to run through a number of macro-cycles to increase the SML of ANC-systems.

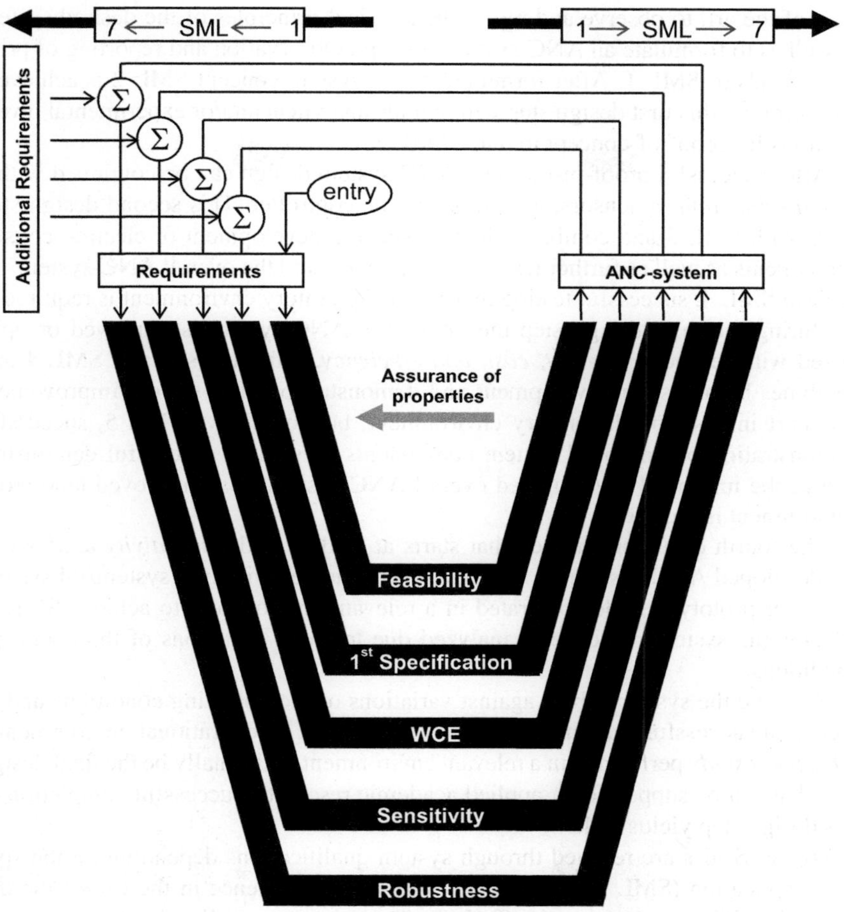

Fig. 8.3 ANC-system design considering multiple design steps modeled by five macro-cycles

The pyramiding of five macro-cycles according to single design steps (such as the feasibility step) is shown in Fig. 8.3. This figure clarifies that the output of each individual design step (modeled by a single V-process) gives an input to the next (superior) macro-cycle.

Furthermore, it has been taken into account that the requirements can change during the design process. An example can be motivated by the total weight, space allocation and/or the power consumption of the ANC-system components. While these parameters are usually not important, if the SML is less or equal to 3, their relevance increases significantly, for higher SMLs.

To analyze the applicability of ANC according to the requirements of a specific application, ANC-system design should start with a well researched *feasibility study*, as shown in Fig. 8.3. Such a study will include applied research, based on the state of the art, to observe and report the physical principles of the desired system as well as to formulate an ANC-system concept. Observation and reporting of principles leads to SML 1. After formulation of a system concept SML 2 is achieved. Furthermore, this first design step will include analytical and/or experimental investigations for proof-of-concept to reach SML 3.

After successful proof-of-concept, ANC-system design can be continued with a *first specification* of sensors, actuators and the controller. This second design step starts with SML 3 and combines domain specific development of electro-acoustic components as well as further tests for components and the overall ANC-system. To achieve SML 4, successful testing in a simple laboratory environment is required.

During the third design step the laboratory ANC-system is improved or optimized with respect to *weight, cost, and efficiency*. This step starts at SML 4 and combines both system development and demonstration. The system improvement can start in a simple laboratory environment, but to achieve SML 5, successful demonstration of improved system components as well as successful demonstration of the improved or optimized overall ANC-system in an improved laboratory environment is required.

The fourth development step that starts at SML 5 is the *sensitivity analysis* of the developed ANC-system. In this design step the behavior of a system/subsystem model or prototype is demonstrated in a relevant environment to achieve SML 6. Hereby, the system response is analyzed due to small variations of the operating conditions.

To make the system robust against variations of the operating conditions and to prepare a successful demonstration of a prototype in an operational environment a *robustness study* performed in a relevant environment will usually be the final design step that can be supported by applied academic research. Successful completion of this design step yields SML 7.

Higher SMLs are reached through system qualifications depending on the specific application (SML 8) and through long term experience in the use of the developed ANC-system in praxis (SML 9). These last two SMLs are determined by industrialization and commercial usage of a certain ANC-system and have therefore not been included into this (academic) discussion.

8.3.3 Matrix Model of the ANC-System Design Process

ANC-system design may require several system design steps, compare Sect. 8.3.2, thereby each design step is subdivided into four system design tasks as discussed in Sect. 8.3.1. In order to process a specific design task, we will have to solve problems on the micro-level. For this purpose we can use the problem-solving cycle that has also been summarized in Sect. 8.1.

In this subsection we will discuss the interaction between the design steps and the design tasks that determine the ANC-system design process. This is, considering the different SMLs, illustrated by Fig. 8.4. In the *matrix model* shown in this figure,

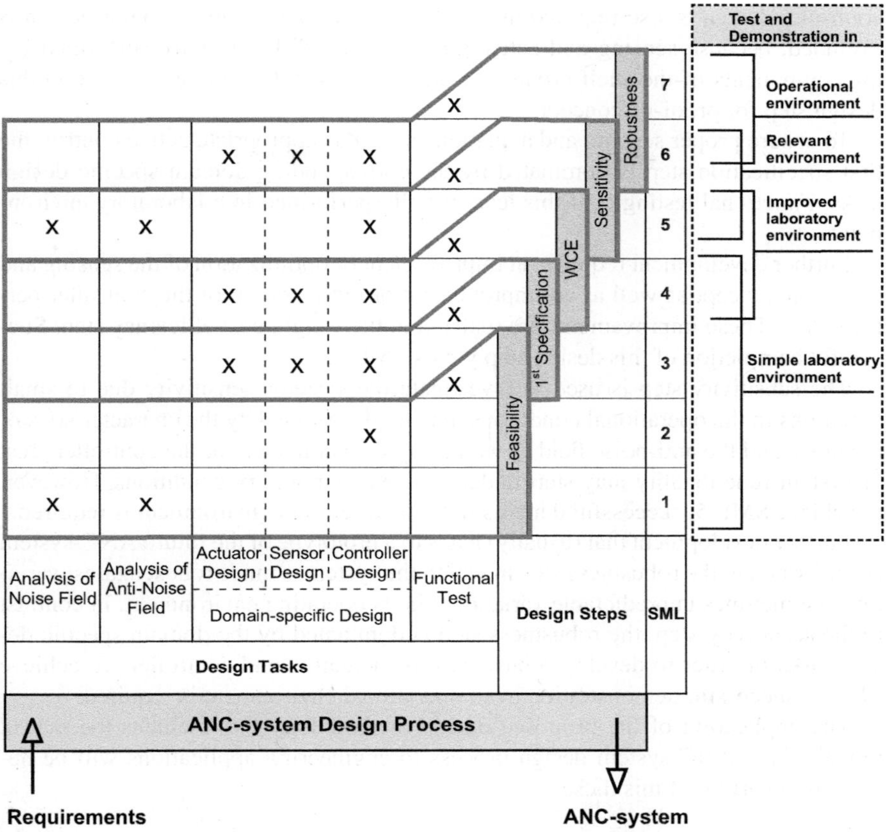

Fig. 8.4 Interaction between design steps and design tasks during ANC-system design process. Development level is given in form of a system maturity level (SML)

an x is used to indicate design tasks which are (usually) the most important ones within a certain design step.

The single design steps are framed by a thick gray line. Black horizontal lines enable the correlation between a certain SML and a certain design step. Because successful functional testing leads to a higher SML, the cells in the sixth column are divided by the gray "design step frame line." This division reflects that the successive design step starts at the SML that was achieved after successful completion of the previous design step.

Figure 8.4 also clarifies that the process input is given by the requirements, while the output of the design process is the ANC-system. Its SML depends on the outcome of a certain design step.

As stated above, every design step requires the handling of specific design tasks. Figure 8.4 shows that the feasibility step (usually) includes both analysis of primary noise field and analysis of anti-noise field. To formulate a system concept, controller design is also required in this design step. Sensor and actuator design is excluded, because sensing and actuation is at these SMLs usually performed using components-of-the-shelf products. Finally, functional testing is included in this design step for proof-of-concept.

To ensure proper sensing and actuation as well as appropriate active control, the first specification step is dominated by the corresponding domain specific design tasks. Functional testing is at this level usually performed in a laboratory environment.

Further development requires an improvement or optimization of the sensing and actuation concept as well as an improvement or optimization of the controller performance. These improvements are realized in the weight cost efficiency step. Successful completion of this design step yields SML 4.

The sensitivity step is used to investigate the systems sensitivity due to small variations in the operational conditions that are determined by the characteristics of the noise and the anti-noise field as well as by the parameters of the controller. Test of system functionality may start under improved laboratory conditions. However, to achieve SML 5, successful demonstration in a relevant environment is required.

Further development that (usually) leads to a prototype of the future ANC-system is performed in the robustness step in which the system is made robust against variations (sometimes unpredictable variations) in its operating environment. In contrast to the sensitivity step, the robustness step is dominated by the domain specific design tasks in order to develop robust sensors, actuators and controller. To achieve SML 7, successful demonstration in an operational environment is required.

The application of the proposed design methodology that includes the matrix model of the ANC-system design process to engineering applications will be described in part IV of this theses.

Part IV
ANC-System Design: Examples

Chapter 9
Active Noise Control in a Semi-closed Interior

Abstract As a first example this chapter presents the application of the general design methodology for ANC-systems (proposed in Chap. 8) to a particular interior noise problem in an aircraft cabin. To show the practical use of the proposed (matrix) design approach, considering the mechatronic background of feed-forward ANC presented in part I as well as the ANC-system design tools discussed in Chap. 7, this chapter reports on five successive design steps that were performed to support the development of a robust active noise system (ANS) for a special working area (WA) of a certain military aircraft (MA). The chapter starts with a motivation and a description of the requirements. Afterwards, the output of all design steps, beginning with the feasibility study and ending with the robustness study, will be summarized. The project was intended to support the activities of Airbus Germany. It was carried out at the professorship for Mechatronics of Helmut-Schmidt-University/University of the Federal Armed Forces Hamburg between 2002 and 2009. This support has been continued in 2010 and in 2011.

9.1 Description of Problem and Requirements

The interior noise field in turboprop aircraft is below 500 Hz dominated by tonal components. The excitation frequencies are given by the blade passage frequency (BPF) of the propeller passing the fuselage in a short distance as shown in Fig. 9.1 and its higher harmonics. The dominating tonal noise levels, measurable at these frequencies, can occur because the blade profiles are only optimized in terms of performance. The acoustic properties are usually not taken into account.

To obtain an acceptable noise level, and to guarantee both work-security and comfort in aircraft interiors, synchro-phasing of engines (Jones and Fuller 1986) as well as passive methods (e.g. adding material that increases the transmission loss of the cabin sidewall) are commonly used. Especially if low frequency noise has to be reduced, the latter solution results in unwanted additional weight.

To avoid this extra weight and to avoid the radiation of sound from the primary structure into the aircraft interior, passive and/or active vibration absorbers (tuned to the BPF tones, see Benassi and Elliott 2003) can be installed as reported in (Breitbach and Sachau 2009).

T. Kletschkowski, *Adaptive Feed-Forward Control of Low Frequency Interior Noise*, Intelligent Systems, Control and Automation: Science and Engineering 56,
DOI 10.1007/978-94-007-2537-9_9,

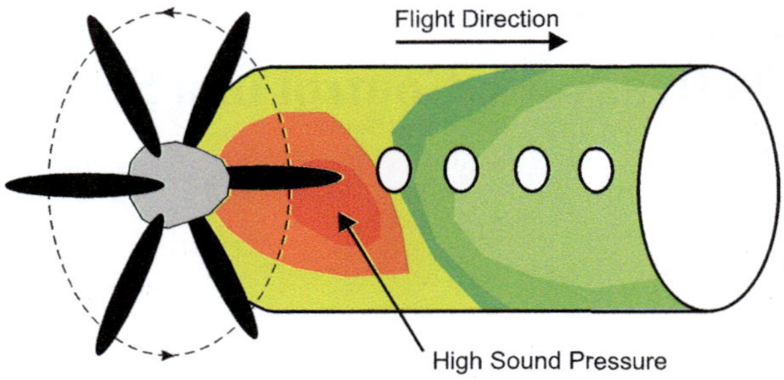

Fig. 9.1 Generation of engine noise caused by propeller rotation (adapted from Johansson 2000). Propeller blades passing the fuselage in a short distance and generate high tonal sound pressures that are localized in the propeller plane

However, increasing the transmission loss of the cabin sidewall by passive and active treatments applied to the structure alone can be not sufficient for those turboprop aircraft that have powerful high performance engines, which consequently induce high noise levels into the aircraft. A solution is to apply ANC systems, which focus on the unwanted tonal noise at low frequencies and provide a good compromise of noise reduction performance and the amount of additional weight.

A typical example for a high speed turboprop aircraft is the A400M, see Fig. 9.2. This aircraft has been designed to provide high strategic mission efficiency and has to meet the requirements of tactical operations. It has an advanced aerodynamic design and is equipped with four high performance turboprops, see Fig. 9.2.

Each engine drives a 8-blade propeller made of carbon fiber reinforced plastic with a diameter of 5.33 m that allows for cruising speed up to Mach 0.72 at 37000 ft. This turboprop aircraft can carry an average payload of 25 t over 3000 nm without stop.

In (Breitbach and Sachau 2009) the authors state "Predictions show that without any noise reduction treatment the maximum sound pressure in the ... cargo hold might reach very high levels exceeding 100 dB(A) as measured in some four-propeller military transport aircraft." Therefore, beside passive methods such as

- Absorbing lining (realized in a transparent manner backed up with glass wool),
- Ice shield (in the area around the propeller plane an ice shield is attached to the fuselage to prevent damages resulting from ice stroke. This ice shield also has

Fig. 9.2 Model of military aircraft

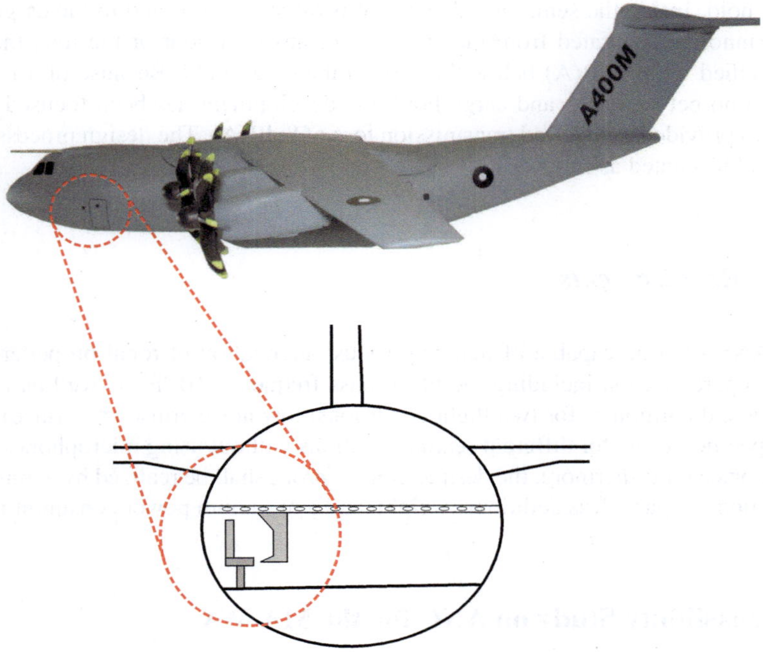

Fig. 9.3 Position of working area in the military aircraft

an acceptable noise transmission loss especially for the higher propeller tones $f > 300$ Hz),

- Primary thermal insulation (made of glass wool) mainly applied to attenuate mid (400 Hz $< f <$ 1 kHz) and high frequency noise ($f >$ 1 kHz),
- Propeller synchro-phasing,

active noise control and active vibration control will be used to reduce the interior noise in the MA. Active vibration control will be applied to attenuate sound radiation at the BPF and its first harmonic from the fuselage into the cargo hold using tunable vibration absorbers (TVA).

Both a decentralized concept (Breitbach et al. 2009) and a centralized approach (Breitbach and Sachau 2009) with a number of approximately 100 TVA units together with remote microphone control (noise is measured remote from the TVA-units) were analyzed.

The noise level in the cargo hold will be reduced to an acceptable level by these noise treatments. But, the situation is more sophisticated in the MA WA. Its position, in front of the cargo hold, is shown in Fig. 9.3. From this position an operator is able to observe the cargo hold including secured payloads, doors and ramps. The operator controls all cargo handling operations, such as parachute drops, loading, unloading etc using panels and displays integrated in the workstation in front of his seat.

To provide both a certain speech intelligibility and comfort during long term flights, the requirements concerning the SPL inside the WA are higher than for the cargo hold. Inside the semi-closed WA that is directly connected to the cargo hold and cannot be separated from the cargo by means of a door or the like, the SPL is specified with 8 dB(A) below the SPL in the cargo hold. Because of a missing separation between WA and cargo hold, the development has been focused on an ANS to provide the required transmission loss of 8 dB(A). The design process input can be formulated as follows.

9.1.1 Requirements

The ANS shall be capable of realizing robust attenuation of tonal propeller noise at the operator ears, including the blade pass frequency (BPF) as well as its first and second harmonic (for two flight conditions: low noise cruise and cruise) down to a specified value for different seating positions without using microphones in the operator seat. Furthermore, the desired control profit shall be realized by a minimum allocation of space, less additional weight, and a minimum power consumption.

9.2 Feasibility Study on ANC for the MA WA

9.2.1 Noise Field Analysis for the MA WA

A first insight into the physics of the acoustic interior in the MA can be obtained by analyzing the energy flow between the cargo hold and the semi-closed WA as shown in (Kochan et al. 2008). In this analysis the special characteristic of the future turboprop aircraft—the coupling between the relatively small volume of the WA and the large cargo hold—was investigated by a simplified model of an enclosed sound field. This model is shown in Fig. 9.4. It can bee seen that the cargo hold is represented by a rectangular room. The WA is attached at the front of the cargo hold. S_{12} is the coupling surface between both rooms. The working area has a slightly curved outer wall which is implied by the chamfer. The average value of the reverberation time inside the volumes are assumed to be comparable to the real aircraft using a measure of $\langle T_{60} \rangle = 0.6$ s.

Tonal excitation ($f < 300$ Hz) of the sound field inside the aircraft is assumed to be primarily through the fuselage structure in the propeller plane on both sides. Because of synchro-phasing of the aircraft propellers, the noise entry Π_{11} and Π_{12} into the aircraft is in phase. Inside the aircraft, the sound waves are partly reflected and absorbed by the interior walls. Hereby the major dissipation effect is induced by the absorption on the boundaries, in comparison to the minor effect of internal friction in the fluid. The primary noise field in the WA is excited via the coupling surface S_{12}. If the ANC-system is switched on, the anti-noise field is superimposed on the primary noise field in the WA. This secondary noise field is generated by the canceling sources at the outer wall and the ceiling of the WA. As shown in

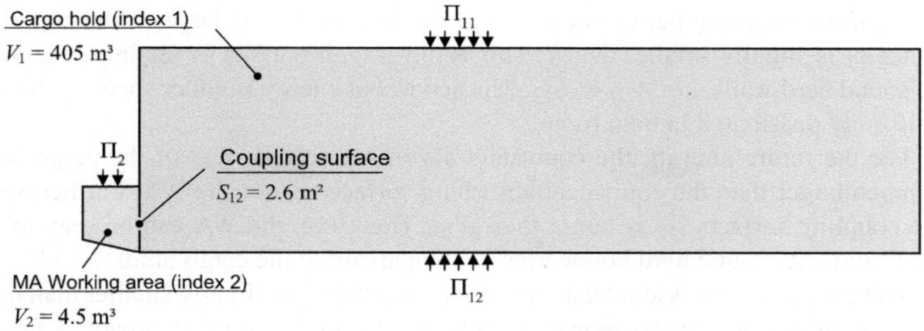

Fig. 9.4 Simplified model of the coupled system working area and cargo hold used to analyze the energy flow between the MA cargo hold and the semi-closed working area (adapted from Kochan et al. 2008). Primary excitations caused by propeller noise are Π_{11} and Π_{12}. Anti-noise caused by canceling sources is indicated by Π_2

(Kochan et al. 2008), the acoustical power balance for the cargo (index 1) and the WA (index 2) can be written as

$$0 = \frac{4}{c}\Pi_1 - A_{11}E_1 + S_{12}E_2 \quad \text{and} \quad 0 = \frac{4}{c}\Pi_2 - A_{22}E_2 + S_{12}E_1. \tag{9.1}$$

E_1 denotes the acoustical energy density in the cargo hold and E_2 denotes the acoustical energy density in the WA. The absorbing surfaces in both rooms can be summarized to the equivalent absorbing surfaces A_{10} and A_{20}. The surface which separates the WA from the cargo hold is S_{12}, compare Fig. 9.4. At this surface, both rooms are able to exchange acoustical energy. $\Pi_1 = \Pi_{11} + \Pi_{12}$ is the acoustical power which is radiated by the primary sources, whereas the power of the secondary sources is denoted by Π_2. $A_{11} = A_{10} + S_{12}$ is the absorbing surface of the cargo hold, and $A_{22} = A_{20} + S_{12}$ is the absorbing surface of the WA.

By applying this energy based analysis method, only the room volumes and the equivalent absorbing surfaces of the model are of interest. This is a strong simplification of the real aircraft. Therefore no exact numerical values and exact sound pressure level distributions are desired, in fact we are more interested in a qualitative analysis of both the primary noise field and the secondary noise field by performing the first two design tasks: *analysis of noise field* and *analysis of anti-noise field*, compare Fig. 8.4.

Noise Field Analysis

For the basic system in which only the primary excitation is present ($\Pi_2 = 0$), the energy density in the cargo is given by

$$E_1 = \frac{4}{c}\frac{\Pi_1}{A_{10} + k_1 A_{20}} \quad \text{with } k_1 = \frac{S_{12}}{A_{22}}. \tag{9.2}$$

The denominator shows that the contribution of the acoustical absorbing surface A_{20} to the energy density E_1 is reduced by the coupling factor k_1. It can be seen that the acoustic coupling between the MA cargo and the WA is large, if the coupling factor k_1 is slightly smaller than 1. This is the case, if the WA is (mainly) enclosed by sound hard walls, i.e. $A_{20} \ll S_{12}$. The acoustical energy is under such conditions uniformly distributed in both rooms.

For the future aircraft, the equivalent absorbing surface A_{10} of the cargo hold is much larger than the equivalent absorbing surface A_{20} of the WA. Furthermore, the coupling surface S_{12} is larger than A_{20}. Therefore, the WA can be seen as an enclosure with sound hard boundaries in comparison to the cargo hold.

Therefore, we can assume that the coupling factor k_1 is slightly smaller than one. Because of this, the energy density in both rooms will be equal, compare (9.1) and (9.2). Hence, neglecting any additional damping in the WA, the coupled rooms can be seen as one room from the view point of the primary excitation.

In view of the ANC-system design process it can be concluded that because of the strong coupling between the MA cargo and the WA high sound pressure levels inside the cargo hold will propagate unimpeded in the WA. Additionally, an amplification of the high SPL can be expected because of local resonance effects.

Anti-noise Field Analysis

In the case of the secondary excitation ($P_1 = 0$), the acoustical power of the system is introduced via the secondary sources and the energy density E_2 can be evaluated using (9.1). This yields

$$E_2 = \frac{4}{c} \frac{\Pi_2}{k_2 A_{10} + A_{20}} \quad \text{with } k_2 = \frac{S_{12}}{A_{11}}. \tag{9.3}$$

If we assume that the absorbing surface in the cargo hold is huge compared to the coupling surface ($A_{10} \gg S_{12}$), the coupling factor k_2 is small and the coupling between the WA and the cargo hold is low. Because of the relatively sound hard WA in comparison with cargo hold, the coupling surface S_{12} can be seen as a fully absorbing surface from the view point of the secondary source. The acoustical energy density in the cargo hold will therefore be smaller than the acoustical energy density inside the WA. For this reason, despite the sound hard material in the WA, the WA can be considered as a damped room. These findings have two consequences for the ANC-system:

1. Damping smooths the transfer behavior between canceling sources and error microphones in the WA resulting in lower resonance peaks and a steady slope of the phase responses. This is advantageous for a robust ANC-system because it reduces the uncertainty in the estimated transfer functions.
2. Because resonance effects are reduced by a higher amount of damping, secondary sources with high performance attributes are required.

Because the physical behavior of the operational environment in which the future ANC-system will operate were analyzed and reported, SML 1 was reached after completion of these first two design tasks.

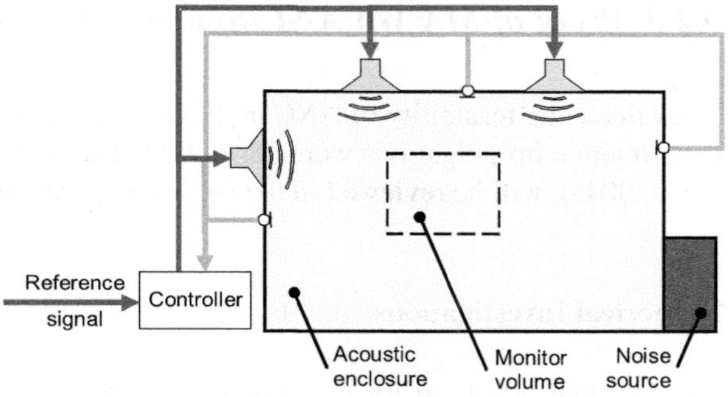

Fig. 9.5 Control task given in the MA WA, see (Kochan and Sachau 2009). The enclosure contains loudspeakers and error microphones located on the interior boundaries. The objective of the active noise control system is to reduce the noise inside the monitor volume, not necessarily at the error microphones

9.2.2 Formulation of ANC-System Concept for the MA WA

A control concept that is capable of meeting the requirements specified in Sect. 9.1 can be based on M loudspeakers and L error microphones that are mounted at the boundaries of the enclosure as shown in Fig. 9.5. The information processing of the ANC-system must be able to control the noise at the error microphones. However, the objective of the active noise control task is to reduce the noise in the center of the enclosure around the head of the sitting operator.

The interior noise will be caused by the engines. For this reason it is possible to design a feed-forward control system using the engine speed to calculate the reference signal by wave form synthesis, see (Kuo and Morgan 1996). Because propeller rotation causes multi-tonal disturbances with high SPLs, we can expect a perfect coherence between the BPF tones that have to be controlled at the L error sensors and the reference signal.

Such a RSC problem, compare Sect. 6.5.2, occurs frequently during the design of ANC-systems inside vehicles or aircrafts. It requires appropriate positions for the canceling loudspeakers and for the error microphones to allow for active noise reduction in the monitor volume. Therefore, it is necessary to use a desired position or to redefine a transducer setting, if a better one is found by numerical or experimental investigations. This is only possible in an early state of the design process. After a design freeze that usually includes a freezing of transducer positions, RSC is only possible with a software solution considering virtual microphones, RSC by noise design at the error microphones or parametric controller design.

However, in the first design step—the *feasibility study*—it is sufficient to show that the SPL at the operator ears can be significantly reduced, if ANC is applied ignoring the constraints that occur, if a desired transducer location is not available. Because an ANC-system concept was formulated, SML 2 was achieved at this point.

9.2.3 Proof of MA WA ANC-System Concept

To evaluate the feasibility of ANC in the MA WA, numerical simulations as well as experimental investigations were performed. The main results, reported in (Gerner et al. 2004a), will be reviewed in the following paragraphs.

Numerical Investigations

The tool that was used for the numerical investigations was the FEM. It was applied to perform time-harmonic simulations. In a first step these investigations were based on a simplified model of the MA cargo hold that was represented by a rectangular enclosure of equivalent volume with sound hard boundaries. The semi-closed WA was simply modeled by a block, integrated at a position equivalent to the WA-position in the MA as shown in Fig. 9.6 (top).

This figure in which all dimensions are given in meters also shows the position of the primary source as well as the position of eight canceling sources inside the WA model. All loudspeakers were modeled as ideal velocity sources with Neumann BCs. According to results of the noise field analysis, the coupling surface between WA and cargo hold was modeled as an internal boundary. The impedance at this surface was therefore set to $\rho_\infty c$.

This simple geometry model was discretized by finite elements. The associated finite element mesh is shown in Fig. 9.6 (below). It was applied to determine the number of actuators necessary to realize the desired noise reduction at the operator ears. For this purpose it was necessary to perform a TPA, compare Sect. 7.3.2, in order to select M_{ANC} actuator positions from $M_{\max} = 108$ possible locations inside the WA for a constant number of $L = 11$ error sensors. One particular set of 8 of loudspeaker locations is shown in Fig. 9.6.

In a first step, the maximum number of $M_{\max}$ actuators was reduced to M by deselecting $M_{\max} - M$ actuators with lowest amplitude in the optimal controlled case. The latter is determined by the minimum of the sum of the squared sound pressure at the $L = 11$ error sensors. This minimum can considering (6.14) as well as (6.84)–(6.89) be expressed as

$$J_{\min} = \mathbf{p}_p^H \left[\mathbf{I} - \mathbf{H}\left(\mathbf{H}^H \mathbf{H}\right)^{-1} \mathbf{H}^H\right] \mathbf{p}_p. \tag{9.4}$$

In a second step the number of actuators was reduced by selecting the best solution out of the M calculated solutions. This procedure was continued with selecting the best solution out of the recalculated remaining $M - 1$ solutions until the desired number of M_{ANC} actuators was reached. Using this approach, it was possible to find 8 actuator positions out of the 108 possible locations that enables the required noise reduction. This is the minimum number of secondary sources, located at optimal positions.

To study the noise reduction potential of the determined transducer set, the primary as well as the controlled sound field in the WA were analyzed using a more

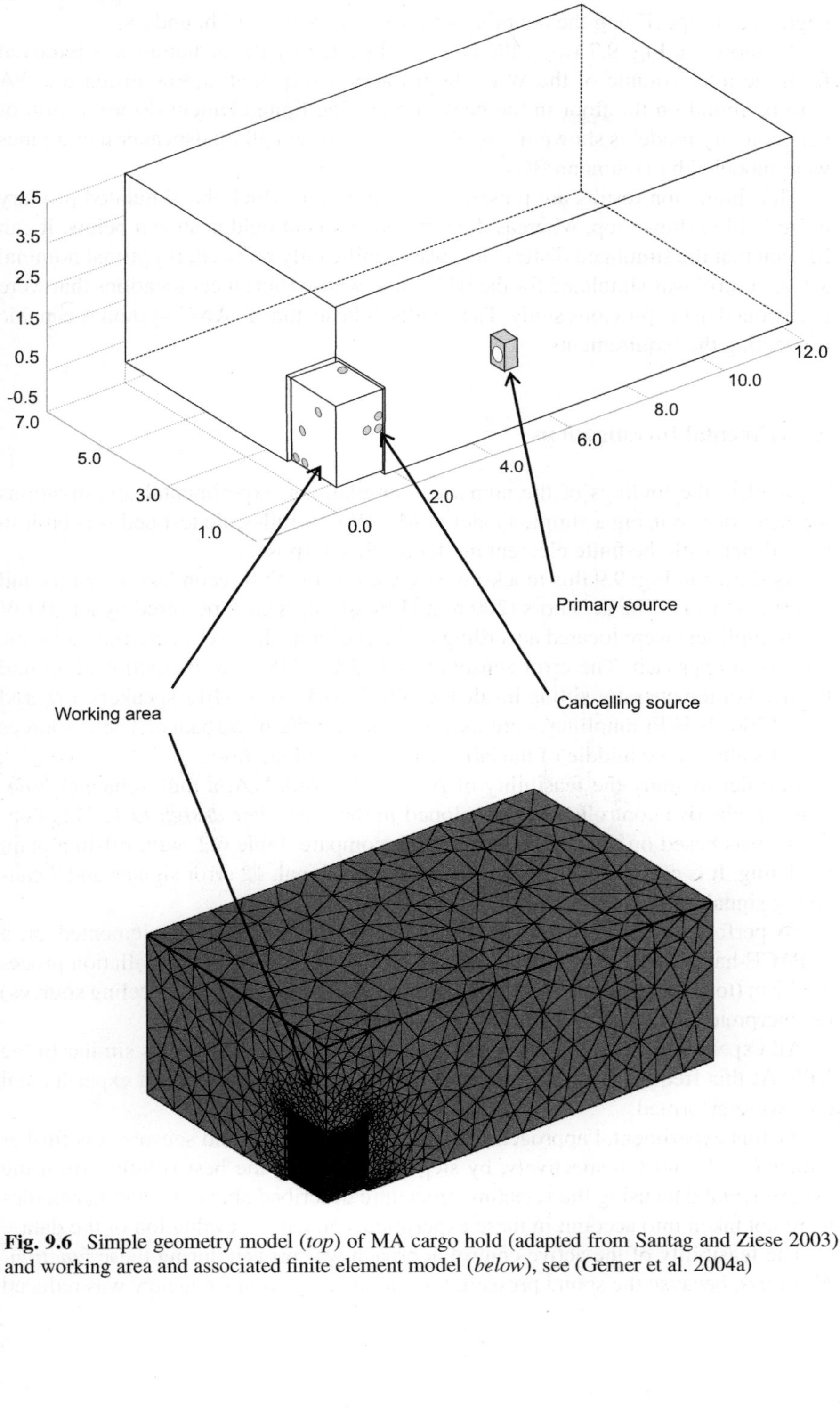

Fig. 9.6 Simple geometry model (*top*) of MA cargo hold (adapted from Santag and Ziese 2003) and working area and associated finite element model (*below*), see (Gerner et al. 2004a)

detailed numerical model in which the coupling between WA and cargo hold was neglected by specifying the coupling surface as a sound hard boundary.

As shown in Fig. 9.7 (top), the volume allocated by the actuators was removed from the total volume of the WA. The primary source that is now inside the WA is to be found on the floor in the back corner. The finite element discretization of the geometry model is shown in Fig. 9.7 (below). Again all loudspeaker membranes were modeled by Neumann BCs.

The simulation results are presented in Fig. 9.8 in which the simulated primary noise field is shown top, whereas the controlled sound field is shown below. It can be seen that the simulated disturbance was significantly reduced, if optimal nominal active control was simulated for the BPF using whose transducer locations that were determined in the previous study. The results indicate that an ANC-system is capable of meeting the requirements.

Experimental Investigations

To validate the findings of the numerical simulations, experimental investigations were performed using a simple model of MA WA. A full-scale test bed was built in accordance with the finite element model for this purpose.

As shown in Fig. 9.9 this mock-up was equipped with 8 secondary actuators and 11 error sensors. The actuators (200 mm HiFi-speakers each powered by an 100 W HiFi-amplifier) were located according to the positions that were determined by the numerical approach. The error sensors (type: B&K 4188) were positioned around the head of the operator sitting inside the WA. Two 380 mm HiFi-speakers powered by a 1200 W HiFi-amplifier were used to generate the disturbance. These sources were located in the middle of the lab room, compare Fig. 9.6.

In order to study the feasibility of ANC in the MA WA, a multi-channel time-domain adaptive controller was developed in the *controller design task*. This controller was based on the FxLMS-algorithm, compare Table 6.2, with off-line plant modeling. It is capable of handling one reference signal, 12 error signals and 8 canceling signals. The controller layout is shown in Fig. 9.10.

To perform *functional testing*, the adaptive controller was implemented on a dSPACE-hardware that was also integrated into the test rig. This installation procedure can (together with the assembling of error microphones and canceling sources) be interpreted as *system integration*.

All experiments were conducted at an excitation frequency that is similar to the BPF. At this frequency an additional optimization that was based on experimental data was performed.

By this experimental approach, the number of actuators and sensors was further reduced to 4, and 6 respectively, by stepwise selecting the best solution from the experimental data using the selecting procedure described above. Higher harmonics were not taken into account in these experiments and at the evaluation of the data.

The feasibility of the active control approach was proven during these *functionality tests*, because the sound pressure level at the excitation frequency was reduced

Fig. 9.7 Model of the MA WA. *Top*: geometry model (adapted from Santag and Ziese 2003). *Below*: finite element model, see (Gerner et al. 2004a)

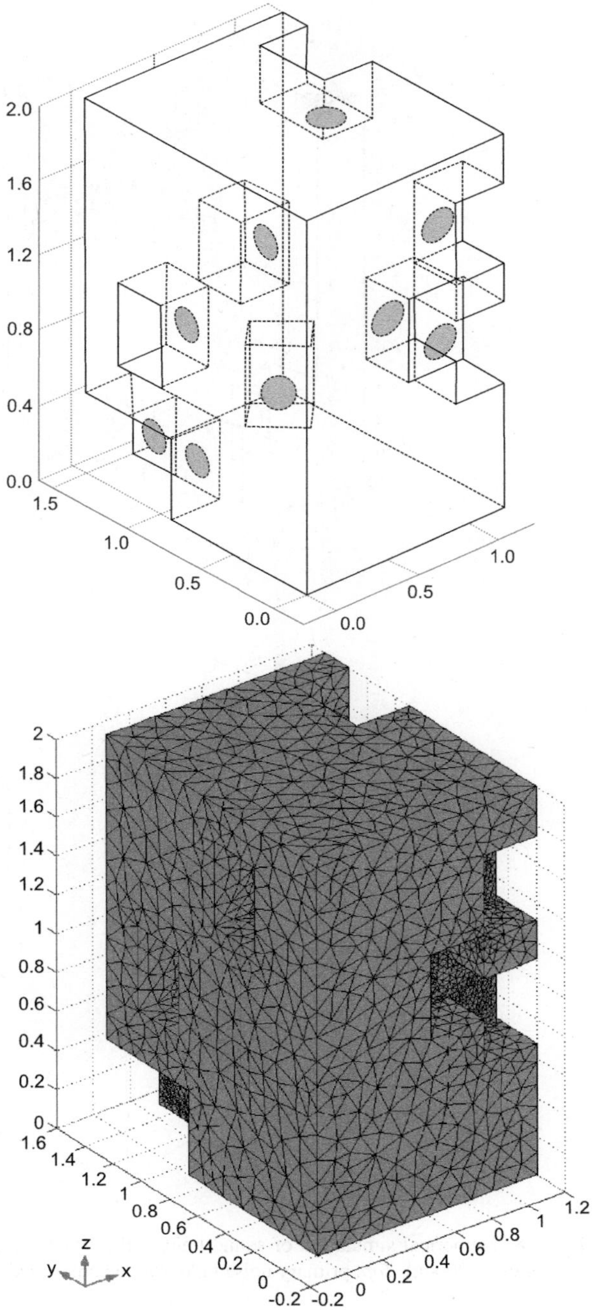

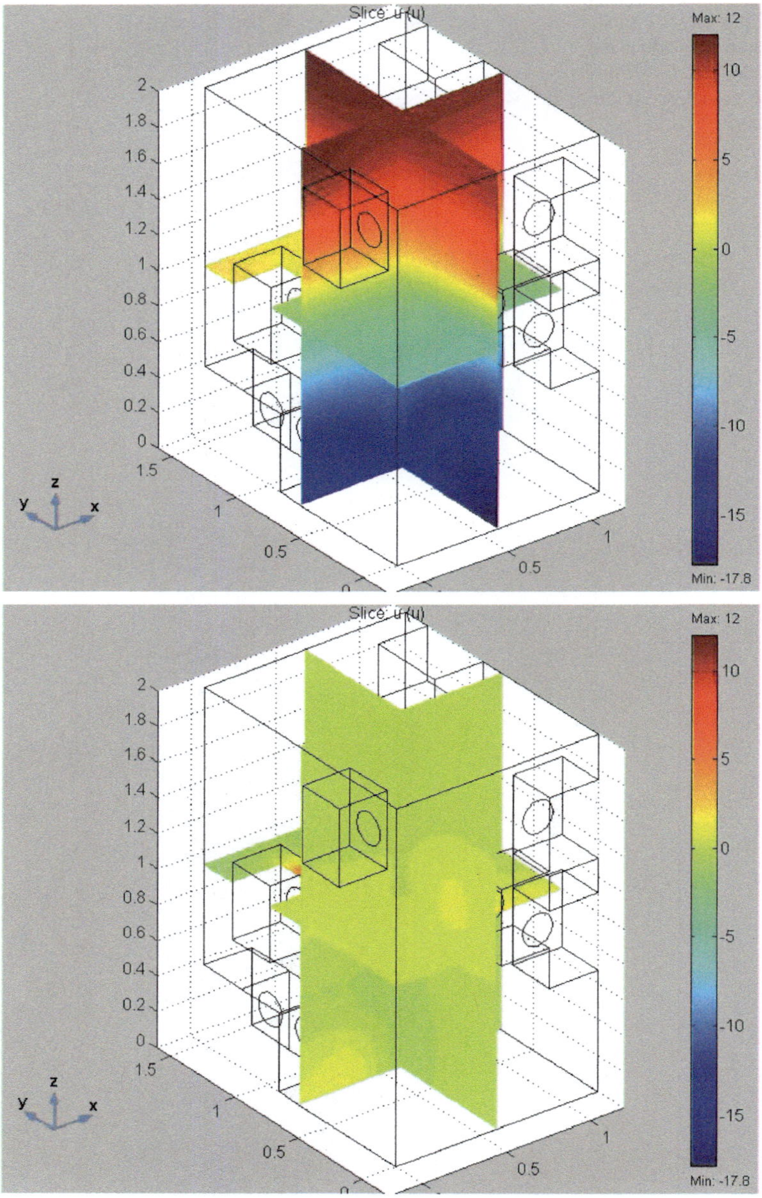

Fig. 9.8 Numerical simulation of sound field in the WA for the blade passage frequency (from Gerner et al. 2004a). *Top*: primary noise field. *Below*: controlled sound field

by 32 dB/37 dB at the left/right ear of a simple operator dummy (compare Fig. 9.9). Therefore, SML 3 was achieved at the end of the first design step.

9.3 First Specification of Active Noise System

In order to compile a first specification of the WA ANS, the second design step—the *first specification*—was processed. It was intended to analyze the performance of commercial off-the-shelf microphones and actuators that were tested stand alone as well as connected with the controller and integrated into the simple mock-up shown in Fig. 9.9.

As described in (Gerner et al. 2004b) and (Sachau and Gerner 2004), these investigations were especially focused on the actuator performance (*actuator design task*), necessary to realize a maximum SPL of 101 dB at the BPF, 90 dB at the first harmonic, and 95 dB at the second harmonic in a distance of 1 m using 4 secondary actuators to meet the **concretized requirement:** *The maximum sound pressure level at the operator ears must not exceed* 86 $dB(A)$ *in the controlled case.*

Furthermore, the controller software was improved to take into account for the contribution of the first and second harmonic of the BPF.

Fig. 9.9 Simple mock-up of MA working area used to evaluate the feasibility of active noise control, see (Gerner et al. 2004a)

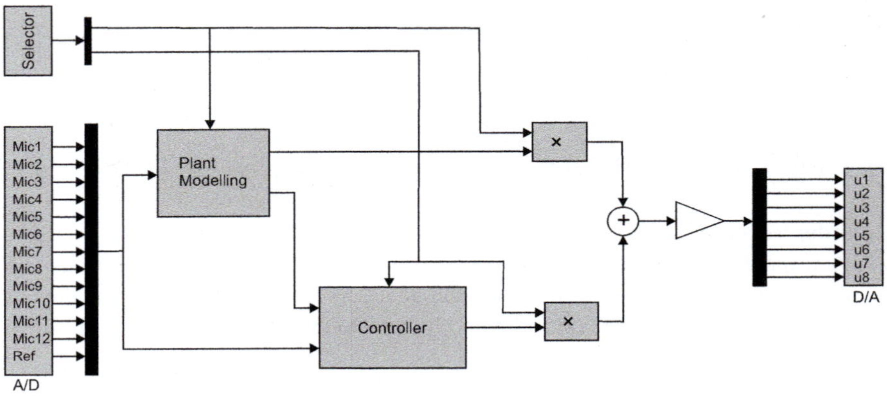

Fig. 9.10 Controller layout for multi-channel adaptive control in time domain. Controller software is capable of handling 1 reference signal, 12 microphone signals (Mic) and 8 control signals (sek)

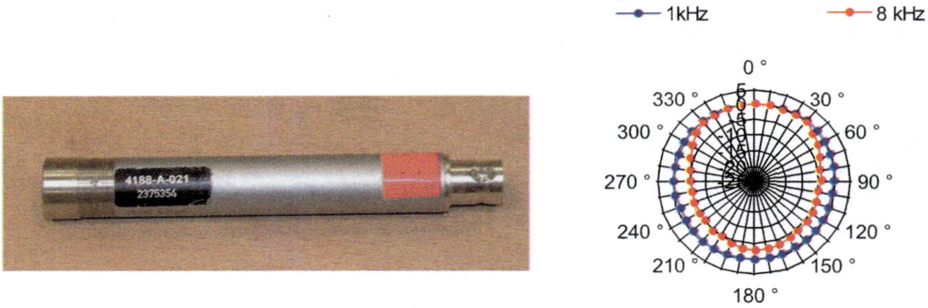

Fig. 9.11 Measurement microphone (type: B&K 4188) (*left*) and directivity of measurement microphone (*right*) determined according to (DIN EN 60268-4 2004)

9.3.1 First Specification of Microphones

Because of their high quality, condenser microphones of type B&K 4188, see Fig. 9.11 (left) were considered as sensors for the first specification of the ANS. For low frequencies $f < 1$ kHz, the microphone acts as an omni-directional transducer, compare see Fig. 9.11 (right). Furthermore, the sensitivity of these microphones is constant in the frequency range of interest as shown in Fig. 9.12 in which the magnitude of the measured frequency response is plotted against the normalized frequency. The latter was calculated as f/BPF.

This is not the case for an ordinary low-price electret microphone that acts like a high-pass filter as also shown in Fig. 9.12. Therefore, without additional effort for the correction of the frequency response, this microphone is not a proper choice for multi-tonal ANC at low frequencies.

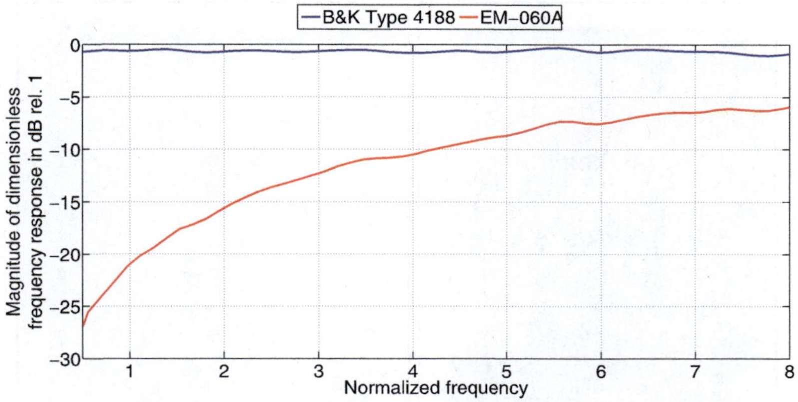

Fig. 9.12 Magnitude of frequency response of a measurement microphone (*blue curve*) and a low-price electret microphone (*red curve*) plotted against frequency that was normalized with respect to the BPF. Measurement data were determined according to (DIN EN 60268-4 2004)

9.3.2 First Specification of Actuators

The analysis of actuators was performed to evaluate the actuator power in view of the concretized requirements. Furthermore, both volume allocation and weight penalty of actuators were evaluated. For this reason various flat panel speakers were analyzed.

It turned out that high power speakers like the IMG Stage Line PAB-8MK2, compare Fig. 9.13 (left), are needed to reach the required SPL at the BPF as well as at the first and second harmonic with normal HiFi-speakers. This loudspeaker type was further on used as reference.

It was also found that flat panel speakers, especially the Honeycomb speaker ML Audio High Gain 60 shown in Fig. 9.13 (right), are capable of generating the required SPL at the first and second harmonic, but not at the BPF. This lower frequency was a problem for all investigated flat panel speakers. The magnitude responses of two actuators are shown in Fig. 9.14. It illustrated the inability of the honeycomb speaker to control the blade pass frequency and the ability of the IMG Stage Line PAB-8MK2 speaker to control all necessary frequencies.

9.3.3 First Specification of Controller Software

Because the controller software developed in the *feasibility study* was only capable of handling a tonal reference, it was necessary to consider the harmonics of the BPF. Therefore, *controller design* was also a part of the *fist specification study*. The layout of the revised controller is shown in Fig. 9.15.

Fig. 9.13 Actuators that were analyzed in the fist specification study. *Left*: HiFi speaker IMG Stage Line PAB-8MK2. *Right*: ML Audio High Gain 60 with two electro-dynamical exciters

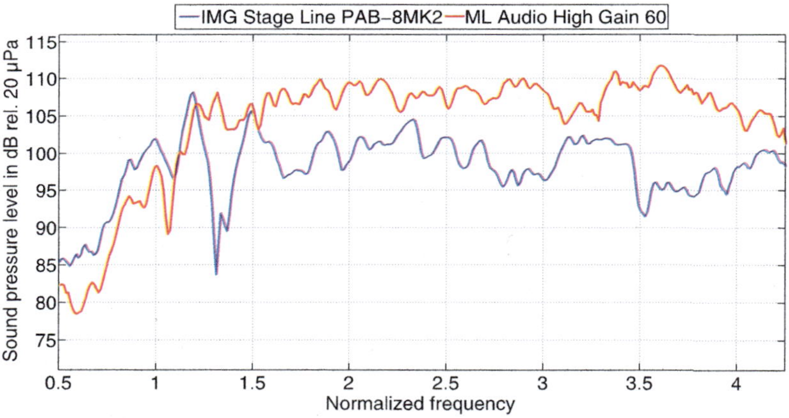

Fig. 9.14 Magnitude response of actuators that were analyzed in the fist specification study. Measurements were performed in an anechoic chamber and oriented on (DIN EN 60268-5 2004). Sound pressure was determined in a distance of 1 m to the center point of the loudspeakers using an actuator input voltage of 15 V. Normalized frequency is calculated as f/BPF

9.3.4 Evaluation of First Specification

Functional testing was performed using condenser microphones (type: B&K 4188), electro-dynamical loudspeakers (type: IMG Stage Line PAB-8MK2) and the improved controller software. All hardware elements were integrated into the simple WA mock-up, compare Fig. 9.9. It turned out that a complete ANS consisting of 11 error microphones, 6 canceling sources of the reference type and 1 dSpace unit

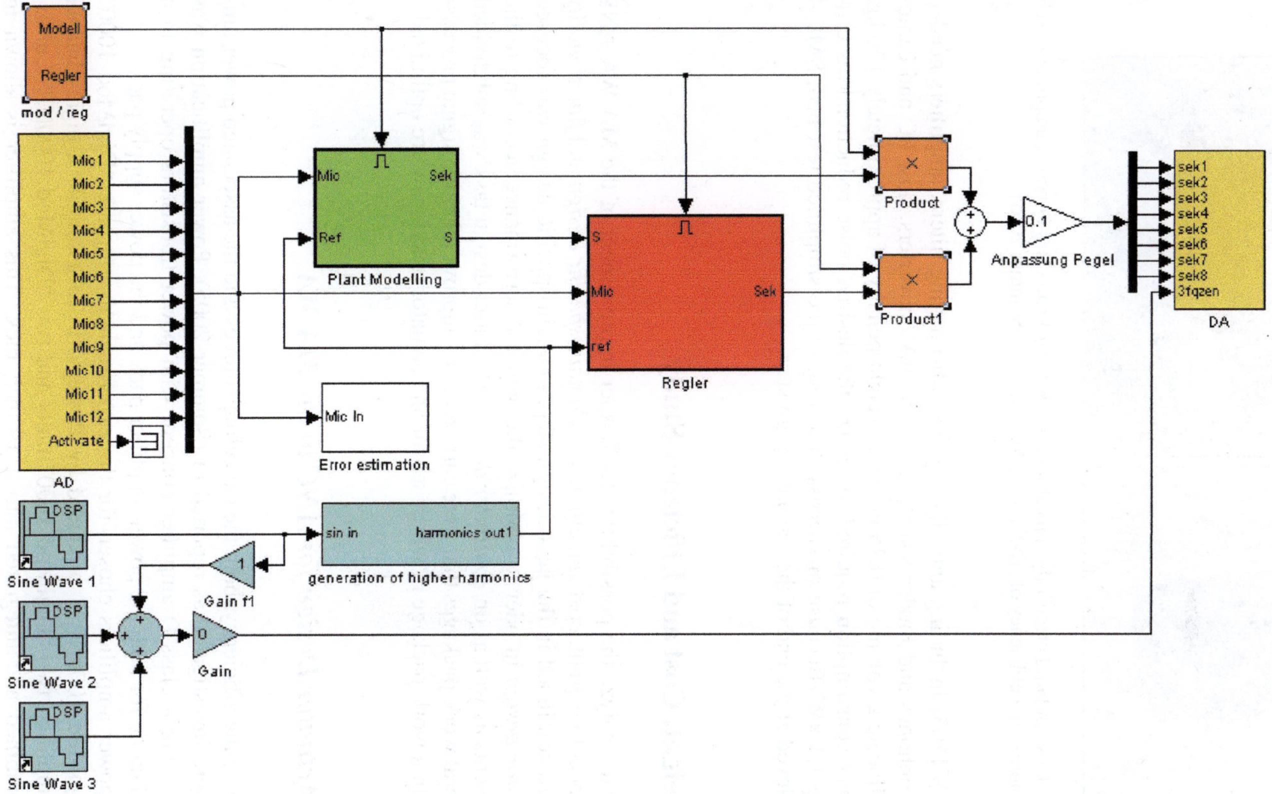

Fig. 9.15 Controller layout for feasibility study (Sommer 2005). Layout contains a plant modeling block (*green*), a control block (*red*), and blocks for generation of disturbance and reference signal (*blue*). Inputs and outputs are represented by *yellow blocks*

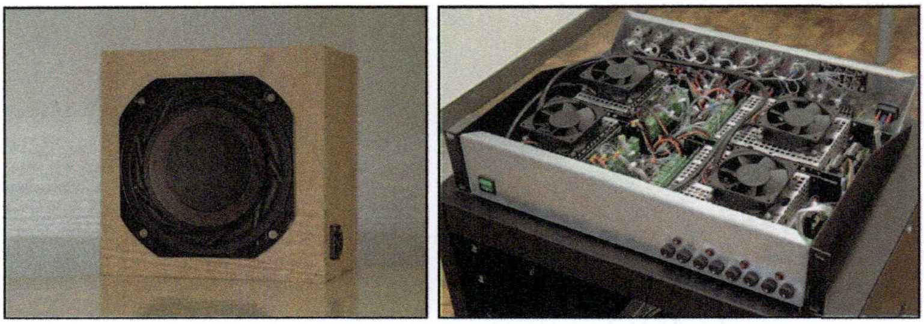

Fig. 9.16 Custom build electro-dynamical loudspeaker (*left*) and class-D power amplifier developed to improve the efficiency of ANS for the MA WA, see (Sommer 2005)

(Type DS 1103) including anti-aliasing and anti-imaging filters, amplifiers needed for microphones and loudspeakers, A/D and D/A converters, host PC and cables would allocate a volume of 0.25 m^3 by a weight penalty of approximately 100 kg. The power consumption required to realize the desired noise reduction was estimated to 1.1 kW. Because of successful testing in the simplified test bed, SML 4 was achieved at the end of the second design step.

9.4 Weight, Cost and Efficiency Study

In order to analyze the possibilities for further improvement of the MA WA ANS with respect to weight, cost and efficiency, *domain-specific design* and *functionality tests* were continued in the next design step. Domain-specific design was focused on *actuator design* in oder to optimize the actuator performance according to the requirements as well as on *controller design*. The latter design task was subdivided into several work-packages that were carried out to improve adaptive signal processing and in a work-package in which sensor and actuator positions were optimized.

9.4.1 Actuator Design for ANC in the MA WA

To improve the efficiency, both the canceling sources and the associated power amplifier were investigated as reported in (Sommer 2005). Power amplification was realized using a class-D amplifier concept with power transistors operating in a switched mode for both the power amplifier pack and the power supply pack.

The power amplifiers chosen for the experimental setup are "DigiMod 1000" and "DigiMod 1000NPS" produced by Powersoft. To assemble a eight-channel power amplifier, two "DigiMod 1000" units and two "DigiMod 1000NPS" units, were combined and integrated into a 19-inch rack. This amplifier is shown in Fig. 9.16 (right). It represents an attempt to achieve a high output at low frequencies with a compact power amplifier.

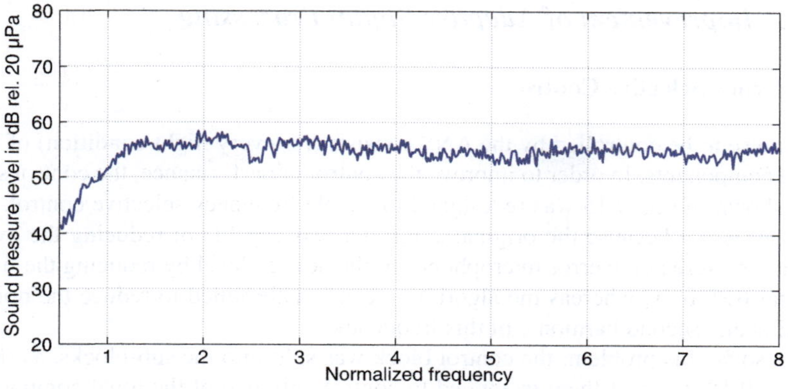

Fig. 9.17 Magnitude response of custom made electro-dynamical loudspeaker. Measurements were performed in an anechoic chamber according to (DIN EN 60268-5 2004). Sound pressure was determined in a distance of 1 m to the center point of the loudspeakers membrane using an input power of 1 W. Normalized frequency is calculated as f/BPF

In total six different chassis were built in a baffle and tested in free field half-space according to (DIN EN 60268-16 2004), to find the best one for the canceling sources.

It turned out that the chassis 6ND430 produced by Eighteen Sound are well suited for the desired application, because of its superior values for the SPL as well as the distortion. This chassis is developed especially for low frequency reproduction of sound. It has a membrane diameter of 152 mm and a robust mechanical design. Furthermore, it has a low weight penalty of only 1.25 kg. The latter is realized by a light neodymium magnet. For the chosen 6ND430 chassis, different cases were designed. Their effects on the sound pressure and the distortion were analyzed for the BPF of different flight conditions. The final design was given by a wooden box with dimensions 210 mm × 210 mm × 126 mm, built from wooden plates with a thickness of 12 mm. Finally, the whole loudspeaker mass, including the chassis, resulted to 2.280 kg. This design, shown in Fig. 9.16 (left), represents an attempt to achieve a high output at low frequencies with a compact box of approximately 4 liter volume.

The magnitude response of the developed electro-dynamical loudspeaker is shown in Fig. 9.17 in which the normalized frequency is calculated as f/BPF. It can be seen that the magnitude response of this actuator is nearly constant in the frequency range of interest. Its natural frequency is close to 125 Hz. However, the decay is moderate. Thus, this actuator is capable of generating the required anti-noise at all frequencies.

9.4.2 Improvement of Adaptive Signal Processing

Frequency Selective Control

The noise to be controlled by the ANS consists (for every flight condition) of three tonal components. In order to improve the controller performance, the control structure shown in Fig. 9.15 was redesigned to enable frequency selective control. This was necessary because the original controller was capable of reducing the overall sound pressure at the error microphones to the desired level by reducing the fundamental BPF only, whereas the algorithm occasionally failed to reduce the noise at the first and second harmonic of this frequency.

To solve this problem, the control block was split into the sub-blocks, as shown in Fig. 9.18. Each of them is applied to control only one of the tonal components. By this approach the overall system performance was significantly improved concerning its noise reduction abilities and the convergence speed of the regulation.

A-Weighting of Microphone signals

To analyze, if the consideration of human *hearing perception* can improve the control success, two approaches for *A-weighting* of microphone signals were investigated.

A according to (DIN EN 61672-1 2003), A-weighting is applied to ensure that the measured loudness corresponds well with the subjectively perceived loudness. For this purpose, the SPL around 3–6 kHz (where the human ear is most sensitive) is emphasized, while the SPL of very high ($f > 6$ kHz) and very low frequencies ($f < 500$ Hz), is attenuated using the frequency dependent function

$$H_A(f) = \frac{12200^2 \cdot f^4 \cdot 1.2589}{(f^2 + 20.6^2)(f^2 + 12200^2)\sqrt{f^2 + 107.7^2}\sqrt{f^2 + 737.9^2}}. \tag{9.5}$$

The first approach was realized by weighting all microphone signals in time domain. The required FIR-filter was realized with linear phase and with an uneven number of coefficients, see (Sommer 2005). This approach can be summarized as follows:

1. Frequency sampling of (9.5):

$$H(k) = H_A\left((k-1)\frac{f_s}{L_{FIR}}\right), \quad k = 1, \ldots, \frac{L_{FIR}+1}{2}. \tag{9.6}$$

2. Crossover mirroring of results:

$$\hat{H}(k) = H(L_{FIR} + 2 - k), \quad k = \frac{L_{FIR}+3}{2}, \ldots, L_{FIR}. \tag{9.7}$$

3. Inverse discrete Fourier transformation:

$$\hat{h}(n) = \frac{1}{L_{FIR}} \sum_{k=1}^{L_{FIR}} \hat{H}(k) e^{j2\pi(k-1)\left(\frac{n-1}{L_{FIR}}\right)}, \quad n = 1, \ldots, L_{FIR}. \tag{9.8}$$

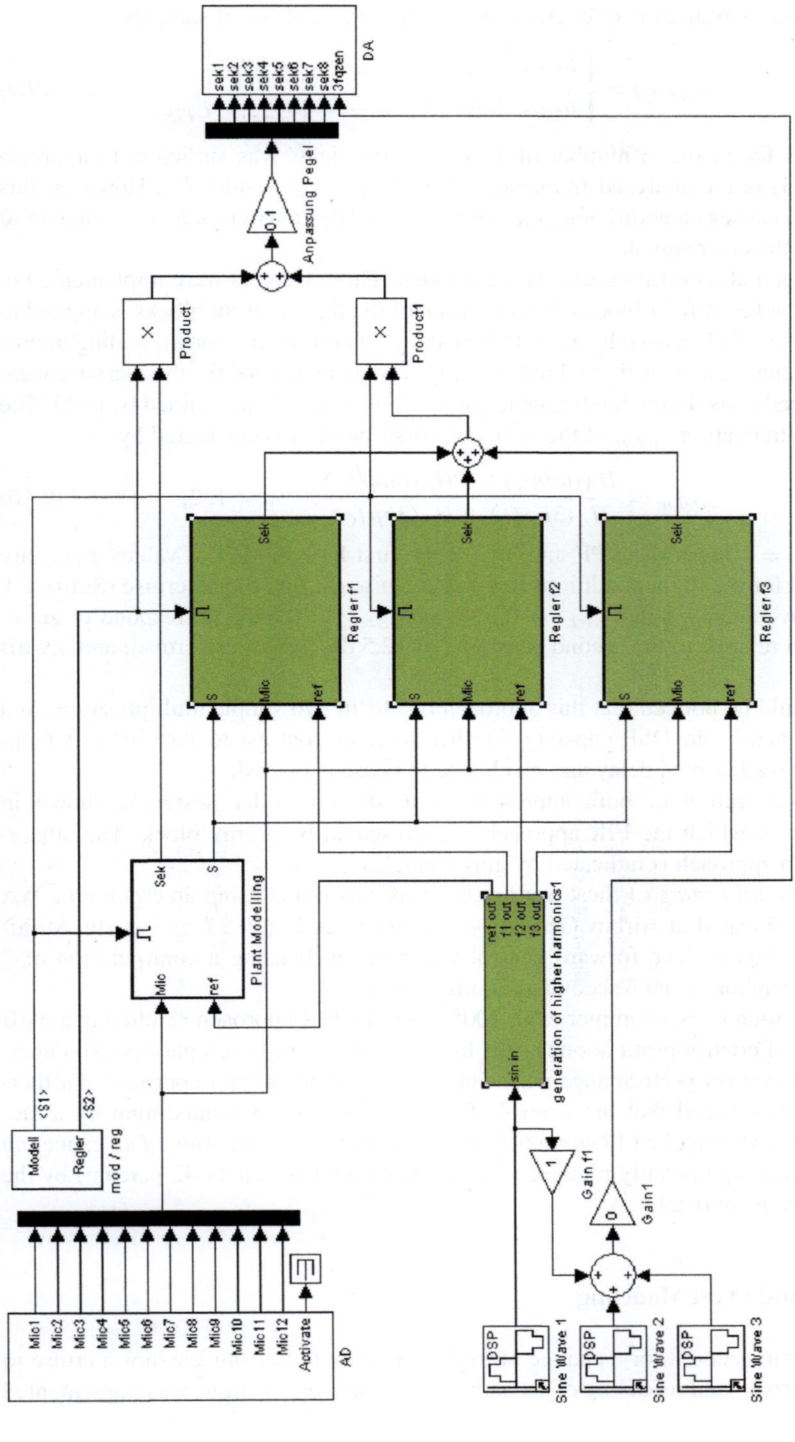

Fig. 9.18 Controller layout for frequency selective control, see (Sommer 2005), containing a plant modeling block and three control blocks (*green*), each used to handle a single frequency

4. Circular permutation of coefficients according to number of samples:

$$h_{FIR}(n) = \begin{cases} \hat{h}(n + \frac{L_{FIR}+1}{2}), & n = 1, \ldots, \frac{L_{FIR}-1}{2} \\ \hat{h}(n - \frac{L_{FIR}-1}{2}), & n = \frac{L_{FIR}-1}{2}, \ldots, L_{FIR}. \end{cases} \tag{9.9}$$

It was found that a number of 19 filter coefficients was sufficient to approximate (9.5) in the analyzed frequency band (50 Hz $< f <$ 400 Hz). However, this approach causes an additional delay in the secondary transfer path and a change of phase in the error signal.

The second approach uses relative weights. These weights were implemented in front of two controller blocks, because each of the three control blocks is applied to attenuate the SPL caused by a single frequency. Therefore, the corresponding microphone signals can be weighted individually. The gains needed for this purpose were determined considering the frequency dependent A-weighting defined by (9.5). The relative attenuation $g_{rel,m}$ of the m-th controller block was calculated by

$$g_{rel,m} = \frac{H_A(mbpf_L) + H_A(mbpf_C)}{H_A(3bpf_L) + H_A(3bpf_C)}, \quad m = 1, 2, \tag{9.10}$$

where $m = 1$ signs the BPF and $m = 2$ the first harmonic. The values $g_{rel,m}$ are averaged for the flight conditions low-noise cruise (Suffix L) and cruise (Suffix C).

The averaged weights $g_{rel,1} = 0.2362$ and $g_{rel,2} = 0.6358$ correspond to an attenuation relative to the second harmonic of 12.5 dB (low-noise cruise) and 3.9 dB (cruise).

It should be noticed that this approach results in two simple multiplications and requires nearly no DSP capacity. Furthermore, in contrast to the FIR-Filter approach, no additional delay and no change in phase is caused.

The integration of both approaches into the controller design is shown in Fig. 9.19 in which the FIR approach is represented by a gray block. The attenuation gain approach is indicated by gray triangles.

Functional testing of these approaches was performed using an ergonomic WA mock-up (located at Airbus Germany—Bremen), see Fig. 9.23, as test rig. Multichannel adaptive feed-forward control was performed using a configuration of 7 error microphones and 6 secondary loudspeakers.

It was found, see (Sommer et al. 2005), that the FIR approach resulted in a 6 dB decrease of control profit at one of the head mounted error microphones. No degradation of control performance was caused by the relative gain approach. Furthermore, it was found that the control effort was increased (in maximum by a factor 10) by applying the FIR-approach, while the power consumption of all canceling sources was significantly reduced (in average by approximately 42 percent) by the relative gain approach.

Broadband Plant Modeling

To take into account for a change of flight condition (e.g. from low-noise cruise to cruise), broadband modeling of secondary-path transfer functions was implemented

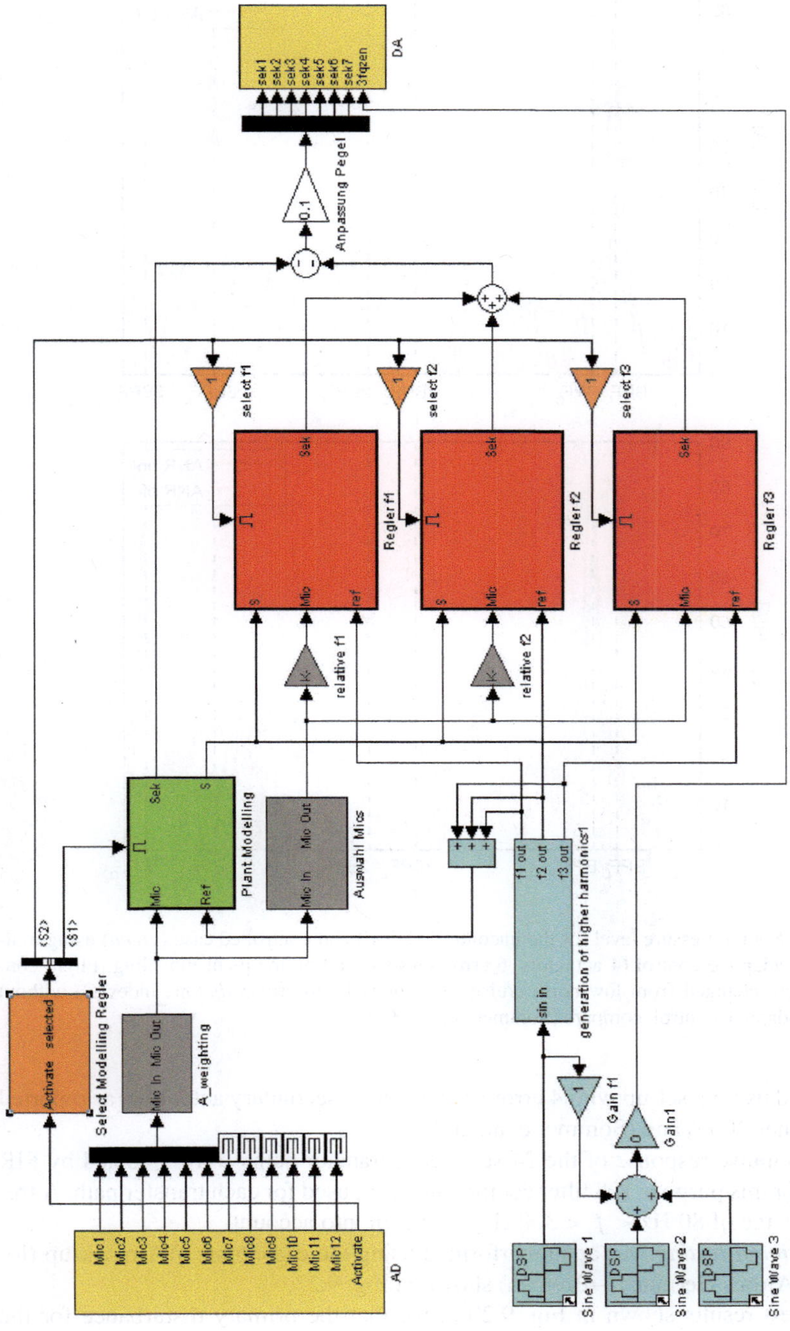

Fig. 9.19 Multi-channel adaptive control system based on the FxLMS algorithm, frequency selective control and A-weighting of error signals (Sommer 2005). Layout contains the FIR approach (*gray colored rectangle*) and the attenuation gain approach (*gray colored triangles*)

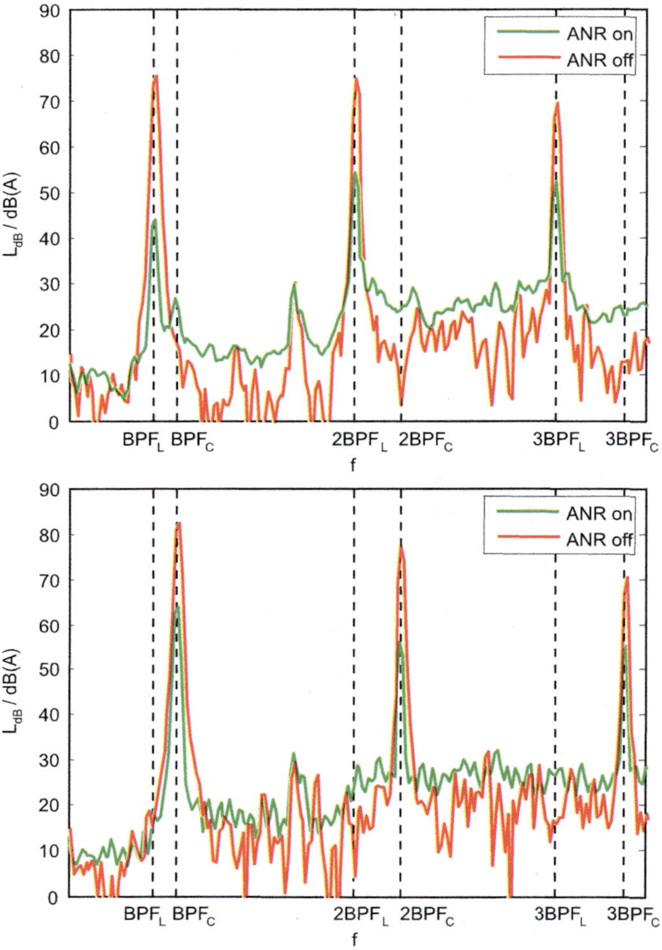

Fig. 9.20 Sound pressure level for the uncontrolled (*red*) and controlled case (*green*) using multi-channel adaptive control (4 actuators, 6 error sensors) and online-plant modeling. Flight conditions were changed from low-noise cruise (*top*, index L) to cruise (*below*, index C) without stopping adaptive control, compare (Sommer et al. 2005)

and tested using a set-up with 4 error sensors and 6 secondary actuators, as reported in (Sommer 2005) and (Sommer et al. 2005).

The impulse response of the 24 secondary transfer paths were modeled by FIR filters. For this purpose 128 filter coefficients were used for each transfer path. A frequency range of 80 Hz $< f <$ 300 Hz was taken into account.

Functional testing was again performed using the ergonomic WA mock-up (located at Airbus Germany—Bremen) shown in Fig. 9.23.

The test results shown in Fig. 9.20 prove that the primary disturbance for the flight condition low-noise cruise (top chart) can be reduced by −30 dB (BPF),

−20 dB (first harmonic) and −17 dB (second harmonic), if the ANS is activated. These reduction correspond to an overall reduction of −19 dB.

After the change from low-noise cruse to cruse (lower chart) the ANS is converged within a few seconds and a new persistent status is reached (graph below, green curve). The single components are now reduced by −18 dB (BPF), −21 dB (first harmonic) and −15 dB (second harmonic). For the second flight condition an overall noise reduction of −18 dB was achieved.

Software-Limitation of Controller Outputs

As described in (Klünder 2004) a software-limiter for the loudspeaker outputs was integrated to avoid nonlinear distortions. This limiter consists of a level estimation followed by a scaling of signals passed to the loudspeakers.

As summarized in (Gerner 2005) limitation of loudspeaker driving signals was implemented separately for each output channel m using q_{peak} as upper bound and $g_m(n)$ as correction factor for the m-th channel at discrete time n.

The sample time T as well as the response time t_{ab} and the return time t_r were used to adjust the software-limiter. This approach can be summarized as follows:

1. Calculate peak value:

$$q_{m,peak}(n) = \begin{cases} \gamma|q_m(n)| + (1-\gamma)q_{m,peak}(n-1) \\ \quad \text{if } |q_m(n)| \geq q_{m,peak}(n-1) \\ (1-\eta)q_{m,peak}(n-1) \\ \quad \text{else } |q_m(n)| < q_{m,peak}(n-1). \end{cases} \tag{9.11}$$

2. Calculate the m-th correction gain:

$$g_m(n) = \begin{cases} 1 & \text{if } q_{m,peak} \leq q_{\max} \\ \frac{q_{\max}}{q_{l,peak}} & \text{else } q_{m,peak} > q_{\max}, \end{cases} \tag{9.12}$$

with

$$\gamma = 1 - e^{-2.2\frac{T}{t_{ab}}}, \tag{9.13}$$

and

$$\eta = 1 - e^{-2.2\frac{T}{t_r}}. \tag{9.14}$$

3. Calculate the m-th output signal:

$$q_m'(n) = q_m(n)g_m(n). \tag{9.15}$$

9.4.3 Optimization of Sensor and Actuator Locations

To identify loudspeaker and microphone positions that enable the ANS to realize a significant reduction of the SPL around the operator head to the limit specified in the concretized requirements, TPA, compare Sect. 7.3.2, in combination with an optimization procedure was applied. These investigations were also performed using the ergonomic WA mock-up that is shown in Fig. 9.23. It was utilized to measure the acoustic transfer behavior between 69 possible actuator positions and 53 sensor positions. The whole optimization procedure was presented in (Gerner 2005). The summary given here is based on (Gerner et al. 2005) and (Kletschkowski et al. 2005). Another approach can be found in (Böhme et al. 2006). Because we will present a frequency domain approach, the arguments of all variables have been omitted for an easy notation, except for those equations in which a more precise notation is necessary.

Formulation of Optimization Problem

According to the requirements the optimization problem can be described as shown in (Gerner et al. 2005):

- Find a set of actuators and sensors, which is able to reduce the sound pressure at the operator ears to a certain value $p_{\max}$.
- Ensure that the input to the actuators does not exceed a certain value $q_{\max}$.
- If there are different possible sets take the one with the minimum weight. That means the one with the minimum number of actuators and sensors, because the weight of the controller is fixed.
- For different sets with the same weight take the one with the minimum actuator input power.

In order to present a mathematical formulation, we have to calculate the controlled sound field. To prepare TPA we have to consider the transducer locations. For this purpose we introduce the $(L \times L)$ sensor matrix $\mathbf{L}$ and the $(M \times M)$ actuator matrix $\mathbf{M}$. As specified in Sect. 7.3.2, the elements of these diagonal matrices are one/zero, if the associated transducer is activated/deactivated. Using the transducer matrices, the $(L \times 1)$ column matrix that contains the sound pressure data measured at the error microphones can be expressed by

$$\mathbf{p} = \mathbf{L}(\mathbf{p}_p + \mathbf{HMq}), \tag{9.16}$$

where $\mathbf{p}_p$ is the $(L \times 1)$ column matrix of the disturbance signals and $\mathbf{H}$ is a $(L \times M)$ matrix that contains the transfer behavior between all actuators and all error sensors. $\mathbf{q}$ is the $(M \times 1)$ column matrix of optimal control signals. The (2×1) column matrix that contains the sound pressure at the operator ears can in direct analogy to (9.16) be expressed as

$$\mathbf{p}_o = \mathbf{p}_{op} + \mathbf{H}_o\mathbf{q}, \tag{9.17}$$

where $\mathbf{p}_{op} = [P_{left}(j\omega)\;\; P_{right}(j\omega)]^T$ is a (2×1) column matrix that contains the complex sound pressure at the left ear $(P_{left}(j\omega))$ and the right ear $(P_{right}(j\omega))$. The $(2 \times M)$ matrix $\mathbf{H}_o$ contains the transfer behavior between each actuator and the operator ear microphones. For different frequencies to be taken into account the matrix of total magnitudes of squared sound pressures at the operator ears is

$$\mathbf{p}_{o,tot} = \sum_{n=1}^{3} \left[\left| P_{left,n}(j\omega_n) \right|^2 \; \left| P_{right,n}(j\omega_n) \right|^2 \right]^T \quad \text{with } \omega_n = 2\pi(n \cdot f_{BPF}). \tag{9.18}$$

The matrix of the total magnitudes of squared control signals is given by

$$\mathbf{q}_{tot} = \sum_{n=1}^{3} \left[\left| Q_{1,n}(j\omega_n) \right|^2 \; \ldots \; \left| Q_{M,n}(j\omega_n) \right|^2 \right]^T \quad \text{with } \omega_n = 2\pi(n \cdot f_{BPF}). \tag{9.19}$$

These signals specify the power consumption of all canceling sources driven to control the BPF as well as its first and second harmonic. The latter can be defined as

$$\Pi_q(\mathbf{L}, \mathbf{M}, \mathbf{q}) = \sum_{n=1}^{3} \mathbf{q}^H(j\omega_n)\mathbf{q}(j\omega_n) \quad \text{with } \omega_n = 2\pi(n \cdot f_{BPF}). \tag{9.20}$$

Based on the description of the optimization problem and the considerations outlined above it is possible to present a mathematical formulation of the optimization problem such as

$$\underset{\mathbf{L},\mathbf{M},\mathbf{q}}{\text{minimize}} \; J_R(\mathbf{L}, \mathbf{M}, \mathbf{q}) \tag{9.21}$$

with

$$J_R(\mathbf{L}, \mathbf{M}, \mathbf{q}) = \alpha_L L_a(\mathbf{L}) + \alpha_M M_a(\mathbf{M}) + \alpha_{\Pi_q} \Pi_q(\mathbf{L}, \mathbf{M}, \mathbf{q}), \tag{9.22}$$

where α_L, α_M and α_P are weighting parameters, L_a/M_a is the number of activated sensors/actuators, and $\Pi_q(\mathbf{L}, \mathbf{M}, \mathbf{q})$ is the actuator input power, by considering the constraints

$$\max(\mathbf{p}_{o,tot}) \leqslant p_{\max}^2 \tag{9.23}$$

and

$$\max(\mathbf{q}_{tot}) \leqslant q_{\max}^2. \tag{9.24}$$

Equation (9.23) limits the SPL at the two operator ears, whereas (9.24) is used to limit the actuator power consumption. Because of the transducer matrices **L** and **M** the problem is a discrete optimization problem. Together with the constraints (9.23) and (9.24) it becomes a restricted discrete optimization problem.

It is possible to transform this restricted discrete optimization problem consisting of the weighted sum of three cost functions into an unrestricted optimization problem. For this purpose a penalty term $j_o(\mathbf{L}, \mathbf{M}, \mathbf{q})$ can be introduced to account for violating the constrains (9.23) and (9.24). The cost function of the resulting unrestricted discrete optimization problem reads

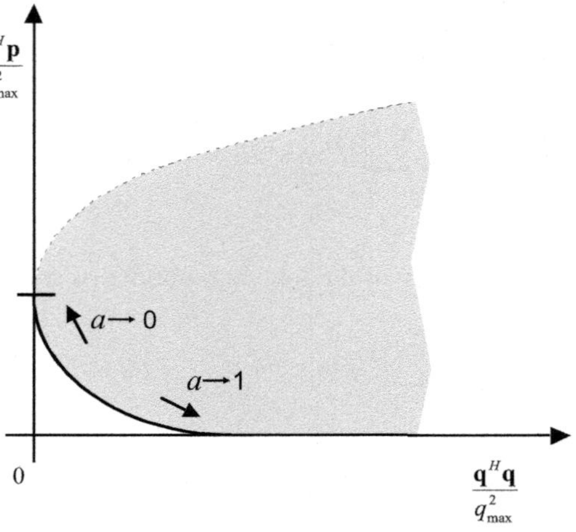

Fig. 9.21 Effect of weighting parameter a on the optimal solution considering one configuration, see (Gerner et al. 2005)

$$J_U(\mathbf{L}, \mathbf{M}, \mathbf{q}) = \alpha_L L_a(\mathbf{L}) + \alpha_M M_a(\mathbf{M}) + \alpha_{\Pi_q} \Pi_q(\mathbf{L}, \mathbf{M}, \mathbf{q}) + j_o(\mathbf{L}, \mathbf{M}, \mathbf{q}). \tag{9.25}$$

Analysis of Single Configurations

To solve the restricted optimization problem (9.21) considering the constraints (9.23) and (9.24), it is necessary to calculate both optimal driving signals for the canceling sources and the controlled sound pressure at the operator ears. The same holds, if (9.25) is to be minimized.

For this purpose we can use the TPA cost function that is defined by (7.58). However to follow *Gerners* approach we have to replace the scalar valued weighting parameter β by

$$\beta = \frac{1-a}{a} \cdot \frac{p_{\max}^2}{q_{\max}^2} \quad \text{with } 0 < a < 1. \tag{9.26}$$

For a given configuration the optimal solution $\mathbf{q}_{opt}$ depends for each frequency on the weighting parameter a. Its effect on the control performance is considering one configuration shown in Fig. 9.21. The gray area is given by (9.16). By minimizing the TPA cost function (7.58) it is reduced to the thick black curve that is given by the optimal controlled sound field $\mathbf{p}(\mathbf{q}_{opt})$. On this curve the decrease/increase of the control profit leads to an increase/decrease of the control effort. Therefore, the curve consists of *Pareto-optimal* points, see (Pareto 1971).

Using the parameter a it is possible to select a certain point on this *Pareto-curve*. In the presented approach this parameter was introduced to find one particular optimal solution that represents the following state

$$\max(\mathbf{p}_{o,tot}) = p_{\max}^2 \quad \text{and} \quad \max(\mathbf{q}_{tot}) \leqslant q_{\max}^2. \tag{9.27}$$

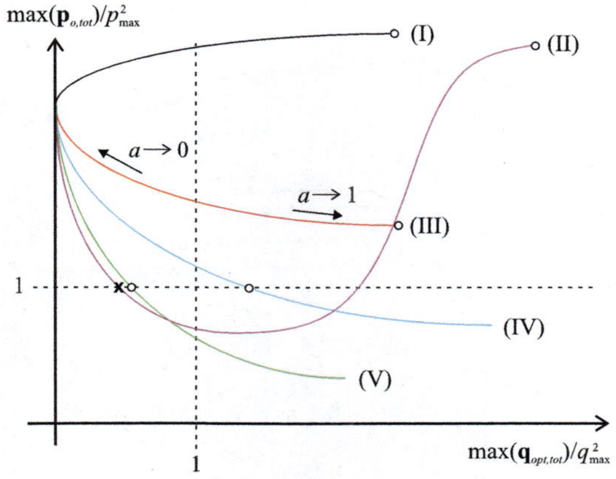

Fig. 9.22 Effect of weighting parameter a used to tune the optimal solution obtained for a certain set of sensor and actuator locations in order to hold the constraints given by a maximum sound pressure at the operator ears $p_{\max}$ and a maximum power consumption $q_{\max}$ of the canceling sources, see (Gerner et al. 2005)

This state is a proper optimum of the optimization problem stated above, because the constraints (9.23) and (9.24) are not violated and the actuator input is minimized.

The effect of tuning the optimal solution of a certain configuration using the weighting parameter a is illustrated in Fig. 9.22 in which the curves (I)–(V) represent 5 different configurations. In this figure the normalized maximum squared sound pressure at the operator ears $\max(\mathbf{p}_{o,tot})/p^2_{\max}$ is plotted against the normalized maximum squared actuator input $\max(\mathbf{q}_{opt,tot})/q^2_{\max}$. These terms represent the constraints (9.23) and (9.24).

For each configuration evaluated during the optimization it can be tested if the configuration is able to satisfy constraint (9.23) for $a \to 1$. If the configuration does so (IV), (V), the minimum necessary actuator input $\mathbf{q}_{\min}$ that is needed to satisfy (9.23) can be calculated by minimizing a until $\max(\mathbf{p}_{o,tot})/p_{\max} = 1$. For the other configurations (I)–(III) $\mathbf{q}_{\min}$ can be calculated for $a \to 1$. Then it is checked if $\mathbf{q}_{\min}$ violates the constraint (9.24).

If one of the constraints is violated (I)–(IV) it is possible to add the penalty term $j_o(\mathbf{L}, \mathbf{M}, \mathbf{q}_{opt})$ to the objective function. This ensures that a configuration violating one of the constraints is always assessed worse than a configuration not doing so (V). At the moment (marked by o) the configuration (II) is assessed to violate both constraints, even it is able to fulfill them (see the marker x) even better than (V) after tuning the weighting parameter a.

Comments on the Optimization Procedure

Using the mathematical formulation outlined above, the optimization problem was solved in two steps. First the necessary number of actuators and sensors was calculated as follows:

1. Calculate the minimum necessary number of sensors by stepwise increasing L_a until (9.23) is fulfilled. For this step all actuators were activated $\mathbf{M} = \mathbf{I}$.

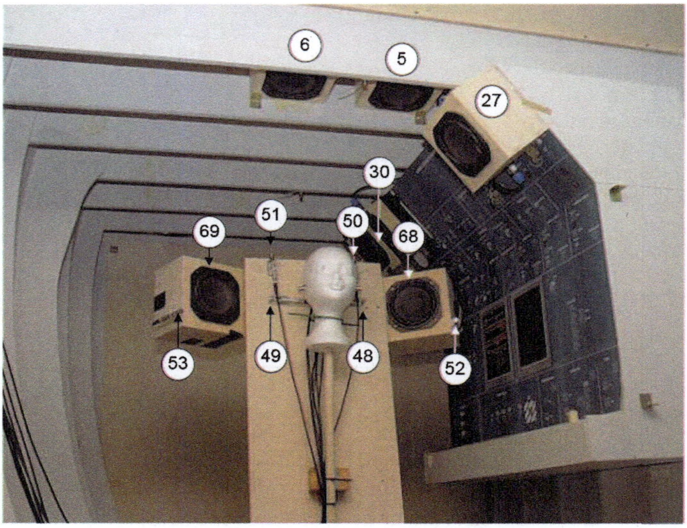

Fig. 9.23 Ergonomic mock-up of MA WA used for optimization of sensor and actuator positions, see (Gerner 2005)

2. Calculate the minimum necessary number of actuators by stepwise increasing M_a until (9.24) is fulfilled. For this purpose, the sensor positions calculated in step 1 were used.

By applying this method is was possible to determine the upper limit of actuators and sensors necessary to fulfill the constraints (9.23) and (9.24). With this upper limit as a fixed number of actuators and sensors the cost function (9.25) reduces to

$$J_{U,red}(\mathbf{L}, \mathbf{M}, \mathbf{q}) = \Pi_q(\mathbf{L}, \mathbf{M}, \mathbf{q}) + j_o(\mathbf{L}, \mathbf{M}, \mathbf{q}). \tag{9.28}$$

Equation (9.28) was used to optimize the L_a sensor positions as well as the M_a actuator positions. For this purpose the COSA-algorithm developed by Wendt (Wendt 1995) was applied. Details about the implementation are to be found in (Gerner 2005). This algorithm combines elements of simulated annealing, a description of this method may be found in (Press et al. 1994), and genetic algorithms, see e.g. (Holland 1975).

The main difference between COSA and simulated annealing is that the change of the configuration—the transition—isn't done randomly but by using other configurations already optimized. To do this a population of different configurations is build like at the genetic algorithms. Within this population mechanisms like recombination and mutation are used. This kind of transition is called cooperative transition. It was adapted for the actual problem.

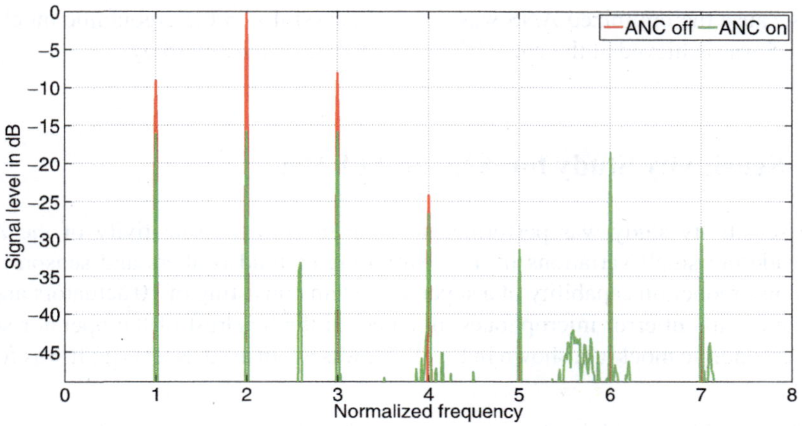

Fig. 9.24 Active noise reduction measured at the right operator ear for an optimal configuration (loudspeaker positions: 5, 6, 27, 30 and microphone positions: 48, 49 defined by Fig. 9.23) that was implemented into the ergonomic mock-up. Low-noise cruse was considered as flight condition. Normalized frequency was calculated as f/BPF. Picture from (Gerner 2005)

Optimization Results

Using the optimization approach outlined above it was possible to find valid sensor and actuator positions. It turned out that the optimization gives a minimum number of 4 actuators and 2 sensors to fulfill the constraints for 2 different seating positions.

It has been shown that such an optimized configuration is connected with a payload of only 20 kg, a volume allocation of only 50 dm^3, and a power consumption of only 200 W. These data represent a reduction in weight, volume and power consumption of approximately 80% compared to the results of the *first specification*. They were used as an indicator for the commercial system.

Functional testing of the optimized ANS was again performed in the ergonomic mock-up. Test results are shown in Fig. 9.24 in which the signal level (or relative SPL) measured at the right operator ear is presented for one optimal configuration. Low-noise cruse was considered as flight condition for this particular test.

Applying frequency selective control, the optimized ANS was capable of meeting the requirements by reducing the total SPL; in minimum −5 dB below the specified limit, in maximum −10 dB below the specified limit.

It should be noticed that the graphs also indicate an amplification at the fourth, fifth and sixth harmonic of the BPF caused by nonlinear distortion of the canceling sources. However, the SPL at these frequency lines is less than the specified limit and can therefore be accepted.

The outcome of this design step was an optimized ANS that was capable of meeting the requirements, except that of excluding microphones from the operator seat. For this reason, further investigations were required to enable RSC without seat mounted microphones. In addition it was necessary to investigate the sensitivity of the optimized ANS.

Because the optimized ANS was tested successfully in the ergonomic mock-up, SML 5 was achieved at the end of the *weight, cost, efficiency study*.

9.5 Sensitivity Study for ANC in MA WA

The sensitivity study was performed to analyze both the sensitivity of the ANS (considering small variations in the positioning of loudspeakers and sensors) and the noise reduction capability of a superset system consisting of 10 actuators and 18 sensors (without error microphones mounted on the headrest of the operator seat). The ergonomic mock-up, shown in Fig. 9.23, was again used as test rig. It was found that:

- The total SPL could, in almost any case, reduced down to $p_{\max}$, if the two microphones at the headrest were utilized alone. Depending on the number of activated secondary actuators greater power reserves did partly exist.
- The positioning of the error sensors has a much greater impact on the performance of the ANS than the positioning of the secondary actuators. The measurement data indicate that the desired actuator positions can be used for the commercial ANS.
- The desired positions for the error microphones along the ceiling and the sidewall were found to be not optimal to meet the requirements. Further experiments were conducted using displaced error sensors. The results of these experiments proved that the displacement of the microphones in most cases led to an improved noise reduction. This strongly indicated that the superset ANS was (in principle) capable of meeting the requirements.

Because *functional testing* was not performed in a relevant environment, the SML was still SML 5 at the end of the *sensitivity study*.

9.6 Robust Control of Sound in the MA WA

The feasibility study as well as the first specification design step made clear that it is possible to design an ANS that is capable of meeting the requirements. The weight cost efficiency step indicated the possibilities for optimizing the ANS. The sensitivity study proved that an optimized system as well as an improved superset system can lead to a successful solution for the ANC task in the WA.

The intention of the robustness study was to improve the RSC performance of the superset system by robust controller design, compare Sect. 6.5.3, using a relevant test environment, to avoid error microphones mounted on the headrest of the operator seat. This study was therefore subdivided into the following parts:

1. Design and construction of a relevant test environment.
2. Robust controller design for RSC without head rest microphones.

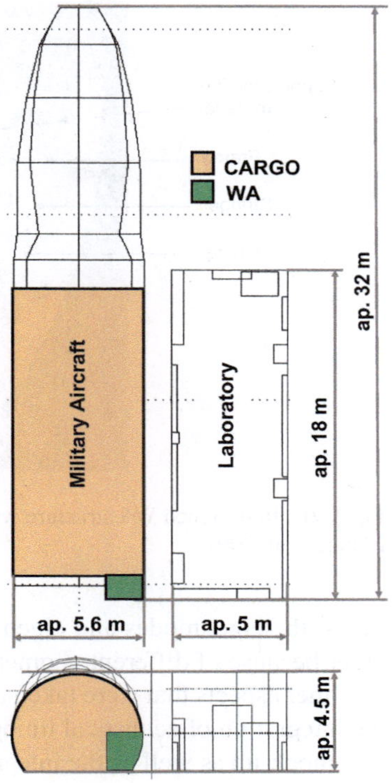

Fig. 9.25 Comparison between dimensions of the MA cargo and the dimensions of the laboratory used for acoustic ground tests (adapted from Böhme et al. 2007)

9.6.1 Design, Construction and Verification of Relevant Test Rig

Preliminary Considerations

The cavity of the WA is not the only air filled volume which was considered for acoustic ground tests. As stated in Sect. 9.2.1 the acoustic coupling of the semi-closed WA to the cargo hold plays a key role for a successful system design. For this reason, the geometry of the surrounding laboratory plays an important role.

In reality the WA is connected to the cargo having an overall length of approximately 32 m. A fraction of 18 m is designed like a cylindrical enclosure with floor. The aft portion of the cargo is cone-shaped.

Taking the operational condition into account, a certain laboratory at the Helmut-Schmidt-University/University of the Federal Armed Forces Hamburg was chosen to simulate the cargo hold. This room has almost the same dimensions as the cylindrical portion of the MA cargo compartment, as illustrated by Fig. 9.25. The side walls and the floor of the laboratory can be assumed to be sound hard in contrast to the ceiling where sound absorbing material is installed.

As reported in (Böhme et al. 2007), an average value of 0.5 s was measured for the reverberation time in the desired laboratory in the frequency range of interest (80 Hz $< f <$ 500 Hz). However, even with comparable volumes and reverberation

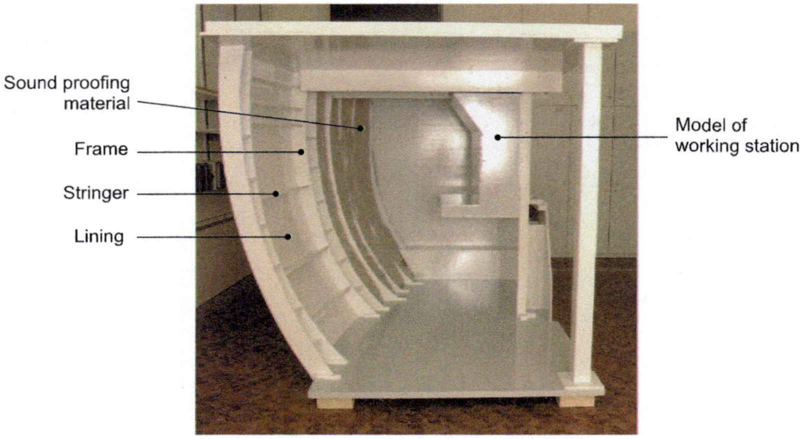

Fig. 9.26 Redesigned WA structure with sound proofing material in the university (adapted from Böhme et al. 2007)

times, the eigenmodes and eigenfrequencies of these two cavities differ from each other because of different geometries and operational conditions.

Other aspects that were taken into account by planning the test environment were the integration of acoustical lining which has an effect on the room acoustics of the WA mock-up as well as the integration of an appropriate operator dummy.

Design and Construction of Mock-up

To simulate the room acoustics of the WA, an accurate model of the semi-closed cavity is required. Therefore, original construction data were extracted from a digital mock-up. In the next phase, the necessary parts, which define the cavity of the WA, were redesigned, to eliminate negligible details of the original set of data.

In order to adapt the design to available material sizes for manufacturing, further simplifications were required. Figure 9.26 shows the assembled mock-up that was made of wood. The reverberation time inside this mock-up was adjusted to be equal to the reverberation time of the supposed aircraft. Therefore, this test facility represents the essential acoustical conditions of the operational environment.

The mock-up was however not capable of representing the vibro-acoustic properties of the operational environment. The eigenfrequencies and eigenmodes will therefore be different to whose of the future aircraft.

System Integration and Verification of Test Rig

The mock-up was equipped with an ANS consisting of $M = 8$ loudspeakers and $L_C = 16$ error microphones (original flight qualified equipment delivered by Ultra

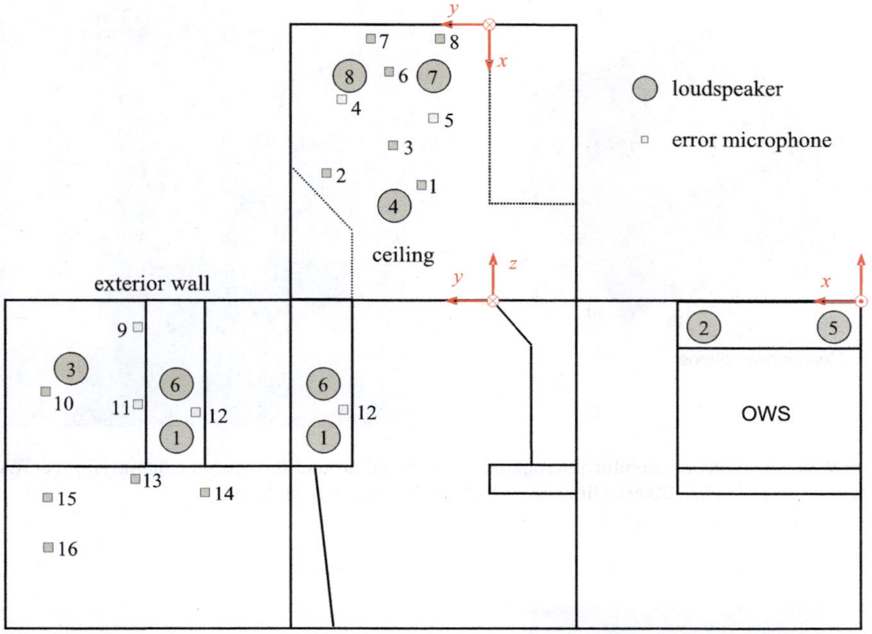

Fig. 9.27 Positions of error microphones and canceling loudspeakers in the WA mock-up, see (Kochan 2009). OWS is the abbreviation for operator working station. Layout is not drawn to scale

Electronics). All loudspeakers and microphones were installed at the outer wall and the ceiling of the WA. The transducers positions shown in Fig. 9.27 were specified by Airbus. These positions were based on the results of the optimization procedure described in Sect. 9.4.3.

To take into account for a working operator as well as to enable aurally compensated pressure signals at the operator ears, the WA mock-up was equipped with an artificial head (type: HEAD Acoustics, HMS III.0) and a wooden torso box (covered with felt) to simulate the acoustic properties of a human body as shown in Fig. 9.28 (right).

Because the objective of the control task is to reduce the sound pressure around the operator head, $L_M = 12$ monitor microphones (type: B&K 4188) were distributes around the artificial head, as also shown in Fig. 9.28 (right). Their signals were analyzed using a multi-channel FFT system. The monitor microphone positions are to be found in Fig. 9.28 (left). To enable ANC, a rapid prototyping board (dSPACE 1103) was integrated in the test rig. This system was connected with both the WA error microphones and the canceling sources via a breakout panel as well as with a second dSpace system (used to generate the primary noise).

The latter connection was established to obtain the reference signal. 4 public address loudspeakers were positioned 6 m away from the mock-up to excite tonal disturbances. The complete installation as well as the information flow are shown in Fig. 9.29.

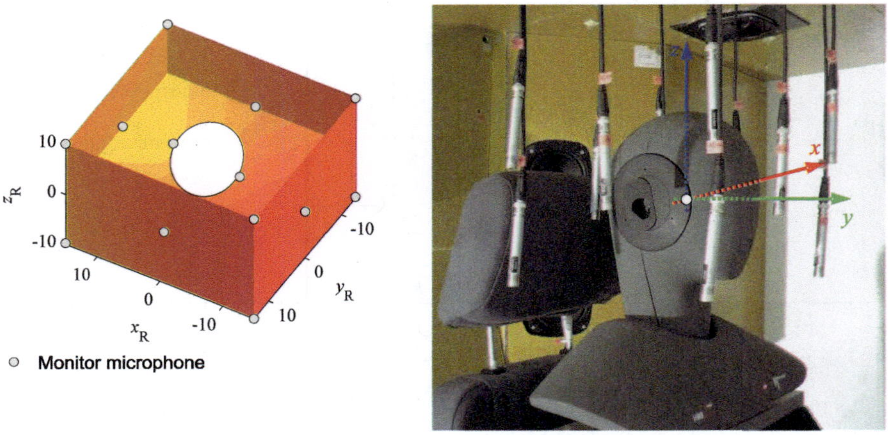

Fig. 9.28 Positions of monitor microphones distributed around the operator dummy in the WA mock-up, see (Kochan 2009). Dimensions in the *left part* are given in cm

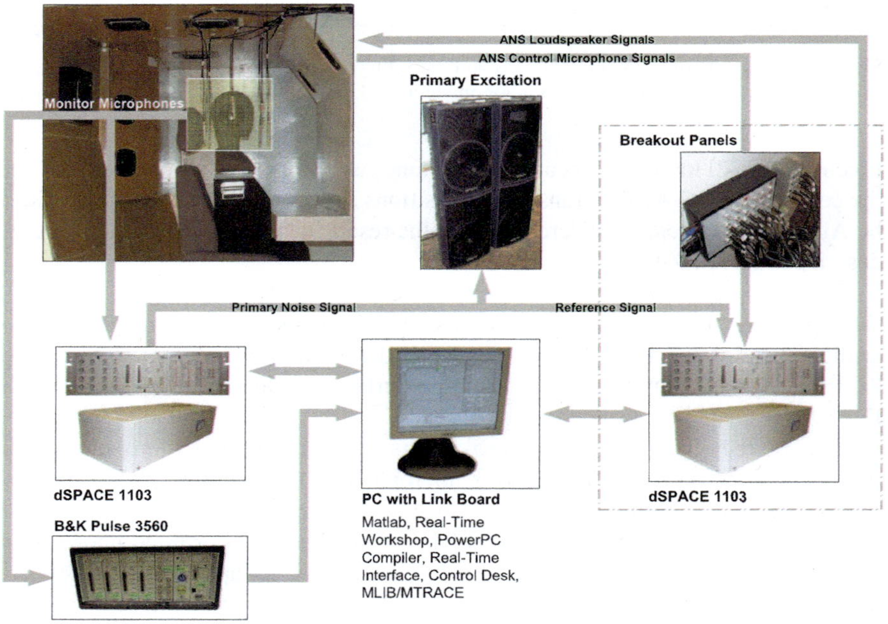

Fig. 9.29 Information flow in test rig used for acoustic ground tests and robust controller design. Test rig is equipped with 16 error microphones, 8 canceling sources, one artificial head and 12 monitor microphones. Noise is generated by public address speakers driven by a dSpace system. A second dSpace system (connected with the ANC components via a breakout panel) was used for active control. Noise reduction around the artificial head was measured using a multi-channel FFT analyzer

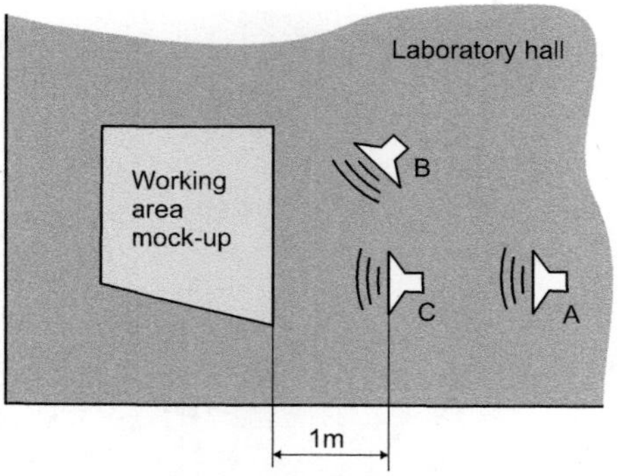

Fig. 9.30 Outline of test idea for verification of primary noise field in the WA mock-up. Positions are as follows: Pos. A—2 m in front of the working area, Pos. B—sideward shifted and 45° rotated; 1 m in front of the working area, Pos. C: lying position; 1 m in front of the working area; 45° tilted. Picture is adapted from (Kochan et al. 2008)

To ensure proper excitation of the WA mock-up both a comparison of the forced vibration using the modal assurance criterion and a principal components analysis of the primary noise field were carried out. For this purpose the WA was excited with a step-sine from $f = 80$ Hz to $f = 200$ Hz. The loudspeaker used for excitation was placed in front of the WA at three different positions (A, B and C) as shown in Fig. 9.30. For more details see (Kochan et al. 2008).

It turned out that the shape of the primary noise field is less sensitive to the three considered excitation cases, and that the shape of the excited primary noise field is correlated over a large frequency range. The geometry of the WA mock-up was most significantly responsible for the shape of the sound field. Therefore, it was concluded that the test rig can give a good (relevant) estimate of the sound field in the WA of the real aircraft.

9.6.2 Determination and Modeling of Uncertainties

As outlined in Sect. 6.5.3, robust controller design requires a measure of the uncertainties in the secondary transfer paths as well as a measure of the uncertainties in the primary disturbance. Because ANC in the WA requires RSC, it was necessary to consider disturbances for 4 variables.

These variables, compare Fig. 6.28, are the column matrix of the primary disturbance measured at the error microphones $\mathbf{p}_{pM}$, the column matrix of the primary disturbance measured at the monitor microphones $\mathbf{p}_{pC}$, the matrix $\mathbf{H}_C$ that contains the transfer behavior between the error microphones and the canceling sources, and the matrix $\mathbf{H}_M$ that contains the transfer behavior between the monitor microphones and the canceling sources. The uncertainties are modeled as (small) variations around the nominal values, compare Sect. 6.5.3, such as

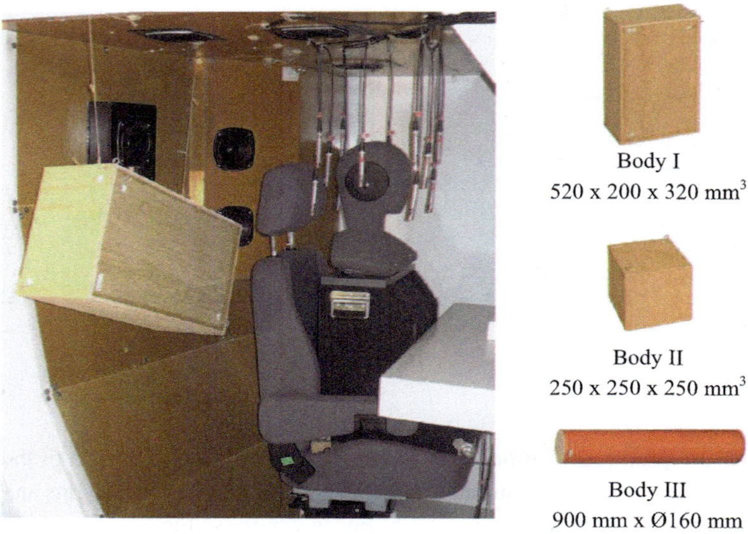

Fig. 9.31 *Left*: WA with diffraction body. *Right*: shape and dimensions of diffraction bodies used to simulate uncertainties in the acoustic transfer paths. Picture is taken from (Kochan et al. 2009a)

$$\begin{aligned}\mathbf{p}_{pC} &= \mathbf{p}_{pC0} + \Delta\mathbf{p}_{pC}\\ \mathbf{p}_{pM} &= \mathbf{p}_{pM0} + \Delta\mathbf{p}_{pM}\\ \mathbf{H}_{C} &= \mathbf{H}_{C0} + \Delta\mathbf{H}_{C}\\ \mathbf{H}_{M} &= \mathbf{H}_{M0} + \Delta\mathbf{H}_{M}.\end{aligned} \tag{9.29}$$

Measurement of Uncertainties

To determine the uncertainties caused by aircraft crew and equipment inside the cabin, the acoustic field in the WA mock-up was altered by diffraction bodies, shown in Fig. 9.31 (right), using three different sound hard diffracting objects suspended from the ceiling of the workstation, as shown in Fig. 9.31 (left).

Two flight conditions (low-noise cruise and cruise) were considered in order to measure the uncertainties for the BPF and the first two higher harmonics. First, the transfer functions and the disturbance noise fields were determined for the nominal state. Then measurements were performed in the disturbed sound fields.

The experimental results are shown in Table 9.1 which contains the uncertainty norms of disturbance noise fields and transfer functions for both flight conditions. For more details about these investigations see (Kochan et al. 2009a), (Weizel 2009) and (Kochan 2009).

Table 9.1 Uncertainty norms of disturbance noise fields and transfer functions according to (Kochan et al. 2009b). For each flight condition the uncertainty norm is given for three frequencies such as $BPF/2 \times BPF/3 \times BPF$

Flight condition:	Low-noise cruise	Cruise
$\frac{\|\Delta \mathbf{H}_C\|_F}{\|\mathbf{H}_{C0}\|_2} \cdot 100$	3.6/12.6/13.9	3.2/11.7/12.2
$\frac{\|\Delta \mathbf{p}_{pC}\|_2}{\|\mathbf{p}_{pC0}\|_2} \cdot 100$	4.7/10.5/14.6	3.7/8.6/8.8
$\frac{\|\Delta \mathbf{H}_M\|_2}{\|\mathbf{H}_{M0}\|_2} \cdot 100$	3.4/4.9/12.1	3.4/5.9/8.6
$\frac{\|\Delta \mathbf{p}_{pM}\|_2}{\|\mathbf{p}_{pM0}\|_2} \cdot 100$	4.8/6.8/10.7	5.0/13.1/9.3

Modeling of Uncertainties

Modeling of the uncertainty matrices $\Delta \mathbf{H}_C$ and $\Delta \mathbf{H}_M$ was based on the unitary matrices $\mathbf{U}_C$ and $\mathbf{V}_C$ as well as $\mathbf{U}_M$ and $\mathbf{V}_M$ that were obtained from the singular value decomposition of the nominal transfer matrices such as

$$\mathbf{H}_{C0} = \mathbf{U}_C \mathbf{\Sigma}_C \mathbf{V}_C^H \quad \text{and} \quad \mathbf{H}_{M0} = \mathbf{U}_M \mathbf{\Sigma}_M \mathbf{V}_M^H. \tag{9.30}$$

As shown in (Elliott 2001) it is possible to decompose the matrix of perturbations in the secondary transfer paths, so that

$$\Delta \mathbf{H}_C = \mathbf{U}_C \mathbf{\Delta\Sigma}_C \mathbf{V}_C^H \quad \text{and} \quad \Delta \mathbf{H}_M = \mathbf{U}_M \mathbf{\Delta\Sigma}_M \mathbf{V}_M^H, \tag{9.31}$$

where $\mathbf{\Delta\Sigma}_C$ and $\mathbf{\Delta\Sigma}_M$ are the singular value perturbation matrices. The approach presented in (Kochan et al. 2009a) and (Kochan 2009) suggests that these matrices can be multiplicatively decomposed into uncertainty norms and structural matrices such as

$$\mathbf{\Delta\Sigma}_C = \|\mathbf{\Delta\Sigma}_C\|_2 \mathbf{N}_C \quad \text{and} \quad \mathbf{\Delta\Sigma}_M = \|\mathbf{\Delta\Sigma}_M\|_2 \mathbf{N}_M. \tag{9.32}$$

The uncertainty norms $\|\mathbf{\Delta\Sigma}_C\|_2$ and $\|\mathbf{\Delta\Sigma}_M\|_2$ were described by the median of the measured norms such as

$$\|\mathbf{\Delta\Sigma}_C\|_2 = \|\mathbf{\Delta}\tilde{\mathbf{H}}_C\|_2 \quad \text{and} \quad \|\mathbf{\Delta\Sigma}_M\|_2 = \|\mathbf{\Delta}\tilde{\mathbf{H}}_M\|_2. \tag{9.33}$$

The median values at the right hand side of (9.33) were determined so that 50% of the measured uncertainties were above and 50% of the measured uncertainty were below these quantities. The structure of the uncertainties were represented by normalized uncertainty structure matrices

$$\begin{aligned} \mathbf{N}_C &= \frac{\hat{\mathbf{N}}_C}{\|\hat{\mathbf{N}}_C\|_2} \quad \text{with } \hat{\mathbf{N}}_C = [\hat{N}_{lcm}] \\ \mathbf{N}_M &= \frac{\hat{\mathbf{N}}_M}{\|\hat{\mathbf{N}}_M\|_2} \quad \text{with } \hat{N}_M = [\hat{N}_{lMm}] \end{aligned} \tag{9.34}$$

that were, according to (Kochan et al. 2009a) and (Kochan 2009), based on the structural models such as

$$\hat{N}_{lCm} = \sqrt{\frac{1}{l_C^2 + m^2}} \quad \text{for } l = 1, \dots, L_C \text{ and } m = 1, \dots, M \tag{9.35}$$

and

$$\hat{N}_{lMm} = \sqrt{\frac{1}{l_M^2 + m^2}} \quad \text{for } l = 1, \dots, L_M \text{ and } m = 1, \dots, M. \tag{9.36}$$

Because no significant structure was found by analyzing the disturbed primary noise fields, the uncertainties in these signals were modeled such as

$$\Delta \mathbf{p}_{pC} = \|\Delta \tilde{\mathbf{p}}_{pC}\|_2 \mathbf{n}_C \quad \text{and} \quad \Delta \mathbf{p}_{pM} = \|\Delta \tilde{\mathbf{p}}_{pM}\|_2 \mathbf{n}_M, \tag{9.37}$$

where $\|\Delta \tilde{\mathbf{p}}_{pC}\|_2$ and $\|\Delta \tilde{\mathbf{p}}_{pM}\|_2$ are the mean values of measured uncertainties. The normalized structural column matrices $\hat{\mathbf{n}}_C$ and $\hat{\mathbf{n}}_M$ are specified by

$$\mathbf{n}_C = \frac{\hat{\mathbf{n}}_C}{\|\hat{\mathbf{n}}_C\|_2} \quad \text{with } \hat{\mathbf{n}}_C = [n_{l_C}] \quad \text{and} \quad \mathbf{n}_M = \frac{\hat{\mathbf{n}}_M}{\|\hat{\mathbf{n}}_M\|_2} \quad \text{with } \hat{\mathbf{n}}_M = [n_{l_M}]. \tag{9.38}$$

$\hat{\mathbf{n}}_C$ is a $(L_C \times 1)$ is column matrix that contains 16 elements equal to 1 and $\hat{\mathbf{n}}_M$ is a $(L_M \times 1)$ column matrix that contains 12 elements equal to 1.

9.6.3 Review of the MA WA Robust Control Approach

Comments on the Frequency Domain Controller Implementation

Because it is possible to compute the reference signal from the engine speed, a feed-forward frequency domain implementation of the steepest-decent algorithm, compare Sect. 6.4.2 was chosen to perform the adaptive control task. This implementation

- Enables frequency selective control that was found to be much more effective than non-selective control (as mentioned in Sect. 9.4.2),
- Results a low computational complexity,
- Allows for simple parameter adjustment.

As reported in (Kochan et al. 2009b) the convergence performance was a minor criterion, because the frequency shifts of the propeller engines will be slow in comparison to the convergence speed of the adaptive algorithm.

Because of the requirements, the information processing must be able to handle the BPF and its two first higher harmonics. The principle scheme of the controller is shown in Fig. 9.32. It can be seen that the continuous-time reference signal $x(t)$ and the continuous-time error signals $\mathbf{p}_M(t)$ (measured at the 16 error sensors) are feed to the controller. Afterwards, the error signals are transformed to the frequency

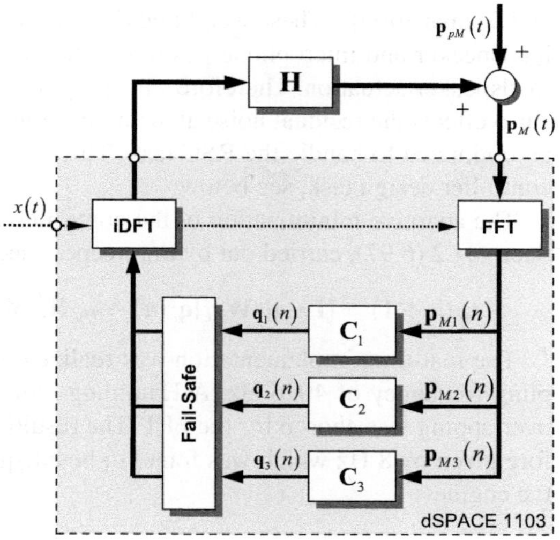

Fig. 9.32 Implementation of frequency domain controller acting on three tonal components, see (Kochan et al. 2009b)

domain using the Fast-Fourier-Transformation[1] (FFT) in which the reference signal is used to discard the idle bins of the complex frequency spectrum.

Three independent controller $\mathbf{C}_1$, $\mathbf{C}_2$, and $\mathbf{C}_3$ are used to calculate three control signals. Each of them is used to reduce one BPF tone. They are passed through the "Fail Safe" block that disables the controller in case of handling errors or instabilities. For this purpose it is verified that the magnitude of the control signals is less that the magnitude of the maximum admissible source strength.

This is a (very) simple but effective implementation of a limiter, compare Sect. 9.4.2. Finally, the resulting complex actuation signals are transformed in the time domain by an inverse FFT.

At each discrete time n the complex error signals according to the BPF ($i = 1$), its first ($i = 2$) and second ($i = 3$) harmonic are grouped in (16×1) column matrices such as

$$\mathbf{p}_{Ci}(n) = \left[P_{Ci}^{1}(n), \ldots, P_{Ci}^{2}(n), \ldots, P_{Ci}^{16}(n)\right]^T \quad \text{with } i = 1, 2, 3, \qquad (9.39)$$

where $P_{Ci}^{l_C}(n)$ is the individual complex error signal according to the i-th frequency measured at the l_C-th error sensor at discrete time n. The three (8×1) column matrices of the complex loudspeaker actuation signals $\mathbf{q}_i$ are calculated in each controller $\mathbf{C}_i$ by minimizing parametric cost functions such as

$$J_{Ci} = \mathbf{p}_{Ci}^H \mathbf{W}_{pi} \mathbf{p}_{Ci} + \mathbf{q}_i^H \mathbf{W}_{qi} \mathbf{q}_i \quad \text{with } i = 1, 2, 3, \qquad (9.40)$$

where the controller weighting matrices $\mathbf{W}_{pi}$ and $\mathbf{W}_{qi}$ are real valued diagonal matrices. They are used to weight the error microphones signals and the loudspeaker actuation signals independently for each excitation frequency. As stated in (Kochan

[1] Details about the algorithmic implementation of the FFT can be found in textbooks such as (Press et al. 1994).

and Sachau 2009): "These weightings have a similar mathematical effect as different loudspeaker and microphone positions and change the phase and amplitude of the loudspeaker actuation. Therefore, changing the weighting matrices," $\mathbf{W}_{pi}$ and $\mathbf{W}_{qi}$, "also effects the residual noise at the monitor microphones." Their elements have to be optimized to handle the RSC task. This optimization was performed within the controller design task, see below.

The adaptive minimization of the controller cost functions (9.40) was, compare Sect. 6.4.2 (6.97), carried out by the steepest-decent algorithm

$$\mathbf{q}_i(n+1) = [\mathbf{I} - \mu_i \mathbf{W}_{qi}]\mathbf{q}_i(n) - \mu_i \mathbf{H}_i^H \mathbf{W}_{pi} \mathbf{p}_{Ci}(n) \quad \text{with } i = 1, 2, 3. \tag{9.41}$$

The real-time implementation was realized on a dSpace hardware using a sampling frequency of 4096 Hz. A Hamming-window with 2048 coefficients and 3/4 overlapping was chosen for the FFT. The resulting controller update rate was therefore given by 8 Hz which was found to be adequate for the slow frequency shift of the engines.

Robust Controller Design

The goal of the control task is to enable robust RSC around the operator head. For this purpose it was necessary to determine the controller weighting matrices in such a way that the lowest residual error at the monitor microphones is achieved. For each frequency i, one set of these weighting matrices was defined. Therefore, the (same) controller design approach was performed for each excitation frequency. Please notice that we will omit the subscript i that was introduced to indicate a single frequencies in the following for an easy notation.

The controlled sound pressure at the 16 error sensors and the positions of the 12 monitor sensors is according to Sect. 6.5.2 specified by (6.110)–(6.115). The uncertainties in the disturbance signals and the uncertainties in secondary transfer paths are described by (9.29). This simplified approach is an approximation for the dynamic behavior of real time-variant systems with weak nonlinearities in which it is assumed that the real physical plant varies around the nominal plant. Furthermore, it is assumed that the uncertainties are arbitrary and can be modeled by their norm, compare (9.30)–(9.38).

This robust RSC problem was formulated as an optimization problem that combines *robust control stability*, see (6.164), and *robust control quality*, see (6.160). The solution of the optimization problem must guarantee the stability and has to achieve the best performance for all norm-bounded uncertain plants. For this reason the objective function and the constraints were evaluated for the worst-case.

As shown in (Kochan et al. 2009b) and outlined in Sect. 6.5.3, compare (6.171)–(6.179), this worst-case approach requires estimates of the worst-case residual noise at the monitor microphones, the magnitude of the maximum control signal, and (to evaluate the robust stability constraint) the eigenvalues of the controller matrix $\mathbf{C}_C = -[\mathbf{H}_C^H \mathbf{W}_p \mathbf{H}_C + \mathbf{W}_q]^{-1} \mathbf{H}_C^H \mathbf{W}_p$ that determines the optimum of (9.40) for one excitation frequency.

According to (6.177), considering the application of RSC, the estimate of the optimal control signal matrix in the worst-case scenario is given by

$$\mathbf{q}_{opt} = \mathbf{q}_{opt,0} + \Delta\mathbf{q}_{opt}, \tag{9.42}$$

with

$$\mathbf{q}_{opt,0} := \mathbf{C}_{C0}\mathbf{p}_{pC0} \quad \text{and} \quad \Delta\mathbf{q}_{opt} := \mathbf{C}_{C0}\Delta\mathbf{p}_{pC} + \mathbf{C}_{C0}\Delta\mathbf{H}_C\mathbf{C}_{C0}\mathbf{p}_{pC0}, \tag{9.43}$$

where

$$\mathbf{C}_{C0} = -\left(\hat{\mathbf{H}}_{C0}^H\mathbf{W}_p\mathbf{H}_{C0} + \mathbf{W}_q\right)^{-1}\hat{\mathbf{H}}_{C0}^H\mathbf{W}_p \tag{9.44}$$

is the matrix of the optimal controller transfer functions in the nominal state. Equations (9.42)–(9.44) enable an approximation of $Q_{opt}^{\max}$ as well as an approximation of the sound pressure at the position of the monitor microphones. The latter is in the worst-case scenario determined by

$$\mathbf{p}_M = \mathbf{p}_{pM0} + \mathbf{H}_M\mathbf{q}_{opt}. \tag{9.45}$$

Based on (9.45) it is, considering the application of RSC, possible to specify the upper bound for the squared sound pressure at the monitor microphones that is according to (6.179) given by

$$\max\left[\mathbf{p}_M^H\mathbf{p}_M\right] \leqslant \left\{\begin{array}{l} \quad\|\mathbf{H}_{M0}\mathbf{q}_{opt,0} + \mathbf{p}_{pM0}\|_2 + \cdots \\ \max\|\Delta\mathbf{H}_M\mathbf{q}_{opt,0}\|_2 + \cdots \\ \max\|\mathbf{H}_{M0}\mathbf{C}_{C0}\Delta\mathbf{p}_{pC}\|_2 + \cdots \\ \max\|\mathbf{H}_{M0}\mathbf{C}_{C0}\Delta\mathbf{H}_C\mathbf{q}_{opt,0}\|_2 + \cdots \\ \max\|\Delta\mathbf{p}_{pM}\|_2 \end{array}\right\}^2 =: p_M^2. \tag{9.46}$$

The uncertainties models that have been specified by (9.30)–(9.38) were applied to evaluate (9.46). This analysis requires the introduction of an upper limit for the *Euclidean* norm of the matrix product $\mathbf{Ab}$, where $\mathbf{A}$ is a $(M \times N)$ matrix with MN elements A_{mn} and $\mathbf{b}$ is a $(N \times 1)$ column matrix that contains N elements b_n. As shown in (Kochan 2009) such a limit can be defined by

$$\|\mathbf{Ab}\|_2^2 = \sum_{m=1}^{M}\left(\sum_{n=1}^{N} A_{mn}b_n\right)^2 \leqslant \sum_{m=1}^{M}\left(\sum_{n=1}^{N} |A_{mn}||b_n|\right)^2 =: \left\||\mathbf{A}|\cdot|\mathbf{b}|\right\|_2^2, \tag{9.47}$$

where the absolute value bars $|.|$ stand for the usage of the absolute values of the individual matrix elements in the calculation. Considering (9.47) it is possible to specify the upper limit of the worst-case approximation (9.46) such as

$$\begin{aligned} p_M^2 &\leqslant \left\{\begin{array}{l} \|\mathbf{H}_0\mathbf{q}_{opt} + \mathbf{p}_{p0}\|_2 + \cdots \\ \|\Delta\tilde{\mathbf{H}}_M\|_2 \cdot \left\|\left|\mathbf{U}_M\mathbf{N}_M\mathbf{V}_M^H\right| \cdot |\mathbf{q}_{opt,0}|\right\|_2 + \cdots \\ \|\Delta\tilde{\mathbf{p}}_{pC}\|_2 \cdot \left\||\mathbf{H}_{M0}\mathbf{C}_{C0}| \cdot |\mathbf{n}_C|\right\|_2 + \cdots \\ \|\Delta\tilde{\mathbf{H}}_C\|_2 \cdot \left\|\left|\mathbf{H}_{M0}\mathbf{C}_{C0}\mathbf{U}_C\mathbf{N}_C\mathbf{V}_C^H\right| \cdot |\mathbf{q}_{opt,0}|\right\|_2 + \cdots \\ \|\Delta\tilde{\mathbf{p}}_{pM}\|_2 \end{array}\right\}^2 \\ &=: \max\left[p_M^2\right]. \end{aligned} \tag{9.48}$$

Minimizing the worst-case approximation leads to the worst-case solution of the robust optimization problem. The latter can in direct analogy to the robust control optimization problem (specified by (6.169) and (6.170)) be formulated such as

$$\underset{\mathbf{W}_p, \mathbf{W}_q}{\text{minimize}} \max\left[p_M^2(\mathbf{q}_{opt})\right] \tag{9.49}$$

by considering the constraints

$$\begin{aligned} C_1&: \quad Q_{adm}^{\max} - Q_{opt}^{\max} > 0 \\ C_2&: \quad \min\left\{\text{Re}\left[\hat{\lambda}_m(\mathbf{W}_p, \mathbf{W}_q)\right]\right\} > 0 \\ C_3&: \quad \lambda_l(\mathbf{W}_p) > 0 \quad \text{for all } l \\ C_4&: \quad \lambda_m(\mathbf{W}_q) > 0 \quad \text{for all } m. \end{aligned} \tag{9.50}$$

C_1 avoids nonlinearities and C_2 ensures that the robust stability criteria (6.167) can be used to specify a step size for which (9.41) is a stable iterative mapping.

C_3 and C_4 are again taken into account to guarantee that the controller design parameter matrices belong to the group of positive definite matrices, necessary to obtain a convex cost function. Please notice that according to (6.165) and neglecting plant modeling errors ($\Delta\hat{\mathbf{H}}_C = \mathbf{0}$), the eigenvalues $\hat{\lambda}_m$ have to be computed from the matrix

$$\hat{\mathbf{A}} \approx \left(\mathbf{H}_{C0}^H \mathbf{W}_p \mathbf{H}_{C0} + \mathbf{W}_q\right) + \mathbf{H}_{C0} \mathbf{W}_p \Delta\mathbf{H}_C, \tag{9.51}$$

where the uncertainty is given by $\Delta\mathbf{H}_C = \mathbf{H}_{C0}\mathbf{W}_p\mathbf{U}_C\|\Delta\tilde{\mathbf{H}}_C\|_2\mathbf{N}_C\mathbf{V}_C^H$. The set of controller weighting matrices $\mathbf{W}_p$ and $\mathbf{W}_q$ were optimized according to the optimization problem described above. For this purpose Matlab with the Genetic Algorithm and Direct Search Toolbox was applied. The nonlinear constraints were included by using the barrier function method, and the uncertainties listed in Table 9.1 were used to specify the deviation from the nominal state. The loudspeaker actuation was limited by $Q_{adm}^{\max} = 2$ V. More details about the implementation and solution of the optimization problem are reported in (Kochan 2009).

As also reported in (Kochan et al. 2009b) and (in more detail) (Kochan 2009), two additional controller designs were used to assess the performance of the robust controller.

In the first design, all uncertainties were neglected. Hence, the transfer functions and disturbance noise were equal to the nominal transfer functions and nominal disturbance noise fields. This leads to the nominal controller design problem, compare (6.153) and (6.154), that can easily deducted from (9.49) and (9.50) by setting all uncertainties to zero.

The second additional controller design in which the weighting matrices were set to $\mathbf{W}_p \equiv (1-\alpha)\mathbf{I}$ and $\mathbf{W}_q \equiv \alpha\mathbf{I}$ was chosen as a reference for the robust design. This simple weighting approach in which all microphones and loudspeakers were equally weighted was used as the reference design. The real valued weighting parameter α was optimized according to the maximum noise reduction in the monitor volume using the *Pareto-curve* approach outlined in Sect. 9.4.3. However, in contrast to *Gerners* optimization approach, additional constraints such as stability and loudspeaker actuation were not taken into account for the reference design.

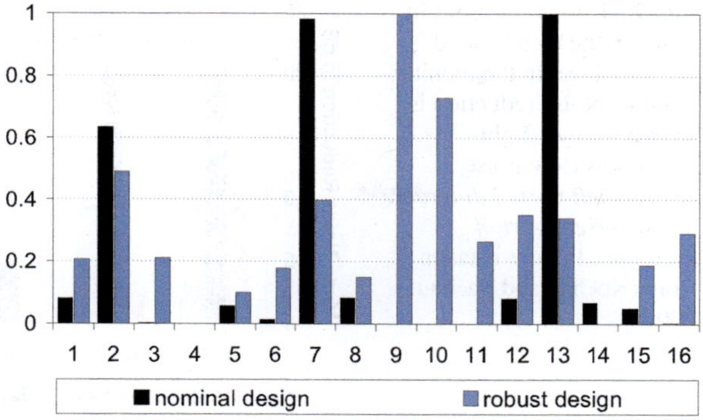

Fig. 9.33 Diagonal elements of the weighting matrix $\mathbf{W}_p$ optimized for the blade pass frequency of load case low-noise cruise. Picture is taken from (Kochan and Sachau 2009)

Some results of the optimization processes are shown in Fig. 9.33 that contains the elements of the weighting matrix $\mathbf{W}_p$ (optimized for the BPF of the load case low-noise cruise) as an illustrative example for the optimized weighting matrices.

This figure clarifies that the nominal design method results in a controller which only uses 8 of the 16 error microphones. The robust design method leads to a controller which uses 14 of 16 error microphones. In comparison to an optimal placement that results from a TPA, the loudspeakers and microphones are weighted differently, not only by setting an activated element to one and a deactivated element to zero. However, it can be seen that the trace of $\mathbf{W}_p$ is dominated by three single elements in the case of nominal design.

The presented results also confirm the outcome of the *weight cost efficiency* study, see Sect. 9.4, which make clear that an optimized system can work successfully with a small number of error sensors.

Such a domination was not found in the case of robust design in which the differences between single weights are smaller. Kochan, see (Kochan and Sachau 2009), concluded that: "In general, the robust design includes more microphones in the controller cost function than the nominal design. This general observation also applies for the weighting matrix where the loudspeaker excitation was more uniformly in the robust design method." For more details about the optimization results, the reader is referred to (Kochan 2009).

Experimental Validation of Controller Performance

The controller performance was validated in the experimental set up shown in Fig. 9.29. Two load cases (low-noise cruise and cruise) were considered in these experiments.

In a first step, the experiments were performed for each frequency individually. The primary noise field was set to at least 75 dB(A) at the ear microphones of the artificial head. The steady state noise reduction at the monitor microphones was according to (6.168) evaluated as the logarithm of the ratio between the mean SPL in the controlled state and the mean SPL in the uncontrolled state.

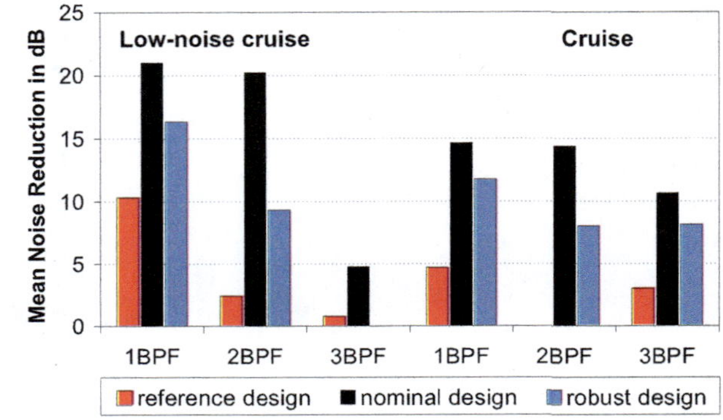

Fig. 9.34 Control profit in terms of the mean sound pressure level in the monitor volume. Noise reduction is shown for two flight conditions (low-noise cruise—*left part of diagram*, cruise—*right part of diagram*). Picture is taken from (Kochan and Sachau 2009)

Some experimental results are shown in Fig. 9.34. In comparison to the simple weighting approach in which the weighting parameter α was chosen according to the maximum noise reduction in the monitor volume, the nominal as well as the robust control approach improved the amount of active noise reduction at the BPF significantly. It can be seen that a reduction of more than 20 dB was achieved at this frequency for both load cases.

As also shown in Fig. 9.34, no noise reduction was achieved at the second harmonic for load case low-noise cruise, if robust control was applied. At all other frequencies, a significant noise reduction was measured.

The weighting of the nominal design method achieved a better noise reduction than the robust design method. However, with uncertainties, the robust weighting will achieve a more reliable noise reduction.

A graphical representation of the mapped sound fields around operator head is shown in Fig. 9.35. The primary noise is shown top, the remaining noise in case of robust control is shown below. The fundamental frequency of the primary noise field was in this experiment set to 92 Hz. In addition, two higher harmonics were excited at 184 Hz and 276 Hz. The sound pressure level at the artificial head was again adjusted to 75 dB(A) for each frequency.

This figure illustrates that an almost uniform noise reduction up to 20 dB can be achieved for 92 Hz and 184 Hz. More experimental results were published in (Kochan 2009).

The outcome of the robust design step was an ANS incorporating original manufacturer equipment that was successful tested in an acoustic mock-up used to simulate relevant conditions. Because it was proven that this system is capable of meeting the requirements, SML 6 was achieved at the end of the *robustness study*.

9.7 Design Process Summary

Noise control in the WA of a high-speed turboprop transport aircraft is an extraordinary challenge, because of the acoustic coupling between the (small) semi-closed

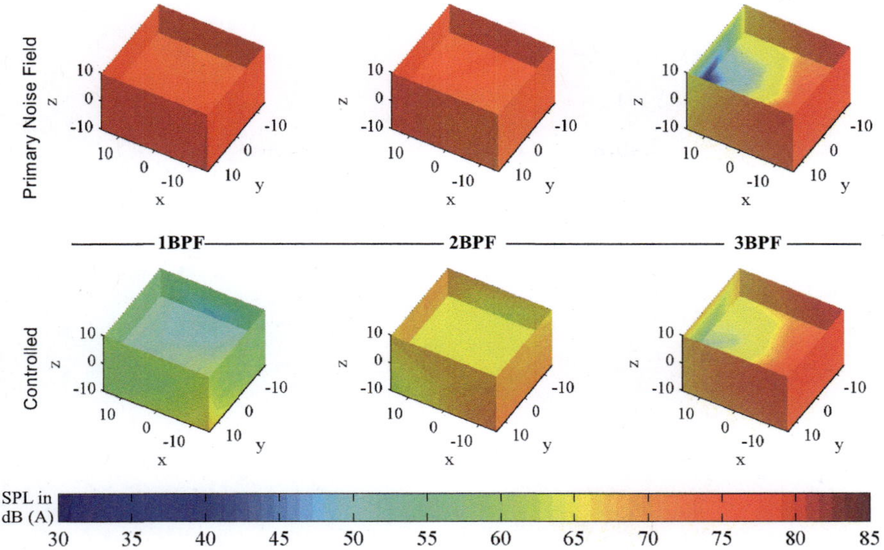

Fig. 9.35 Map of sound field around operator head for the controlled and uncontrolled case, if robust control is applied (adapted from Breitbach and Sachau 2009)

WA and the (large) cargo hold. In consequence of weight issues and the demanding noise requirements the development of possible solutions has been focused on active noise control.

A summary of five successive design steps—beginning with the *feasibility study* and ending with the *robustness study*—that were performed to enable robust control of tonal noise (associated to the BPF and the first two higher harmonics) in the MA WA has been presented in this chapter.

It has been shown that the design process led to a local feed-forward ANS with 16 error microphones and 8 canceling sources that is capable of reducing the sound pressure around the operator head down to the specified limits considering a limitation of the actuator driving signals.

Because it is required that all actuators and sensors have to be located away from the operator seat, it was necessary to design an ANS that is capable of handling RSC. For this reason, the suggested approach is based on optimized transducer locations as well as on a parametric controller software. The latter uses two controller weighting matrices to balance the acoustic potential energy measured at the error sensors and the control effort required to drive 8 canceling sources. The elements of these matrices were optimized to meet both the robust performance criteria and the robust stability requirements of ANS considering a worst-case performance scenario caused by uncertainties in the disturbances and the secondary transfer paths. The performance of the ANS incorporating original manufacturer equipment was successfully tested in a relevant mock-up representing the essential acoustic properties of the operational environment. For this reason, SML 6 was achieved at the end of the *robustness study*.

Chapter 10
A Sound Intensity Probe with Active Free Field

Abstract To avoid the time and cost consuming process of realizing artificial free field conditions by introducing passive damping to the interior (which in addition changes the global characteristic of the investigated enclosure) or the need of sophisticated numerical models required to apply inverse noise source localization techniques such as the IFEM, compare Sect. 7.2 and Chap. 13, a novel sound intensity probe with an active free field (SIAF) was proposed in (Sachau et al. in Schallintensitätsdetektor sowie Verfahren zum Messen der Schallintensität. Patent. DE102004009644A1 22.09.2005, 2005a). This chapter reports on the application of the design methodology for ANC-systems proposed in Chap. 8 to the SIAF design process in which a first specification of the novel sound intensity probe was compiled in two successive design steps. The chapter also includes comments on the functional testing of the first SIAF realization. These experiments were especially focused on free field calibration considering the sound intensity measurements errors discussed in Sect. 4.2.2.

10.1 Feasibility Study on an Active Intensity Probe

The input to the SIAF design process is given by the following **requirements**: *The SIAF shall be capable of identifying acoustic hot spots at the interior wall by measuring the local sound intensity in front of vibrating panels.*

In order to verify, if these requirements can be fulfilled with an ANC-system, the design process started according to Sect. 8.3 with the *feasibility study*. Because it is intended to use a SIAF in various interior noise fields that can be excited by harmonic as well as by stochastic disturbances, the analysis of the primary noise field that is usually the first design task was omitted. For this reason the first design step included the analysis of the controlled sound field using the FEM tool, the formulation of a system concept as well as the proof-of-concept.

10.1.1 Analysis of the Controlled Sound Field

Preliminary Considerations

In order to measure the local sound intensity in front of vibrating panels it is necessary to channelize the sound radiation, e.g. by using a tube with a sound hard lateral

T. Kletschkowski, *Adaptive Feed-Forward Control of Low Frequency Interior Noise*, Intelligent Systems, Control and Automation: Science and Engineering 56, DOI 10.1007/978-94-007-2537-9_10, © Springer Science+Business Media B.V. 2012

area. One end of such a tube must be open to enable sound radiation into the tube. The second termination must be a boundary with absorbing properties to enable wave propagation inside the tube.

As shown in Sect. 3.4.2, see (3.58), the wave field in such an one-dimensional enclosure can be decomposed into a standing wave and into a traveling wave. The time-averaged sound intensity is determined by the acoustic absorption, compare (3.59), that enables the energy transport. Unfortunately, the traveling wave and consequently the time-averaged sound intensity vanish in a sound field without absorption ($\alpha = 0$) that only consists of a standing wave, compare also (Crocker and Arenas 2003). For this reason it is essential to tune the impedance of the second termination by an active controller—in best case to that of a fully absorbing boundary ($\alpha = 1$). This best case results in free field conditions inside the probe.

A conventional sound tube (without free field) that is placed directly in front of the interior wall would change the impedance of investigated surface. The resulting impedance could be determined by standard methods, see (EN ISO 10534-2 2001), but this value would vary from point to point.

In contrast to this unwanted situation, a SIAF enables both the energy transport from the source into the mechanical device as well as its quantification under comparable boundary conditions inside the mechanical device. Because of the free field inside the probe, the SIAF acts as a local sound absorber that enables the measurement of the local sound intensity that would be radiated from the vibrating panel into a free field. A SIAF can therefore be applied for sound source localization, especially in weakly damped interior noise fields.

Numerical Simulations

To study the qualitative behavior of the future ANC-system in an operational environment, numerical simulation were performed in frequency domain using the FEM, compare Sect. 7.1.2. As shown in Fig. 10.1 (top) the calculation domain was given by a circle having a diameter of 1.6 m. The outer boundary of this domain was modeled as a sound hard boundary, whereas the inner boundary was modeled as a velocity boundary at which a uniform normal velocity was prescribed.

The simple cavity model of the SIAF is also shown in Fig. 10.1. It represents a tube with sound hard lateral area. Its open end is modeled by an internal boundary (no change of impedance is seen by the fluid), whereas the closed end is modeled as a sound hard boundary (inactive SIAF) or as an absorbing boundary with free field impedance $Z = 428.75\ \mathrm{kg\,m^{-2}\,s^{-1}}$ (active SIAF).

The calculation domain was discretized using more than 6000 triangular elements with quadratic shape functions resulting in more than 12000 degrees of freedom. All time-harmonic simulations were performed using an excitation frequency of $f = 300$ Hz.

The simulation results are shown in Fig. 10.1 in which the (qualitative) sound pressure distribution is implied using the same scale for each charts, whereas the intensity flow is represented by black arrows. The results shown in the top chart of

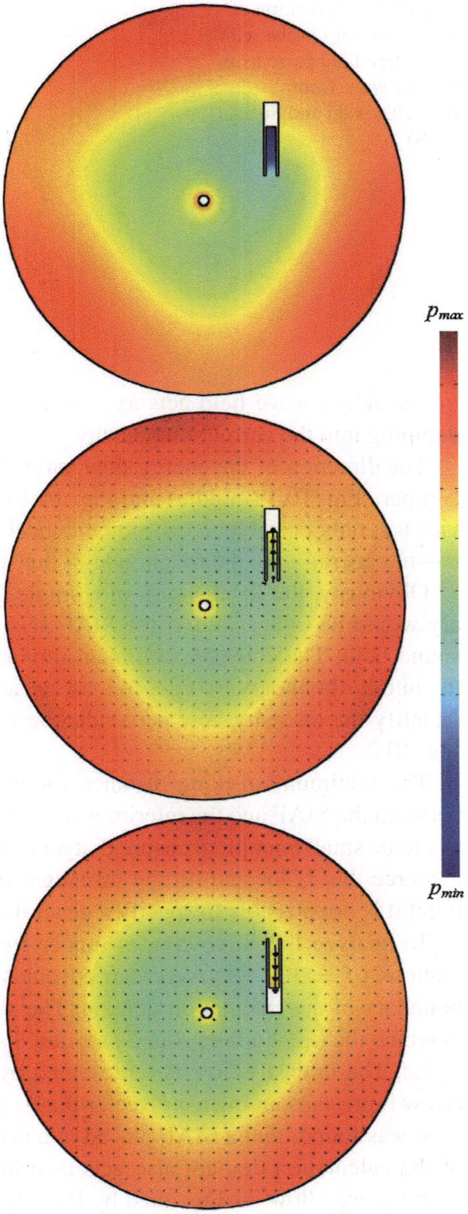

Fig. 10.1 Standing wave field with inactive SIAF (*top*) and intensity flow into active SIAF (*middle* and *below*). Intensity vector is represented by arrows (length is proportional to magnitude)

this figure prove that no intensity flow can be measured in the purely reactive sound field, if the SIAF is inactive. If the inner termination of the SIAF is described by the free field impedance, an energy flow into the mechanical device is enabled as shown in Fig. 10.1 (middle) and Fig. 10.1 (below). Hence, an activated SIAF placed

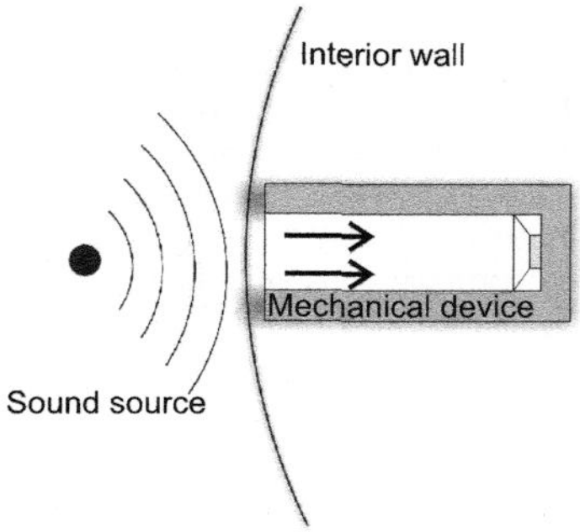

Fig. 10.2 Application of sound intensity probe with active free field close to an interior wall (from Kletschkowski and Sachau 2008)

in a standing wave field acts as a local sound absorber that also introduces some damping into the surrounding cavity.

The direction of the energy flow varies with the spatial orientation of the probe, compare Fig. 10.1 (middle) and Fig. 10.1 (below). Thus, inside the mechanical device it is only possible to measure the local intensity in longitudinal direction which is—in general—not the magnitude of the spatial intensity vector.

Obviously, the localization of a single acoustic hot spot on the interior boundary would be impossible, if the SIAF acts as a local absorber in the middle of the sound field. The converse argument implies that the SIAF has to be used close to the interior boundary in order to localize acoustic sources at the interior wall and to quantify the intensity flow perpendicular to a sound radiating surface, as shown in Fig. 10.2.

The minimum working distance should be chosen in such a way that contact between the SIAF and the interior wall is excluded. The maximum working distance has to be small enough to ensure that an energy flow into the SIAF results only from a source that is located on the investigated part of the boundary. However at this point it is not possible to specify a general rule for the maximum working distance.

To analyze the SIAF-concept under free field conditions, time-harmonic simulations ($f = 300$ Hz) were carried out using free field conditions for the outer boundary of the calculation domain. The simulation results are shown in Fig. 10.3 in which the (qualitative) sound pressure distribution is implied using different gray styles. The intensity flow is represented by black arrows. Please notice that every arrow represents an intensity vector that is normalized to its magnitude.

It was found, that a SIAF placed in a certain distance to the source in the center of the calculation domain also acts as a local absorber that changes the direction of the energy flow as illustrated by Fig. 10.3 (top). If two sources are present, and, if the SIAF is not collocated to one of these sources, the time average of the local intensity in longitudinal probe direction is measured, as shown in Fig. 10.3 (middle). This quantity is determined by the sound field generated by the interfering sound sources. If the SIAF is placed between two sources with equal strength, as illustrated by Fig. 10.3 (below), the unidirectional intensity component of the source facing the open end of the sound tube is measured.

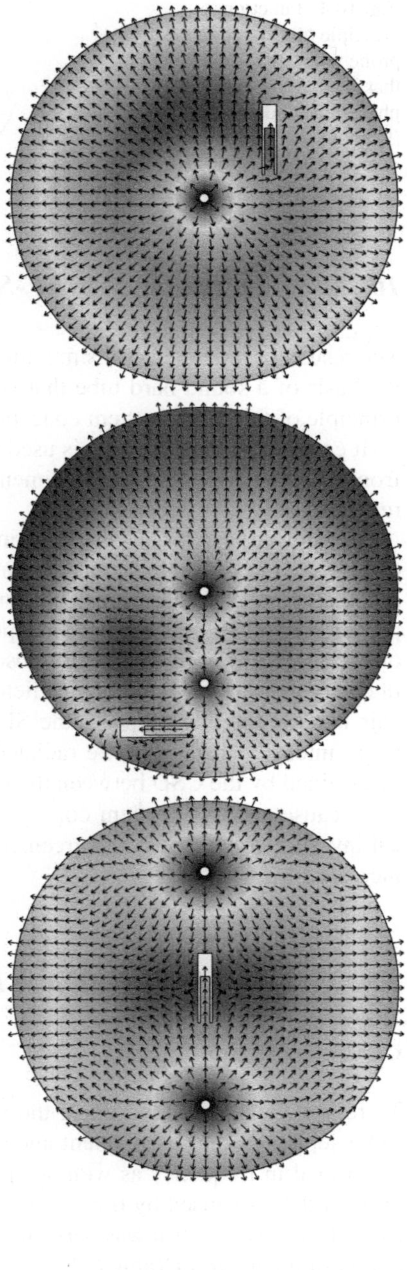

Fig. 10.3 Qualitative plot of pressure distribution and intensity flow in a free field with one (*top*) and two sources (*middle, below*). Normalized intensity vector is represented by *arrows*. Normalization is performed with respect to magnitude of each vector (adapted from Kletschkowski and Sachau 2008)

Because the qualitative behavior of a SIAF was analyzed by time-harmonic finite element simulations considering the application in a standing wave field as well as the application under free field conditions, SML 1 was achieved after processing the first design task.

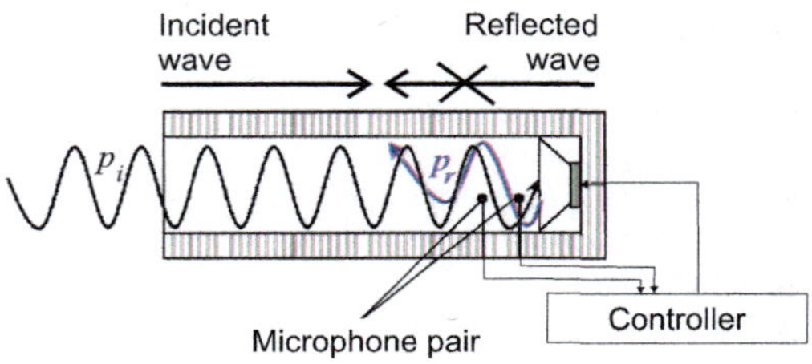

Fig. 10.4 Functional principle of a sound intensity probe with active free field that is based on two phase-matched microphones (from Kletschkowski and Sachau 2008)

10.1.2 Formulation of SIAF-System Concept

The results of the numerical simulations made clear that a SIAF can be designed on the basis of a sound hard tube that is terminated by a loudspeaker. The functional principle of this ANC-system concept is shown in Fig. 10.4.

It can be seen that the tube is used to reduce the three dimensional sound field in front of the interior wall to one dimension. The loudspeaker is needed to cancel the reflected wave component.

For this purpose it is necessary to integrate sensors such as a microphone pair that can be used to decompose the standing wave field in the tube into its incident and reflected part. The latter must be canceled by the active controller that calculates proper driving signals for the loudspeaker. This control strategy is an application of the maximization of acoustic absorption, compare Sect. 6.3. It enables sound intensity measurement under free field conditions inside the tube.[1] If a microphone pair is used for this purpose, the SIAF can be seen as an active p-p probe. The mean intensity that would be radiated from a vibrating surface into a free field is determined by the CSD between the two microphones as shown in Sect. 4.2.1.

Because an ANC-system concept was formulated (on the basis of the numerical investigations) that is in agreement with the initial requirements, the SML was increased from 1 to 2.

10.1.3 Controller Design and Proof of SIAF-System Concept

Controller Design

To realize free field conditions, the sound field inside the mechanical device has to be separated into its incident and reflected components. This separation can be performed in frequency as well as in time domain. However, in order to avoid additional delays caused by time to frequency conversion of signals, wave separation as well as active control was performed in time-domain using the FxLMS algorithm that is summarized in Table 6.2.

[1] This is the main difference between the SIAF approach and active control of duct noise, compare e.g. (Freienstein and Guicking 1996) and (Utsumi 1985), or common sound tubes (with standing wave fields) that are applied to determine material properties, see (EN ISO 10534-2 2001).

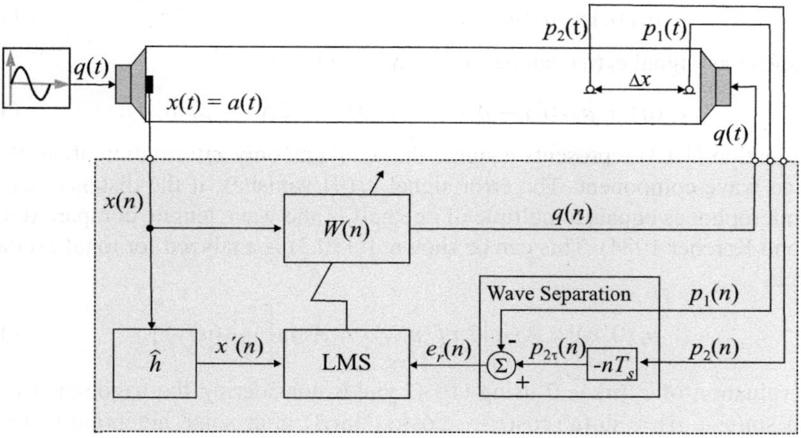

Fig. 10.5 Block diagram of SIAF-control approach (from Kletschkowski and Sachau 2008)

As shown in Fig. 10.5, the control approach is based on one reference signal ($K = 1$), two microphone signals, and one control signal ($M = 1$). The two microphone signals are used to determine one error signal ($L = 1$). For this reason it is possible to use the single-channel version of the FxLMS algorithm for the adaptive control task. The reference signal is obtained from the normal acceleration of the primary noise source that is in the simplest case represented by the membrane of a loudspeaker used to generate the disturbance. Off-line plant modeling based on the common least mean square algorithm, compare Sect. 6.6.2 (6.212), was applied to identify the secondary path model $\hat{h}$.

Wave separation was performed according to (Guicking and Karcher 1984). As shown in this reference, the reflected wave component of a one-dimensional standing wave field can be determined by employing two microphones and a time-delay, if one of the SIAF microphones is placed in front of the canceling loudspeaker at $x = 0$ and the other at a distance Δx. Assuming plane wave propagation, the pressure picked up by the microphones at discrete time n is given by

$$\begin{aligned} p_1(n) &= p_i(0, n) + p_r(0, n) \\ p_2(n) &= p_i(\Delta x, n) + p_r(\Delta x, n) = p_i(0, n + N) + p_r(0, n - N). \end{aligned} \tag{10.1}$$

The continuous time delay $\tau = \Delta x / c$ between the microphones is given by the ratio between the separation distance Δx and the speed of sound c. In order to perform discrete-time adaptive control, this delay has to be approximated by the product of sample time[2] T_s and a number of delays N such as $\tau \approx T_s \cdot N$. If the discrete-time signal $p_2(n)$ is delayed by N time steps, the delayed sound pressure is defined as

---

[2]In order to avoid mistakes, the subscript s is used to indicate the sample time, whereas the periodic time is indicated by T.

$$p_{2\tau}(n) = p_2(n - N) = p_i(0, n) + p_r(0, n - 2N) \tag{10.2}$$

and the error signal $e_r(n)$ can be calculated as follows

$$e_r(n) = p_{2\tau}(n) - p_1(n) = p_r(0, n - 2N) - p_r(0, n). \tag{10.3}$$

Equation (10.3) represents a signal that contains only information about the reflected wave component. The error signal $e_r(n)$ vanishes, if the distance between the microphones equals a multiple of one half of the wave length, compare (Guicking and Karcher 1984). This can be shown, if (10.3) is analyzed for tonal excitation such as

$$p_r(0, n) = A \sin\left[2\pi f(nT_s)\right] = A \sin\left[\frac{2\pi}{T}(nT_s)\right]. \tag{10.4}$$

Evaluation of $e_r(n) = 0$ using (10.4) yields considering the trigonometric formula $\sin(\alpha - \beta) = \sin(\alpha)\cos(\alpha) - \cos(\alpha)\sin(\beta)$ after some algebraic manipulations

$$N = k \cdot \frac{T}{2T_s} = k \cdot \frac{f_s}{2f} \quad \text{with } k = 0, 1, 2, \ldots. \tag{10.5}$$

Equation (10.5) clarifies that the delay number N must not equal a multiple of the (limit) ratio that is defined by *Shannon's* law. If we apply this rule, it is—as also shown in (Kletschkowski and Sachau 2007b)—possible to specify both a lower and an upper time delay limit such as

$$1 \leqslant N < \frac{T_{\min}}{2T_s}, \tag{10.6}$$

where $T_{\min}$ represents the periodic time of the highest frequency of interest $f_{\max}$. If the relation

$$N = \frac{\tau}{T_s} = \frac{\Delta x}{cT_s} \tag{10.7}$$

is used to represent the time delay N, equating of (10.5) and (10.7) yields the "forbidden" separation distance specified in (Guicking and Karcher 1984)

$$\Delta x = k \cdot \frac{\lambda}{2} \quad \text{with } \lambda = \frac{c}{f} \text{ and } k = 0, 1, 2, \ldots. \tag{10.8}$$

Proof-of-Concept

To test the broadband performance of the SIAF approach, real time experiments were performed using the sound tube shown in Fig. 10.7. This tube has a square cross-section (edge length 92 mm) and enables plane wave propagation up to 1864 Hz (compare (EN ISO 10534-2 2001), (2)). It was terminated by electrodynamical loudspeakers (type: SPEAKA MT 60/80) that were used as noise and anti-noise source.

Three microphones (type: Ono Sokki MI-1233) were used for the experiments. According to Fig. 10.5, microphone 1 was placed in front of the canceling loudspeaker. Microphone 2 was placed in a distance of $\Delta x = 13.7$ cm. This separation

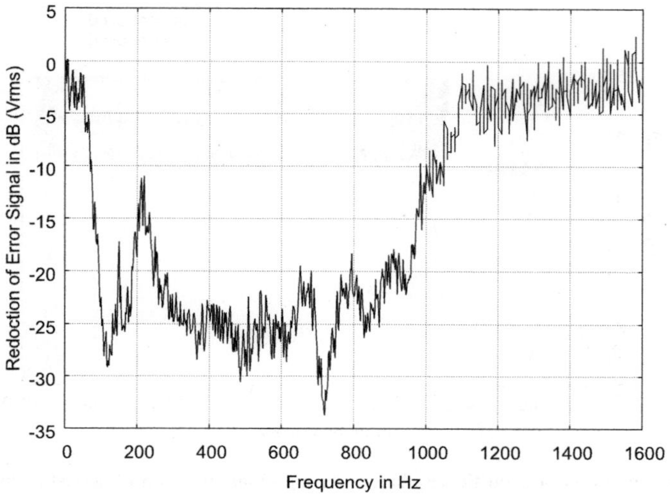

Fig. 10.6 Broadband noise reduction (i.e. reduction of reflected wave component) with the SIAF approach (from Kletschkowski and Sachau 2008)

distance allows the determination of acoustic material properties (such as the absorption coefficient) up to 1152 Hz, compare (EN ISO 10534-2 2001), (4). The third microphone was placed in front of the primary source.

The control algorithm was implemented on a real time processor (type: dSpace DS1103). The sampling frequency was set to $f_s = 5000$ Hz. Considering a speed of sound of $c = 340\ \mathrm{m\,s^{-1}}$ as well as the specified values for f_s and Δx, evaluation of (10.7) yields 2.01. The time delay number was therefore set to $N = 2$. This choice was in agreement with (10.6), because the highest frequency of interest is given by $f_{\max} = 1$ kHz.

An accelerometer (type: B&K 4374) that was amplified by a signal conditioner (type: B&K Nexus 2692) was used to detect the acceleration at the membrane of the primary noise source. This signal was used as reference signal.

Furthermore, several analogue high- and low-pass filters respectively (type: Kemo VBF21) were applied to avoid aliasing and imagine effects. A multi-channel FFT-analyzer (type: Ono Sokki DS2100) was used for data analysis.

The reduction of the reflected wave components was analyzed for broadband disturbances. The system was excited with white noise (50 Hz $\leqslant f \leqslant$ 1 kHz). Therefore, a filter length of $I = 128$ was used for the adaptive controller as well as for the secondary path model.

Experimental results are shown in Fig. 10.6 and Fig. 10.7. The first figure in which the NR, calculated as $NR_{e_r} = 20\log_{10}(e_{r,RMS}^{controlled}/e_{r,RMS}^{uncontrolled})$, is plotted versus frequency clarifies that the controller was capable of reducing the reflected wave significantly. The SIAF-control-strategy was applied with success, because a total reduction of -16.2 dB was achieved in the analyzed frequency band.

The magnitude of the ratio between the complex sound pressure detected by microphone 3 and the driving signal of the primary source (plotted against frequency)

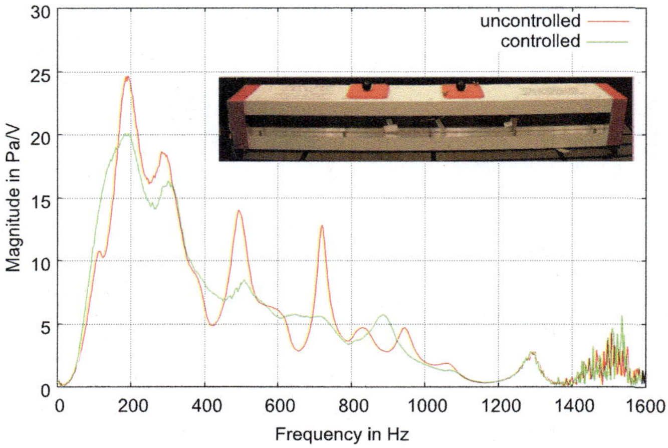

Fig. 10.7 Suppression of acoustic resonances through maximization of acoustic absorption in a sound tube. Magnitude of ratio between complex sound pressure (measured in front of the primary source) and driving signal of primary source is plotted versus frequency

is shown in Fig. 10.7. This figure proves that the applied control strategy was capable of suppressing acoustic resonances that can (considering the length of the sound tube) be approximated by the eigenfrequencies of a sound hard terminated duct, see Table 3.1. Furthermore, it can be seen that the maximization of the acoustic absorption results in an amplification at the anti-resonances of the uncontrolled system as described in Sect. 6.3.

Because the SIAF-concept was tested successfully in a simple laboratory environment, SML 3 was achieved at the end of the *feasibility study*.

10.2 First Specification of an Active Intensity Probe

10.2.1 First Specification of System Hardware

A first realization of a SIAF was proposed in (Kletschkowski and Sachau 2008). It consists of a sound source (type: B&K 4295), compare Fig. 10.8 (left), and a self made cylindrical adapter with inner diameter 38 mm. The latter is shown in Fig. 10.8 (right).

Two phase-matched microphones (type: B&K 4939) were integrated into this adapter. The microphone spacing was set to $\Delta x = 10$ cm. According to (4) in (EN ISO 10534-2 2001), the upper frequency limit for acoustic material testing was therefore given by 1543 Hz. A portable FFT-analyzer (type: B&K: 3560B) with a phase match of $\pm 0.017°$ at 50 Hz using B&K sound intensity probes was used for pre-amplification of the SIAF microphones.

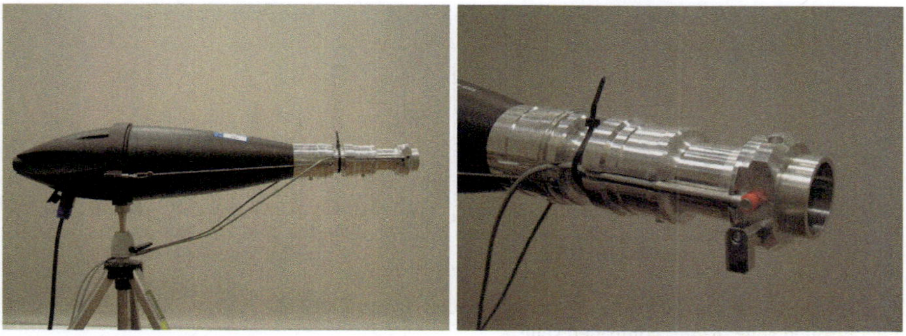

Fig. 10.8 First realization of SIAF (*left*) and detailed view of self made adapter with integrated microphones (*right*)

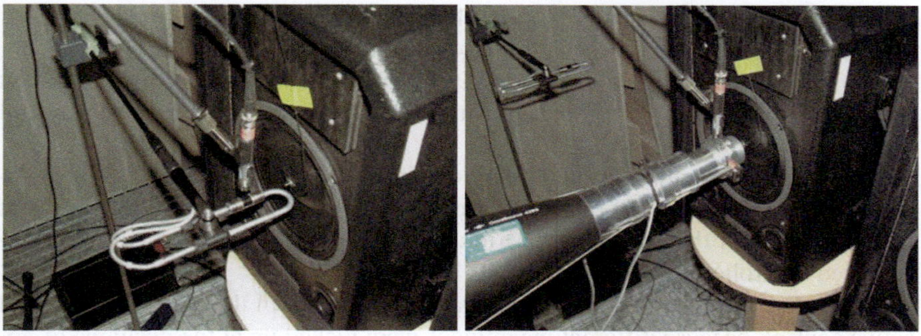

Fig. 10.9 Arrangement for sound intensity measurements using a conventional p-p probe (*left*), and using the novel sound intensity probe with active free field (*right*)

10.2.2 Test of First SIAF-Specification

Calibration Tests in an Anechoic Camber

An anechoic chamber was used to test the possibility of calibrating the first SIAF realization. An electro-dynamical loudspeaker (type: PAB-8MK2) was used as acoustic source, and a measurement microphone (type: B&K 4188) was applied to measure the SPL at a fixed reference point. An accelerometer (type: B&K 4374), amplified by a signal conditioner (type: B&K Nexus 2692), was used to assure a precise reproduction of the excitation and to detect the reference signal. The radiated mean sound intensity was measured for the third octave band center frequencies between 80 Hz and 500 Hz.

First reference data were collected by a conventional sound intensity probe (type: B&K 3595 with a microphone pair of type: B&K 4197 using a spacer of 12 mm) as shown in Fig. 10.9 (left). The measurements were then repeated using the first SIAF realization, compare in Fig. 10.9 (right). Signal processing was performed on

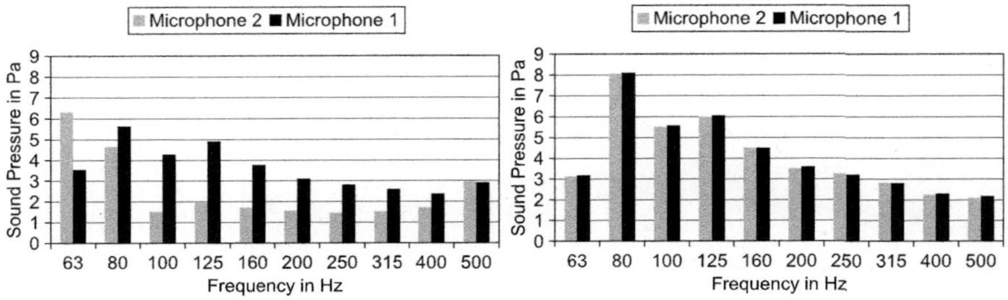

Fig. 10.10 Sound pressure at the SIAF microphones (from Kletschkowski and Sachau 2008). *Left*: uncontrolled state. *Right*: controlled state

a power computer (type: dSpace DS1103), and a filter length of $I = 4$ was used for the control filter as well as for the secondary path model. The latter was remodeled for every excitation frequency. The sampling frequency was set to $f_s = 10$ kHz.

Considering a speed of sound of $c = 340\ \mathrm{m\,s^{-1}}$ as well as the specified values for f_s and Δx, evaluation of (10.7) yields 2.9. However, the time delay number was set to $N = 2$. This choice is in agreement with (10.6), because the highest frequency of interest was given by $f_{\max} = 500$ Hz.

At first it was verified, if the controller was able to maximize the acoustic absorption. The magnitude of the sound pressure measured at the SIAF microphones was compared for this purpose. Experimental results that were recorded for the deactivated controller are shown in Fig. 10.10 (left).

It can be concluded that a standing wave was present in the SIAF tube without active control, because different magnitudes were picked up at the positions of microphone 1 and microphone 2. In contrast to these findings, the results shown in Fig. 10.10 (right) prove that the controller was capable of reducing the reflected wave for all analyzed frequencies. It can be seen that only minor sound pressure deviations were measured between the SIAF microphones, if the controller was activated.

In a second post-processing step the mean sound intensity measured with the conventional probe was compared with the mean sound intensity measured with the activated SIAF. Because the novel intensity probe is an active p-p probe, the mean intensity measured with the SIAF was calculated from the cross spectrum between the SIAF microphones using (4.9). This procedure enables a direct comparison of the "SIAF results" with the results determined by the conventional p-p probe.

The measured results were corrected considering the high frequency error E_{HF} (specified by (4.14)), the low frequency error E_{LF} (specified by (4.24)), and the near field error (specified by (4.35)). The latter was evaluated because the SIAF tube channelizes the propagating spherical wave. This results in plane wave propagation in the SIAF tube that can yield an overestimation of the true mean intensity.

Because the experiments were performed in an anechoic chamber, and because of the verified ability to cancel the reflected wave component inside the SIAF tube, compare Fig. 10.10 (right), the low frequency error formula (4.24) was applied assuming (ideal) free field conditions ($\alpha = 1$).

Table 10.1 Mean intensity measured with a conventional intensity probe in an anechoic chamber

f in Hz	$\bar{I}_{conv}$ in dB	E_{HF} in dB	E_{LF} in dB	E_{NF} in dB	$\bar{I}_{conv}^{corr}$ in dB
80	99	0.0	1.1	0.1	97.8
100	97	0.0	0.9	0.1	96.0
125	100	0.0	0.8	0.1	99.2
160	97	0.0	0.6	0.1	96.3
200	96	0.0	0.5	0.1	95.4
250	96	0.0	0.4	0.1	95.5
315	95	0.0	0.3	0.1	94.6
400	94	0.0	0.2	0.1	93.7
500	94	0.0	0.2	0.1	93.7

Taking into account that both p-p probes were equipped with phase matched microphones, the microphone phase difference was assumed to be $\Delta\varphi_H = 0.3°$ for both intensity sensors.[3]

In order to evaluate the near field error, a microphone separation of $\Delta r = 1.2$ cm and a mean radius, compare Fig. 4.4, of $r = 5$ cm was taken into account for the conventional intensity probe. A microphone separation of $\Delta r = 10$ cm and a mean radius of $r = 7.6$ cm was taken into account for the SIAF. The (uncorrected) experimental results, the calculated errors, and the corrected sound intensity data are listed in Table 10.1 and in Table 10.2. The latter were calculated such as $\bar{I}^{corr} = \bar{I} - E_{HF} - E_{LF} - E_{NF}$. Please notice that measured data are given without decimal points, whereas calculation results are rounded to the first decimal point.

It can be seen that the low frequency error as well as a small near field error were taken into account to correct the mean intensity measured with the conventional p-p probe.[4] In contrast this, the SIAF error is below 200 Hz dominated by the near field error. Above this frequency, the high frequency error increases, because of the 10 cm microphone spacing.

The deviation between the mean intensity measured with these probes is shown in Table 10.3. It can be seen that sound intensity measurement with the actively controlled SIAF results under free field conditions in an overestimation. Without correction, a maximum deviation of 5 dB was determined for $f = 100$ Hz, whereas a minimum deviation of 2 dB was detected for $f = 125$ Hz.

[3]Using a 12 mm spacer for the conventional intensity probe and a spacing of 10 cm for the SIAF, the change of phase over the microphone separation distance was given by $\pm 1°$ for the conventional probe and by $\pm 8.4°$ for the SIAF, considering the lowest analyzed frequency. These phase differences are greater than the (assumed) phase mismatch between the microphones of the p-p probes. However, if the SIAF is equipped with a microphone-analyzer combination that is not perfectly calibrated in phase, phase calibration (e.g. by a procedure based on switching the microphones as described in EN ISO 10534-2 2001) is required.

[4]According to (Gade 1982) this sound intensity probe can because of the 12 mm spacer be used in the frequency range 125 Hz $< f <$ 5 kHz without causing an error magnitude greater than 1 dB.

Table 10.2 Mean intensity measured with the actively controlled SIAF in an anechoic chamber

f in Hz	$\bar{I}_{SIAF}$ in dB	E_{HF} in dB	E_{LF} in dB	E_{NF} in dB	$\bar{I}^{corr}_{SIAF}$ in dB
80	103	0.0	0.2	2.4	100.4
100	102	0.0	0.1	2.4	99.4
125	102	0.0	0.1	2.4	99.4
160	101	−0.1	0.1	2.4	98.6
200	99	−0.1	0.1	2.4	96.6
250	99	−0.2	0.0	2.3	97.0
315	98	−0.2	0.0	2.2	96.0
400	97	−0.4	0.0	2.1	95.3
500	97	−0.6	0.0	1.8	95.8

Table 10.3 Deviation of mean intensity measured in an anechoic chamber using the first SIAF realization and conventional sound intensity probe

f in Hz	$\bar{I}_{SIAF} - \bar{I}_{conv}$ in dB	$\bar{I}^{corr}_{SIAF} - \bar{I}^{corr}_{conv}$ in dB
80	4.0	2.6
100	5.0	3.4
125	2.0	0.2
160	3.0	2.3
200	3.0	1.2
250	3.0	1.5
315	3.0	1.4
400	3.0	1.6
500	3.0	2.1
$\lvert\bar{\Delta}_I\rvert$ in dB	3.4	1.9

$$\lvert\bar{\Delta}_I\rvert := 10\log_{10}\left(\frac{1}{N_f}\sum_{i=1}^{N_f}\frac{10^{\lvert\Delta I_i\rvert/10}}{10^{-12}\,\mathrm{W\,m^{-2}}}\right). \tag{10.9}$$

The mean deviation was calculated according to (10.9), where N_f represents the number of analyzed frequencies. Without correction this quantity was given by 3.4 dB. Considering the corrections, the maximum deviation (still found at $f = 100$ Hz) was reduced to 3.4 dB, and a minimum deviation of 0.2 dB was determined for $f = 125$ Hz after correction. The mean deviation was reduced to 1.9 dB. These results prove that the SIAF can be calibrated successfully under well defined boundary conditions. For practical use, however, a calibrator that can easily be attached and coupled to a well defined sound source (e.g., a pistonphone) is not yet developed.

Table 10.4 Reverberation times and average absorption coefficients measured in a laboratory

f in Hz	Reverberation time in ms	Absorption coefficient
100	674	0.30
125	688	0.30
250	591	0.34
500	543	0.37

Evaluation in a Standing Wave Field

Using the arrangement consisting of loudspeaker, accelerometer and SIAF, compare Fig. 10.9 (right), the SIAF realization was also tested in a typical lab. The reverberation time of this room varies between a maximum value of 0.84 s at 125 Hz and a minimum value of 0.5 s at 400 Hz. The arithmetic mean in the frequency band between 100 Hz and 5 kHz is given by 0.64 s and corresponds to values desired for rooms of medium size (250 m^3–5000 m^3), see (DIN 18041 2004).

The reverberation times measured at the excitation frequencies are listed in Table 10.4. These data have also been used to approximate the average absorption coefficient of the laboratory using the *Sabine equation* (3.73). A room volume of 693 m^3 and an absorbing surface of 545 m^2 were taken into account for these calculations. The results are also listed in Table 10.4.

The experiments were performed separately for each excitation frequency. For every tested frequency an energy flow from the loudspeaker into the SIAF was measured, if active noise control based on wave separation in time domain was applied. This indicates that a SIAF can be used to detect acoustic hot spots in an enclosed sound field as demanded by the requirements.

The mean sound intensities measured with the conventional sound intensity probe—without the surrounding sound tube and without active control—are listed in Table 10.5. The "SIAF-results" are listed in Table 10.6. As in the previous paragraph, these tables also contain the error data as well as the corrected sound intensities. Please notice that the distance between the loudspeaker membrane and the center point of the conventional sound intensity probe was adjusted to equal the distance between the loudspeaker membrane and the rearmost SIAF microphone.

Because the conventional probe was used in a typical lab environment, the average absorption coefficients listed in Table 10.4 were used to calculate the low frequency error (specified by (4.24)) for the conventional p-p probe. The low fre-

Table 10.5 Mean intensity measured with a conventional intensity probe in a laboratory

f in Hz	$\bar{I}_{conv}$ in dB	E_{HF} in dB	E_{LF} in dB	E_{NF} in dB	$\bar{I}_{conv}^{corr}$ in dB
100	111	0.0	3.7	0.1	107.2
125	104	0.0	3.2	0.1	100.7
250	108	0.0	1.6	0.1	106.3
500	109	0.0	0.8	0.1	108.1

Table 10.6 Mean intensity measured with the actively controlled SIAF in a laboratory

f in Hz	$\bar{I}_{SIAF}$ in dB	E_{HF} in dB	E_{LF} in dB	E_{NF} in dB	$\bar{I}_{SIAF}^{corr}$ in dB
100	107	0.0	0.1	0.0	106.9
125	106	0.0	0.1	0.0	105.9
250	105	−0.2	0.0	0.0	105.2
500	103	−0.6	0.0	0.0	103.6

Table 10.7 Deviation of mean intensity measured in a laboratory using the first SIAF realization and a conventional sound intensity probe

f in Hz	$\bar{I}_{SIAF} - \bar{I}_{conv}$ in dB	$\bar{I}_{SIAF}^{corr} - \bar{I}_{conv}^{corr}$ in dB
100	−4.0	−0.3
125	2.0	5.2
250	−3.0	−1.1
500	−6.0	−4.5
$\vert\bar{\Delta}_I\vert$ in dB	4.0	3.3

quency error of the actively controlled SIAF was however calculated for the case of perfect absorption ($\alpha = 1$), because of the free field conditions inside the SIAF tube. As in the previous paragraph a phase mismatch of 0.3° was considered for the microphone pairs of both p-p probes.

The near field error was not taken into account to correct the mean intensity measured with the SIAF. This represents the assumptions that a spherical wave was (without reflections) propagating in the anechoic chamber, whereas the sound field in the lab was assumed to be dominated by standing wave components that differ significantly from a propagating spherical wave. Hence, it was assumed that overestimation of the mean intensity measured with the SIAF—caused by channelizing a spherical wave propagating in free field and resulting in plane wave propagation inside the SIAF tube—can be excluded from the measurement errors.

In contrast to the calibration tests that were performed under idealized conditions, the magnitudes of $\bar{I}_{SIAF}$ (with control) and $\bar{I}_{conv}$ differ significantly as shown in Table 10.7. It can be seen that the application of the first SIAF realization leads in comparison to the results obtained from measurements with the conventional p-p probe to an underestimation of the mean intensity, except for $f = 125$ Hz. This results from the fact that $\bar{I}_{SIAF}$ quantifies the energy transport of traveling plane waves that are absorbed inside the SIAF, whereas $\bar{I}_{conv}$ quantifies the energy flow in a partly active and partly reactive sound field.

Taking into account that the experiments were performed in an enclosure (without free field), it is allowed to assume that the measurement at 125 Hz was performed by placing the conventional p-p probe on a nodal line (that was of course not existing in the tube of the actively controlled SIAF) to explain the results for this particular frequency. However, applying the corrections, it is possible to reduce the magnitude of the maximum deviation from 6.0 dB to 5.3 dB. The magnitude of the minimum deviation is reduced from 2.0 dB at $f = 125$ Hz to 0.3 dB at $f = 100$ Hz. Furthermore, it can be seen that the mean deviation is reduced from 4.0 dB to 3.3 dB, if the corrections are applied.

10.3 Design Process Summary and Outlook

The functional principle as well as a first realization of a new sound intensity probe was tested with success. It was found that a free field can be realized inside this probe by active noise control, if a microphone pair in combination with a wave separation approach is used to cancel the reflected wave components. The FxLMS-algorithm was used for the signal control task using an accelerometer as reference sensor. This approach will be sufficient, if access to a signal which is linearly related to the disturbance is given (e.g. in an acoustic test center or during acoustic ground tests of new aircrafts).

It was shown that a sound intensity probe with an active free field can be calibrated successfully inside an anechoic chamber, if high frequency errors that appear during numerical calculation of the pressure gradient based on a two microphone technique, the low frequency error, and the deviations between plane wave propagation in one dimension and spherical wave propagation in three dimensions are taken into account. It was also found that, because of the active free field inside the probe, a SIAF is capable of detecting an acoustic source in a standing wave field that typically appears in weakly damped interiors. This result is in agreement with the requirements. For this reason SML 4 was achieved at the end of the *first specification step*.

Future work (to be performed in a *weight, cost and efficiency study*) shall be focused on the improvement of the SIAF concept. Such a study shall start with experimental investigations that are used to compare the quality of sound intensity measurements using a SIAF with the quality obtained by the application of conventional sound intensity probes. Based on the results, it will be possible to improve the system design as well as to improve the domain specific design of the SIAF. As mentioned in (Kletschkowski and Sachau 2008) it will be necessary to develop a calibration procedure that allows calibration outside an anechoic chamber. In order to be independent of a reference signal that is well correlated to the disturbance, the application of a robust feedback controller shall be investigated. Furthermore, it should be analyzed, if it is advantageous to redesign the shape of the open end of the SIAF in order to smooth the change of impedance along the inlet of the probe by using a conical inlet or an inlet that has the shape of an exponential horn.

Chapter 11
Active Noise Control Around Human Head

Abstract Increasing industrialization leads to growing noise pollution of the environment. Continuous noise cannot only make a person deaf, it also acts as a factor for stress and can raise systolic blood pressure. One particular problem is noise that keeps a person off from restorative sleep. The application of ANC might be a proper approach to reduce such disturbances, and is motivated by the request to sleep with open windows even in areas with a high volume of traffic. One solution is to create a (at least local) zone of silence around the head of a sleeping person by (at least local) ANC. A particular approach is discussed in the upcoming chapter. It reports on the first design step of a local ANC-system for adaptive control of broadband disturbances around human head. This is another example for the application of the ANC-system design methodology (proposed in Chap. 8) and the usage of the ANC-system design tools (discussed in Chap. 7). The approach is basically similar to other investigations, compare e.g. (Rafaely et al. in J. Acoust. Soc. Am. 116(2):787–793, 1999). However, the long-term objective is to develop a simple, small, portable, and highly effective ANC-system to be utilized in a multitude of situations with various noise exposures. The project described in this chapter was carried out at the Helmut-Schmidt-University/University of the Federal Armed Forces Hamburg between September 2006 and March 2008, see (Kletschkowski et al. in Aktive Lärmminderung in Wohn- und Schlafräumen. Final Report for a Research Project (Az: 24483)—financed by the Deutsche Bundesstiftung Umwelt, Helmut-Schmidt-University/University of the Federal Armed Forces Hamburg, 2008). Financial support by the Deutsche Bundesstiftung Umwelt is gratefully acknowledged.

11.1 Feasibility Study on Local ANC in Bedrooms

In order to evaluate the feasibility of preventing serious diseases caused by continuous noise exposure during sleep, the performance of a local ANC-system was tested. The latter was designed to meet the **requirement**: *Reduce broadband disturbances around human head to the physical limits of the test environment*. The system design process contained only the *feasibility study* in which *noise field analysis*, *anti-noise field analysis*, *domain-specific design* as well as *system integration* and *functional testing* was processed.

T. Kletschkowski, *Adaptive Feed-Forward Control of Low Frequency Interior Noise*, Intelligent Systems, Control and Automation: Science and Engineering 56,
DOI 10.1007/978-94-007-2537-9_11, © Springer Science+Business Media B.V. 2012

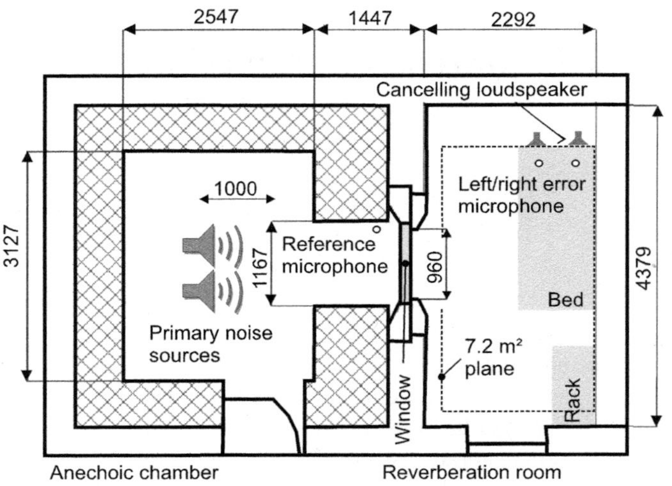

Fig. 11.1 Dimension of transmission test rig given in mm and position of test equipment

11.1.1 Description of Bedroom Test Rig

The plant in which disturbing noise has to be reduced will in practice be given by a bedroom with an open window. This situation was simulated using a transmission test rig that consists of an anechoic chamber and a reverberation room equipped with a conventional bed. The dimensions of this test rig are shown in Fig. 11.1.

In a preliminary test the reverberation time of the reverberation room was determined for the one-third octave band center frequencies between 100 Hz and 4 kHz. A minimum value of 0.5 s was found for $f = 400$ Hz, and a maximum value of 0.84 s was found for $f = 125$ Hz. A mean value of 0.64 s was determined for the analyzed frequency range as reported in (Sachau et al. 2008a).

Acoustic coupling between both rooms was realized by a tilted window, as shown in Fig. 11.2. Primary noise was generated in the anechoic chamber using four electro-dynamical loudspeakers (type: PAB 515/BL). Two active-speakers (type: Genelec 8020) were mounted at the bed, to generate the anti-noise field. Analogue filters (type: Kemo VBF21) were used for band-pass filtering of actuator driving signals. A condenser microphone (type: B&K 4188) was installed in the anechoic chamber to pick up a reference signal. Two microphones of the same type were used to measure the error signals. These error sensors were integrated in the pillow.

The complex pressures P_L and P_R at the left and right error microphone are determined by the transfer behavior of the two primary paths H_{LX} and H_{RX} as well as by the transfer behavior of the four secondary paths, compare (11.1). The latter are represented by H_{LL}, H_{LR}, H_{RL} and H_{RR},

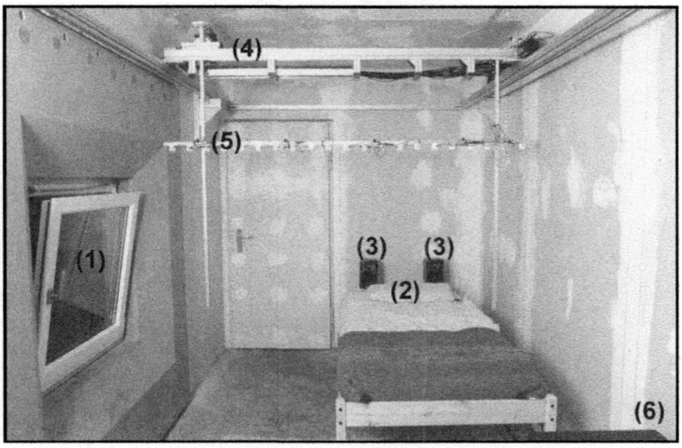

Fig. 11.2 Receiving room of transmission test rig used to simulate a bedroom: (*1*) tilted window, (*2*) pillow with integrated microphones, (*3*) canceling loudspeakers, (*4*) positioning system, (*5*) microphone array, (*6*) rack with electronic equipment

$$\begin{aligned} P_L(j\omega) &= H_{LX}(j\omega)Q_p(j\omega) + H_{LL}(j\omega)Q_{Ls}(j\omega) \\ &\quad + H_{LR}(j\omega)Q_{Rs}(j\omega) \\ P_R(j\omega) &= H_{RX}(j\omega)Q_p(j\omega) + H_{RL}(j\omega)Q_{Ls}(j\omega) \\ &\quad + H_{RR}(j\omega)Q_{Rs}(j\omega). \end{aligned} \tag{11.1}$$

The error sensors were also used to measure the ambient noise in the reverberation room. In the frequency range between 0 Hz and 1.6 kHz a total SPL of 40 dB(A) was determined for the background noise. The window was closed for these preliminary tests.

Because a system concept (one reference microphone located in the anechoic camber, two error microphones integrated in the pillow, and two canceling sources mounted at the bed) was considered when preparing the transmission test rig, this design process started at SML 2.

11.1.2 Noise Field Analysis in the Transmission Test Rig

To get some information about the acoustic behavior of the test rig and to estimate the physical limit of active noise reduction, the primary disturbance measured with the reference microphone and with the error microphones was analyzed. The primary sources were driven by white noise (80 Hz $\leqslant f \leqslant$ 480 Hz) for these measurements. The primary noise was adjusted to a total value of about 70 dB(A) at the left error microphone. The transfer path was given by the tilted window.

The auto spectra caused by the primary disturbance are shown in Fig. 11.3. It can be seen that the former white noise is now colored by the modal characteristics of the test rig. Please notice that a transmission loss of about 20 dB between the error microphones and the reference sensor was measured.

Without changing the primary source strength, the total SPL at the error microphones was attenuated additionally by approximately 20 dB, if the window was closed.

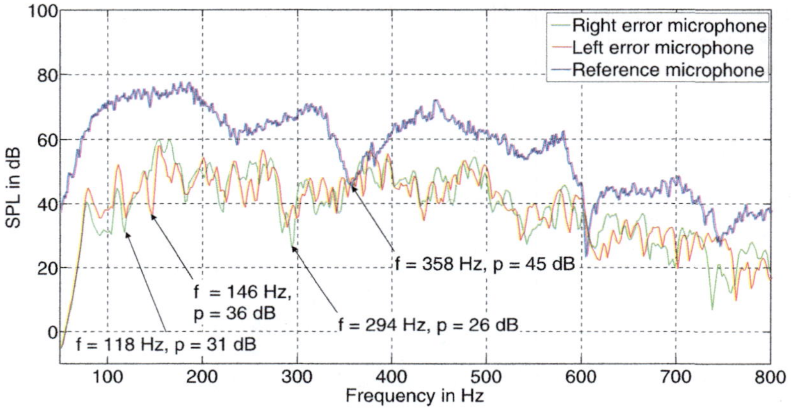

Fig. 11.3 Primary disturbance measured at the reference microphone and two error microphones

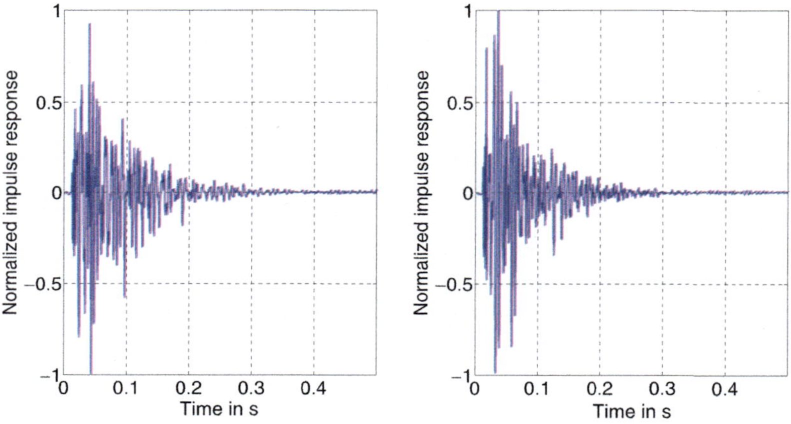

Fig. 11.4 Normalized impulse responses measured for the primary paths in the transmission test rig. Impulse response between the driving signal of primary sources and left error microphone is shown *left*. Impulse response between the driving signal of primary sources and right error microphone is shown *right*. Normalization is performed with respect to the maximum value

Figure 11.4 contains the normalized impulse responses between the driving signal of primary sources and left/right error microphone measured with a sampling frequency of 4096 Hz. Normalization is performed with respect to the maximum value. It can be seen that both transfer paths contain a time delay of about 0.01 s. It can also be seen that the maximum absolute value 1 is reduced below 0.1 for $t > 0.2$ s.

This represents a decay of more than 20 dB in a time span of approximately $\Delta t = 0.2$ s. Considering a sampling frequency of $f_s = 4096$ Hz, a number of approximately $\Delta t \cdot f_s = 820$ filter coefficients will therefore be required to model the main part of such a "long" transfer path.

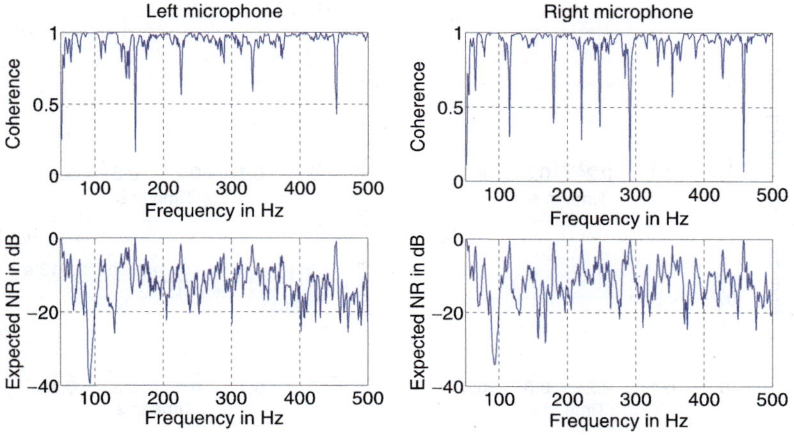

Fig. 11.5 Coherence between error signals and the reference microphone signal (*top*) and expected noise reduction at the error microphones (*below*)

As shown in Sect. 7.3.1, the performance of feed-forward ANC can be estimated by coherence analysis. Equation (7.54) that describes the single-channel case (one reference signal one error signal) was applied to determine an IPE. Using a sample frequency of 4096 Hz and white noise (80 Hz $\leqslant f \leqslant$ 480 Hz) as driving signal for the primary noise sources, CA was performed for both error microphones. As in the desired application, the signal obtained by the reference microphone has been used as reference signal.

The magnitude of the coherence between these sensors is shown in Fig. 11.5 (top). The expected NR, calculated from (7.54), is presented in Fig. 11.5 (below). It is obvious that a low value for the coherence (that can be caused by a low level of the reference signal (compare Fig. 11.3—$f = 358$ Hz) as well as by a low level of the error signal (compare Fig. 11.3—$f = 118$ Hz, $f = 146$ Hz, and also $f = 294$ Hz)) results in a low value for the expected NR.

By analyzing the total levels, it was found that an NR of −17 dB/−18 dB can be expected at the position of the left/right error microphone in the analyzed frequency band, if the sensor arrangement shown in Fig. 11.1 is used for ANC.

11.1.3 Anti-noise Field Analysis in the Transmission Test Rig

Anti-noise field analysis was performed to analyze the transfer behavior of the four secondary paths as well as to evaluate the effect of the secondary sources on the reference sensor. Each canceling source was (successively) driven by white noise (80 Hz $\leqslant f \leqslant$ 480 Hz) for these measurements. The anti-noise was adjusted to a total value of about 70 dB(A) at the left error microphone. The window in the transfer path was again tilted.

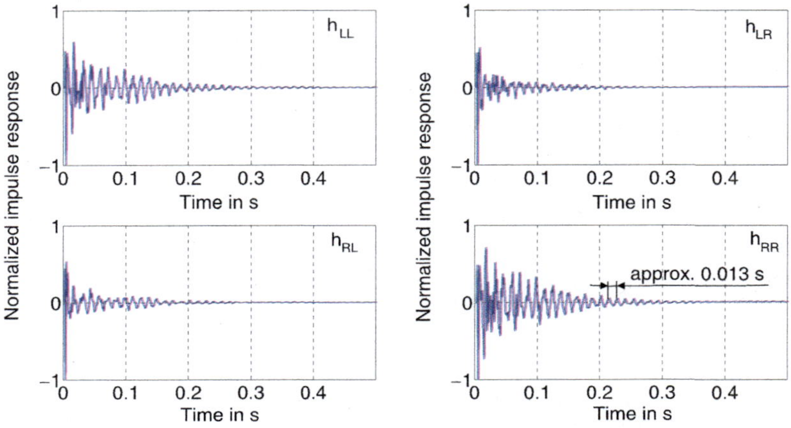

Fig. 11.6 Normalized impulse responses measured for the secondary paths in the transmission test rig. Impulse response between the driving signal of left secondary source and the error microphones is shown *left*. Impulse response between the driving signal of right secondary source and the error microphones is shown *right*. *Top/Below*: response of left/right microphone. Normalization is performed with respect to the maximum value

Figure 11.6 contains the normalized impulse response functions of the secondary transfer paths. A sampling frequency of 4096 Hz was used for these measurements. Normalization is again performed with respect to the maximum value. It was found that all secondary transfer paths contain a time delay of approximately 0.001 s. This is the tenth part of the delay that was determined for the primary transfer paths. Hence, causality can be ensured for the active control task.

The damping of the cross transfer paths $h_{LR} \circ\!\!-\!\!\bullet H_{LR}$ and $h_{RL} \circ\!\!-\!\!\bullet H_{RL}$ is significantly greater than the damping of the direct paths $h_{LL} \circ\!\!-\!\!\bullet H_{LL}$ and $h_{RR} \circ\!\!-\!\!\bullet H_{RR}$. It can be seen that the maximum absolute value 1 of the cross transfer paths is reduced below 0.1 for $t > 0.1$ s. This represents a decay of more than 20 dB in a time span of approximately $\Delta t = 0.1$ s. Considering a sampling frequency of $f_s = 4096$ Hz, a number of $\Delta t \cdot f_s \approx 410$ filter coefficients will therefore be required to model the main part of the cross paths.

The situation is different for the direct paths as also shown in Fig. 11.6. It can be seen that h_{LL} and h_{RR} are (especially after 0.1 s) dominated by a harmonic component with a periodic time of about 0.013 s, i.e. a frequency of about 77 Hz.

Considering the dimensions of the reverberation room, compare Fig. 11.1, this excitation frequency corresponds to the second natural frequency in longitudinal direction, if the formula for sound hard boundaries listed in Table 3.1 is evaluated.

The findings of the impulse response analysis were confirmed by the results of transfer functions measurements shown in Fig. 11.7. This figure contains the magnitudes of the FRFs that were determined between each sensor and the driving signal of the right canceling source. It can be seen that the magnitude of the direct path H_{RR} is greater than the magnitude of the cross path H_{LR}. Furthermore, it can be seen that the FRF indeed peaks at the acoustic resonance found at $f = 76$ Hz.

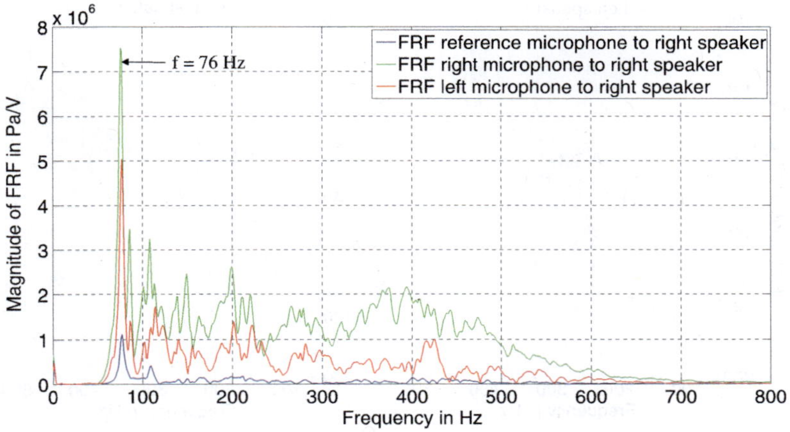

Fig. 11.7 Magnitude of frequency response functions measured in the reverberation room

It should also be noticed that the natural frequency of the secondary sources also appears around 76 Hz (Genelec 2010). Thus, considering that the lower corner frequency of the analogue low pass filter was 80 Hz, the reverberation room was efficiently excited at this frequency. The results also indicate that the reference sensor is (physically) decoupled from the canceling sources. Hence, the effect of ANC in the reverberation room on the reference signal can be neglected.

The number of error microphones equals the number of canceling loudspeakers. Therefore, it is possible to find optimal control signals for the left and for the right actuator that will (in theory) produce zero sound pressure at each error microphone for every tonal excitation in the analyzed frequency band. The auto spectral density of these optimal driving signals are shown in Fig. 11.8. The curves indicate that the control effort has to be increased for frequencies above 250 Hz, if the disturbing noise shall be completely rejected.

11.1.4 Comments on Domain-Specific Design for Broadband ANC

To perform adaptive feed-forward control, a $1 \times 2 \times 2$-implementation (one reference signal, two error signals, two adaptive controller) of the FEFxLMS-algorithm (summarized in Table 6.5) using a power normalized step size $\tilde{\mu}(n)$—as specified by (6.217)—was implemented on a rapid prototyping system (Type: dSpace 1005). This hardware is equipped with a PowerPC running at 1 GHz and multiple fast A/D- and D/A-converters with a hardware latency of approximately 5 μs.

Real time testing requires appropriate values for the control filter length and the step size of the adaptive algorithm. The first parameter is linked to the case of optimal control that is realized, if the (remaining) error signal is uncorrelated to the filtered reference signal. To achieve this goal, it is necessary to specify an optimal control filter length. This was done by IRA, compare Sect. 7.3.4.

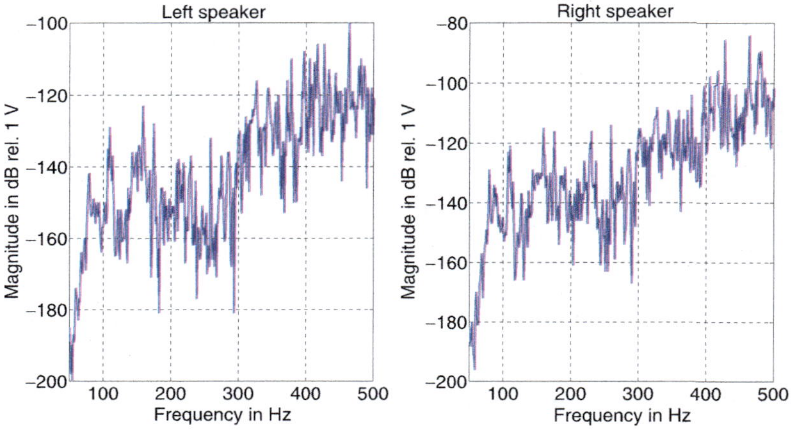

Fig. 11.8 Magnitude of optimal driving signals calculated to cancel the sound pressure at the error microphones at each tonal excitation frequency

Please notice that both reference signal and secondary path impulse response functions were already measured (with a sampling frequency of 4 kHz) during the *noise field analysis* task and the *anti-noise field analysis* task and could therefore be used for IRA. Details about the calculations were reported in (Holters and Zölzer 2008). The IRA results are shown in Fig. 11.9. It can be seen that the attenuation increases nonlinear with an increase of the control filter length. However, saturation seems to be achieved for a number of about 1000 control filter coefficients. This results confirms the initial considerations about the length of the primary path impulse responses, compare Sect. 11.1.2.

The expected attenuation at the error microphones using 1000 control filter coefficients is greater than the amount of NR predicted by CA, compare Sect. 11.1.2. However, by comparing the results it must be taken into account that IRA was performed to evaluate the performance of an optimal $1 \times 2 \times 2$ ANC-system, whereas CA was applied to predict the NR of a $1 \times 1 \times 1$ ANC-system.

Furthermore it must be considered that CA was performed to calculate a performance estimate for the frequency range 50 Hz $< f <$ 500 Hz, whereas the "complete" impulse response was considered during IRA.

For this reason the set of control filter coefficients was optimized to realize an attenuation for all frequency components that contributed to the measured impulse response functions. Based on these findings the controller length was set to $I = 1024$ coefficients for both adaptive filters.

The second parameter is linked to the stability criteria. In the present case, (6.216) was applied to determine a step size that ensures stable iteration. Please notice that no leakage factor ($\beta = 0$) was taken into account. Numerical simulations were performed to calculate the eigenvalues λ_m of the $(2 \cdot 1 \cdot 1024 \times 2 \cdot 1 \cdot 1024)$ matrix $E[\hat{\mathbf{X}}^T(n)\mathbf{X}(n)]$, compare Sect. 6.6.2, on the basis of measured data. Considering the system size ($K = 1$, $M = 2$, $L = 2$, and $I = 1024$), $E[\hat{\mathbf{X}}^T(n)\mathbf{X}(n)]$ was in this particular case composed of the $(2 \times 2 \cdot 1 \cdot 1024)$ reference signal matrix

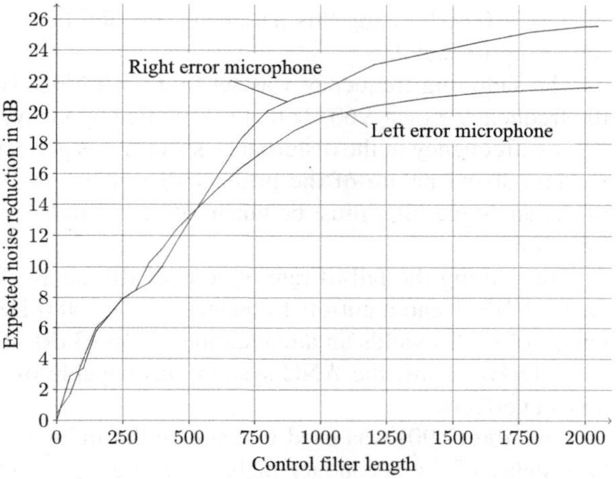

Fig. 11.9 Noise reduction predicted by IRA. Original picture is taken from (Holters and Zölzer 2008)

$$\mathbf{X}(n) = \begin{bmatrix} \mathbf{x}_1^T(n) & \mathbf{x}_1^T(n-1) & \dots & \mathbf{x}_1^T(n-I+1) \\ \mathbf{x}_2^T(n) & \mathbf{x}_2^T(n-1) & \dots & \mathbf{x}_2^T(n-I+1) \end{bmatrix}, \tag{11.2}$$

with reference signals arranged in $(2 \cdot 1 \times 1)$ column matrices

$$\mathbf{x}_l(n) = \left[x_{l1}(n)\ x_{l2}(n)\right]^T, \quad l = 1, 2 \tag{11.3}$$

and of the $(2 \times 2 \cdot 1 \cdot 1024)$ filtered reference signal matrix

$$\hat{\mathbf{X}}(n) = \begin{bmatrix} \hat{\mathbf{x}}_1^T(n) & \hat{\mathbf{x}}_1^T(n-1) & \dots & \hat{\mathbf{x}}_1^T(n-I+1) \\ \hat{\mathbf{x}}_2^T(n) & \hat{\mathbf{x}}_2^T(n-1) & \dots & \hat{\mathbf{x}}_2^T(n-I+1) \end{bmatrix}, \tag{11.4}$$

with reference signals arranged in $(2 \cdot 1 \times 1)$ column matrices

$$\hat{\mathbf{x}}_l(n) = \left[x'_{l1}(n)\ x'_{l2}(n)\right]^T, \quad l = 1, 2. \tag{11.5}$$

Each column matrix $\mathbf{x}_l(n)$ contains two reference signals that are filtered by h_{ml}. The latter variable represents the (physical) impulse response measured between the l-th error microphone and the m-th canceling loudspeaker. As shown by (11.5), $\hat{\mathbf{x}}_l(n)$ is structured in a similar manner. However, filtering is now performed using $\hat{h}_{ml}$—the FIR filter models of the physical plant. 256 filter coefficients were taken into account for each of the four secondary path models.

This choice can be interpreted as a worst case approach, because a number of about 410 filter coefficients would give a better approximation as mentioned in Sect. 11.1.3. However, if the adaptive algorithm is stable for a very simple plant model, it will of course also be stable for a very accurate plant model. Furthermore, a minimum signal power of $\Pi_{x'_{\min}} = 0.02\ \mathrm{V}^2$ was taken into account for the filtered reference to evaluate (6.216).

An upper bound of $\tilde{\mu}_{\max} = 0.02$ was determined in this analysis. By introducing a safety factor of 0.1, such an approach was also proposed in (Kochan et al. 2008), the normalized convergence coefficient used for the real time experiments was set to $\tilde{\mu}_{\text{exp}} = 0.002$. Using this parameter, instabilities were not observed during real time experiments.

The sampling frequency was set to $f_s = 4$ kHz. This value is about four times the frequency $f_{NS} \geq 960$ Hz that is specified by the *Nyquist-Shannon* criteria, if the highest frequency in the disturbance spectrum is given by $f_{\text{max}} = 480$ Hz. However, the broadband nature of the primary disturbance as well as the finite fall-of rate of the analogue filter must be taken into account by specifying f_s as mentioned in Sect. 7.4.

Considering the fall-of rate of $R = 45$ dB/octave for the applied filter of type: Kemo VBF21 and a cut-off frequency of $f_c = 480$ Hz for low-pass filtering, evaluation of (7.89) yields an attenuation of $A \approx 93$ dB, if sampling is performed with $f_s = 4$ kHz. Thus, the ANC-system was capable of suppressing all imaging and aliasing effects.

Equation (7.90) was used to evaluate the influence of the sampling rate on the time delay of the secondary paths. Considering a sampling rate of 4 kHz, a cut-off frequency of $f_c = 480$ Hz and a number of $N_p = 16$ poles, a delay of about $\tau_w = 0.0045$ s was predicted. Please remember that a delay of about $\tau_p = 0.01$ s was found in the primary transfer paths, whereas a delay of about $\tau_s = 0.001$ s was detected in the secondary transfer paths. The difference $\tau_p - \tau_s = 0.009$ s is about two times the delay caused by the analogue filter. Thus, causality was ensured.

11.1.5 System Integration and Functional Testing of Local ANC

Experimental Set-up After System Integration

Functional testing of the local ANC-system was performed in the transmission test rig. The experimental set-up (after integration of the adaptive controller) and the information flow in the ANC-system is illustrated by Fig. 11.10. It can be seen that the primary noise was generated in the anechoic chamber.

Anti-noise was generated in the reverberation room using the two canceling sources. The latter were driven by the multi-channel adaptive controller. The error signals were picked up by two pillow integrated microphones, and the reference signal was measured in front of the window in the anechoic chamber.

Adaptive Control of Tonal Noise

Despite the fact that the ANC-system was designed to cancel broadband disturbances, the effect of ANC on tonal disturbances was analyzed at first. These investigations were performed for $f = 100$ Hz, $f = 200$ Hz, $f = 400$ Hz as reported in (Sachau et al. 2008a). For the tilted window the primary noise was adjusted to 70 dB(A) at the left error microphone. After applying active control, the SPL was at the error microphones for all investigated excitation frequencies reduced to the background noise.

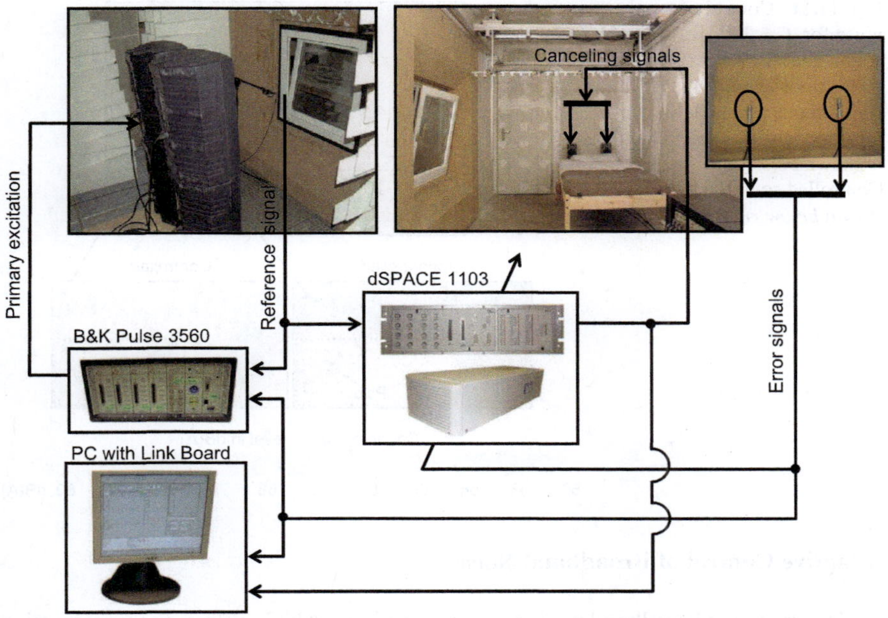

Fig. 11.10 Experimental set-up for real-time testing of local ANC in the transmission test rig

Both the primary and the controlled sound field were monitored in three planes (0.76 m, 1.26 m and 1.76 m above the floor), to determine the global effect of ANC on the sound field in the reverberation room. The microphone array in the reverberation room was used to map an area of 7.2 m^2. In each plane the sound pressure was measured at 777 points, according to a microphone spacing of 10 cm and an equivalent spacing in cross-direction. The measurements were performed using a resolution of 400 lines in the frequency range between 0 Hz and 1.6 kHz.

These investigations were performed for a tonal disturbance with $f = 200$ Hz. For the tilted window the primary noise was again adjusted to 70 dB(A) at the left error microphone.

Results are shown in Fig. 11.11 in which a map of the A-weighted SPL level is shown between 50 dB(A) and 83 dB(A) using different colors in 3 dB-steps. Please notice that unfilled areas represent sound pressure levels below 50 dB(A). The results prove that the active control system was capable of generating a sufficient zone of quiet around the error microphones for a tonal disturbance of $f = 200$ Hz. However, the results shown in this figure also clarify that the control success is limited to the positions of the error sensors, because the SPL was significantly increased in that corner of the transmission test rig that is opposite to the position of the canceling sources.

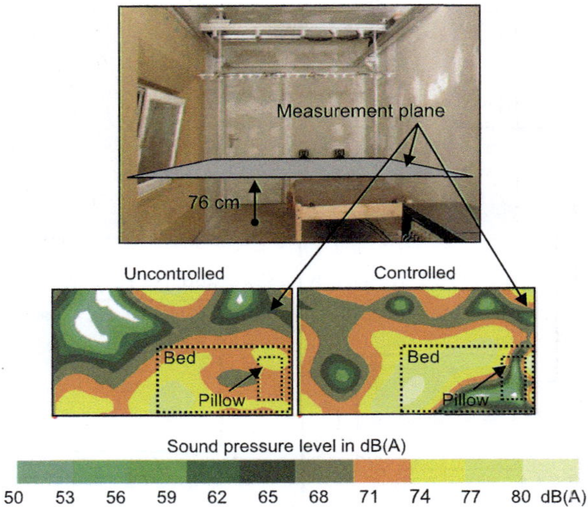

Fig. 11.11 Control of tonal noise for $f = 200$ Hz. Map of sound pressure level (0.76 cm above floor level). Measurement plane is shown *top*. Uncontrolled sound field is shown *below left*. Controlled sound field is shown *below right*

Adaptive Control of Broadband Noise

To determine the broadband performance of the local ANC-system, the transmission test rig was excited by white noise (80 Hz $\leq f \leq$ 480 Hz). For the tilted window the primary noise was adjusted to 70 dB(A) at the left error microphone. All secondary paths were modeled with FIR filters using 256 filter coefficients for each model as in the controller design, compare Sect. 11.1.4.

As shown in Fig. 11.12, the primary disturbance was significantly reduced by adaptive control. As also reported in (Kletschkowski and Sachau 2009), a total NR of -16 dB (calculated for the frequency range specified by 0 Hz $< f <$ 1.6 kHz) was obtained at the left error microphone and a total NR of -18 dB at the right error microphone. Comparing these results with the values estimated by CA, it can be concluded that the physical limits of active NR were achieved.

To give an impression on the effect of ANC in the neighborhood of the error microphones, the sound field was mapped 8 cm above the pillow (84 cm above floor level) using a microphone array with a spacing of 10 cm between each sensor. The NR measured at 32 positions is shown in Fig. 11.13 using a linear interpolation between the measurement points. The results indicate that the disturbance was also reduced away from the two pillow integrated microphones. However, for practical use, the -10 dB zone of quiet (as realized for the area BC-L1R4) must be enlarged.

To achieve this goal it is possible to include more reference signals in the ANC-system. This would lead to an improved coherence that is required to improve the control profit at the error sensors. Furthermore, it can be advantageous to control both pressure and pressure gradient, as proposed in (Elliott and Garcia-Bonito 1995), to smooth the pressure gradient around the error sensors.

To validate the prediction of control filter length, the dependence of the control profit on control filter length (used in real-time experiments) was analyzed. The control filter length was increased from $I_{min} = 128$ to $I_{max} = 2048$ for this purpose.

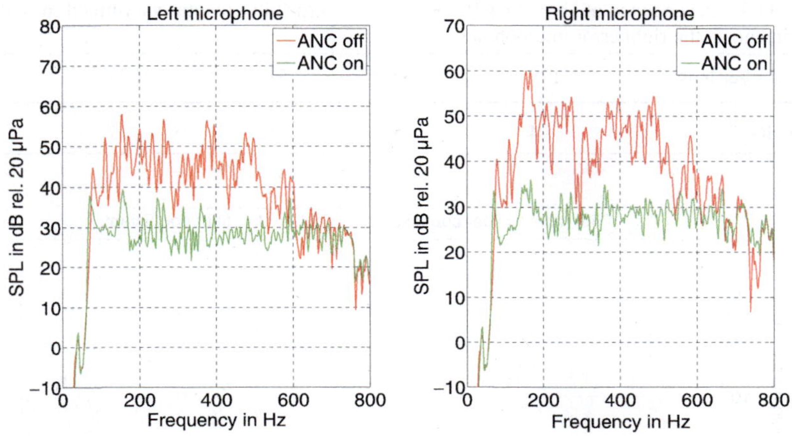

Fig. 11.12 Local NR of broadband disturbances in the transmission test rig at error microphones

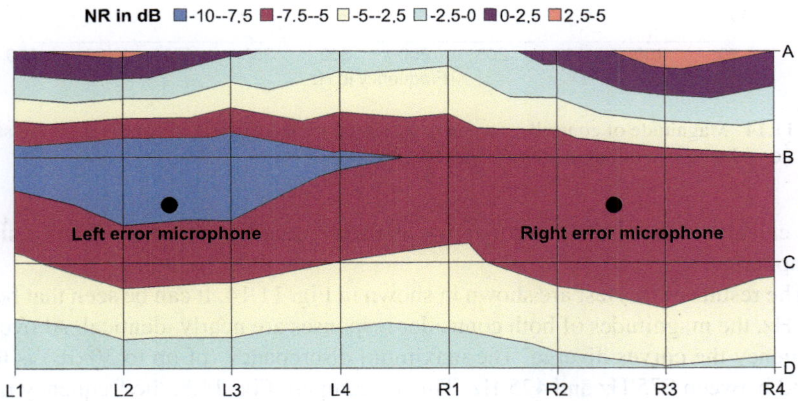

Fig. 11.13 Map of noise reduction due to active control measured 8 cm above the pillow (84 cm above floor level) for broadband excitation (80 Hz $\leq f \leq$ 480 Hz). Distance between the grid points is 10 cm. Error microphones were not used for interpolating of data. Color of legend indicates a certain noise reduction (e.g. *blue*: -10 to -7.5 dB)

The experimental results (measured at the right error microphone) are summarized in Table 11.1. It can be seen that realizing an active noise reduction of at least -10 dB requires a control filter length of 512 taps. Increasing the control filter length from $I = 512$ to $I = 1024$ results in a significant improvement of the controller performance. A further increase up to $I = 2048$ results only in minor changes (as predicted by IRA) and is therefore not necessary.

Finally, to understand, how a reduced set of control filter coefficients influences the controller performance, the steady state controller response was compared for two control filter lengths. For this purpose the frequency response of the controller

Table 11.1 Dependence of (broadband) NR on the control filter length determined in real-time experiments at the right error microphone

Control filter length	128	265	512	1024	2048
NR in dB	−4	−5	−12	−17	−17

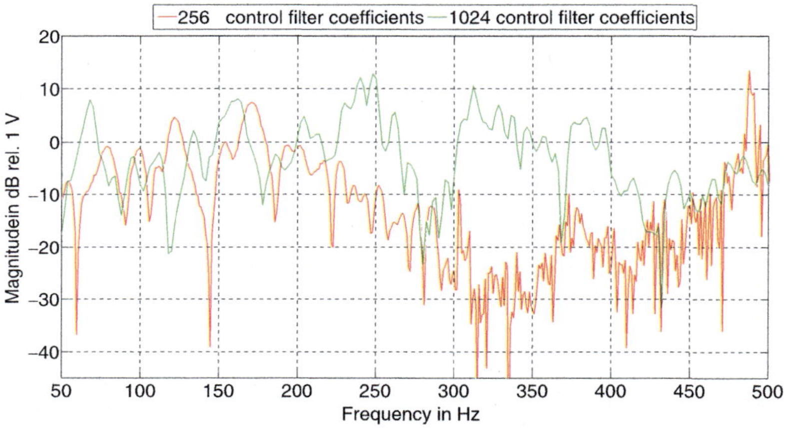

Fig. 11.14 Magnitude of controller response measured for the converged controller in the steady state considering two different control filter lengths

was calculated using the control signal as output quantity and the reference signal as input quantity.

The results of this test are shown in shown in Fig. 11.14. It can be seen that below 175 Hz, the magnitudes of both controller responses are nearly identical. Above this frequency the curves diverge. The maximum discrepancy (of up to 30 dB) is to be found between 275 Hz and 425 Hz. This is, compare Fig. 11.8, the frequency range in which the magnitude of the optimal control strength was increased up to 30 dB.

Active Control of Pseudo-transient Noise

To get a first impression on the capability of the developed ANC-system to attenuate transient noise in the steady state of the controller, active control was applied to a disturbance generated by time recorded data of a passing propeller driven aircraft (type: Cessna) that was repeated continuously. Thus, pseudo-transient excitation was realized.

Before applying active control to this excitation, the developed ANC-system was trained with white noise (80 Hz $\leq f \leq$ 480 Hz) until the controller reached the steady state providing a NR of −17 dB at the right error microphone. Then adaption of filter coefficients was stopped by setting the step size to zero and the "Cessna-signal" was used for excitation.

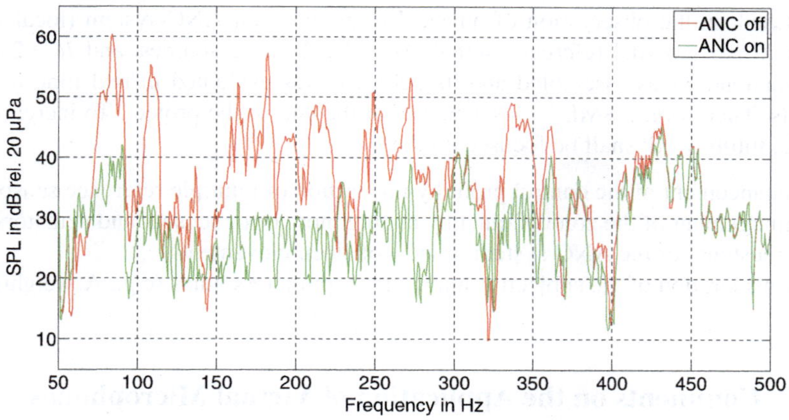

Fig. 11.15 Active suppression of pseudo-transient noise with a converged ANC-system

The experimental results obtained in this test are shown in Fig. 11.15. The power spectra shown for the right error microphone result from time averaging over 1000 data blocks using a sampling frequency of $f_s = 4$ kHz and a frequency resolution of $\Delta f = 1$ Hz. It can be seen that the converged steady controller was capable of reducing pseudo-transient noise. A time averaged reduction of the total sound pressure level of -11 dB was measured at the right error microphone between 80 Hz $\leq f \leq$ 480 Hz.

11.1.6 Summary of Feasibility Study

In order to evaluate the feasibility of lowering the noise exposure during sleep, the concept of ANC was applied to bedrooms. To determine the physical limits in a particular situation, a transmission test rig was used to simulate real world conditions. A conventional bed was placed in the reverberation room, and two error microphones (integrated in the pillow) were used as error sensors. One reference microphone was placed in the anechoic chamber in front of a tilted window that was mounted into the transmission path. The FEFxLMS algorithm was implemented on a real time hardware to perform multi-channel feed-forward control.

At first the physical limits of the control profit were determined by CA. Then a proper filter length was calculated by IRA. This procedure was based on measured data. Finally the noise reductions at the error microphones were measured for broadband disturbances (80 Hz $< f <$ 480 Hz). It turned out that the control profit of up to 18 dB (realized in real time experiments) was equivalent to the noise reduction predicted by CA. It was also found that the active noise control system was capable of generating a zone of quiet in the neighborhood of the error microphones.

Based on the observation of physical limitations, an ANC-system (local ANC with $K = 1$ external reference signal, $M = 2$ anti-noise sources, and $L = 2$ error microphones) was assembled and its behavior was evaluated in real time experiments. Therefore, a SML 3 was achieved at the end of the project. To increase the SML, future work shall be focused on the

- Enhancement of the control profit by integration of multiple reference sensors,
- Enlargement of the zone of quiet by adding more error sensors and actuators,
- Robustness of the ANC-system in view of sleep movements,
- Consideration of the subjective human perception of sounds (e.g., A-weighting).

11.2 Comments on the Application of Virtual Microphones

Because the movements of a sleeping person will have an influence on the error sensors, the performance of an ANC based on virtual microphones, described in Sect. 6.5.2, was analyzed for tonal noise. For this purpose, a conventional bed with dummy head was placed in a typical laboratory.[1] The dummy head was again used to model the sleeping person. Its microphones were used as monitor sensors for RSC. Two microphones (integrated in the mattress) were used as error microphones. Two self made canceling sources[2] were attached to the frame of the bed. This arrangement is shown in Fig. 11.16.

RSC was based on the FxLMS-algorithm (summarized in Table 6.2) considering one internal reference ($K = 1$), and (because of the experimental set-up) two control signals ($M = 2$) as well as two error signals ($L = 2$).

Three test series were performed. At first ANC was directly applied to the error microphones in the mattress, and the NR at the dummy head microphones were simply monitored to measure reference data. Afterwards RSC based on virtual microphones was applied considering both the simplified approach (compare (6.128)) and the advanced virtual microphone technique (compare (6.120)). In the latter case the sound pressure at the monitor microphones was for every discrete time step n according to (6.120) calculated as

$$\begin{aligned} p_{M1}(n) &= \hat{t}_{11} * \left[p_{C1}(n) - \sum_{m=1}^{2} \hat{h}_{C1m} * q_m(n) \right] + \sum_{m=1}^{2} \hat{h}_{M1m} * q_m(n) \\ p_{M2}(n) &= \hat{t}_{22} * \left[p_{C2}(n) - \sum_{m=1}^{2} \hat{h}_{C2m} * q_m(n) \right] + \sum_{m=1}^{2} \hat{h}_{M2m} * q_m(n), \end{aligned} \tag{11.6}$$

where p_{M1} and p_{M2} represent the sound pressure data calculated for the positions of the monitor microphones, whereas p_{C1} and p_{C2} represent the sound pressure data measured at the error microphones. Control signals are indicated by q_m. $\hat{t}_{11}$ and $\hat{t}_{22}$

[1]The acoustic properties of this laboratory were described in Sect. 9.6.1.

[2]This actuator was described in Sect. 9.4.

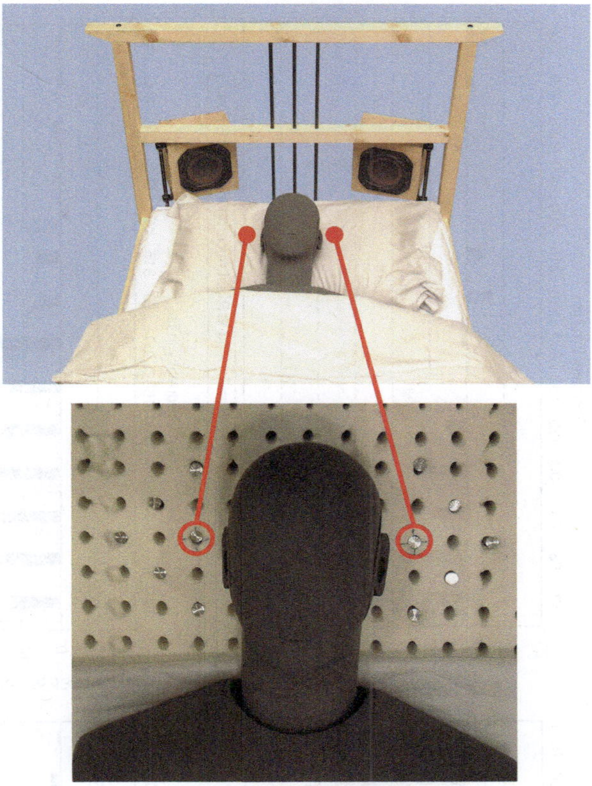

Fig. 11.16 Test bed used to analyze the capability of RSC with virtual microphones using a conventional bed with error sensor in the mattress and a dummy head to monitor the control profit

are the FIR filter models of the transmissibilities between the left/right monitor microphone and the left/right error microphone $(11/22)$. The secondary transfer paths between controller and monitor microphones are described by h_{Cl_Cm} with $l_C = 1, 2$, whereas the secondary transfer paths between controller and error microphones are described by h_{Ml_Mm} with $l_M = 1, 2$. Four filter coefficients were used for each of the six FIR filter that were remodeled for every tested frequency. Please notice that $\hat{t}_{11}$ and $\hat{t}_{22}$ were set to [1 0 0 0] to test the simplified virtual microphone approach.

Experimental results are shown in Fig. 11.17. It is obvious that the highest NR was for all excitation frequencies obtained, if ANC was directly applied to the error microphones, compare Fig. 11.17 (top left). The sound pressure measured at the dummy head microphones was in this case also reduced for frequencies below 200 Hz as shown in Fig. 11.17 (top right). The simplified virtual microphone approach was not capable to improve the control profit at the monitor microphones as indicated by the results presented in Fig. 11.17 (below left). It can be seen that a significant NR at the dummy head sensors was only achieved below 160 Hz.

As shown in Fig. 11.17 (below right), is was possible to improve the control success at the monitor microphones, if the (full) virtual microphone approach—based on (11.6)—was applied. It can be seen that the maximum NR (more than −35 dB at the left dummy head microphone) was achieved for 125 Hz. Between

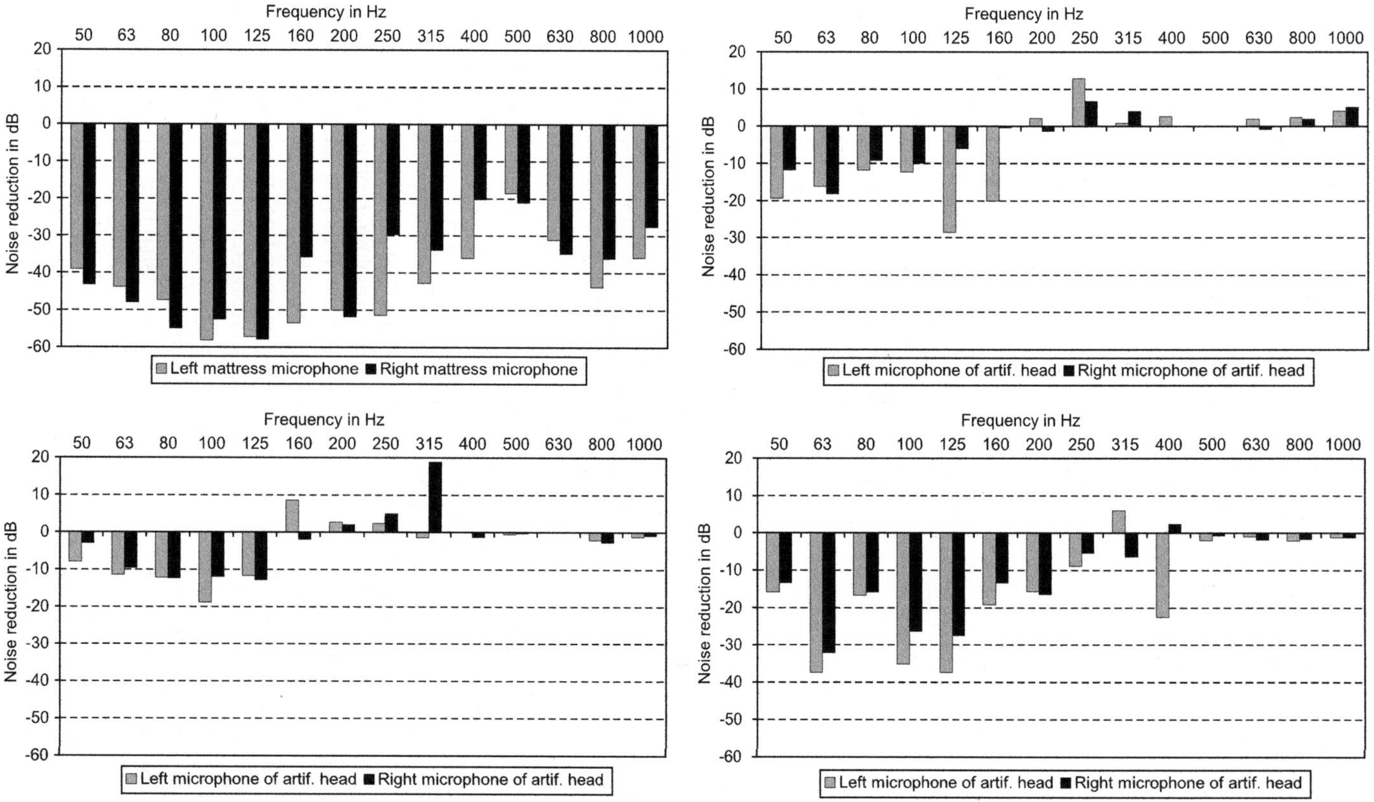

Fig. 11.17 *Top*: NR at mattress microphones (*left*) and dummy head (*right*) without virtual sensing. *Below*: NR at dummy head for simplified virtual sensing (*left*) and full virtual sensing (*right*)

63 Hz and 250 Hz, the NR realized by the (full) virtual microphone approach exceed the NR measured in the reference tests. The results of these investigations prove that RSC based on the virtual microphones offers the possibility to reject disturbing noise around the head of a sleeping person.

Chapter 12
Active Noise Control and Audio Entertainment

Abstract Active systems for noise reduction are especially of interest when considering applications in which low frequency noise is a main source of disturbance and only limited amounts of installation space and payload are available. This makes the adaptation and implementation of such systems plausible in vehicles such as automobiles and aircraft where passive reduction methods are restricted. Taking into account that both automobiles and aircraft are (usually) equipped with an audio entertainment system, it is self-evident to analyze an active control concept that uses the loudspeakers of the entertainment system as canceling sources driven by a combination of audio signal and control signal. As another example for the application of the proposed design methodology (compare Chap. 8) and the usage of the ANC-system design tools (discussed in Chap. 7), the upcoming chapter reports therefore on a project that was carried out to evaluate the feasibility of broadband ANC combined with in-flight entertainment in a very light jet aircraft cabin. Various problems in design, test and optimization such as optimal transducer placement and choice of adaptive control algorithm were taken into account as reported in (Sachau et al. in Audio Interior für Kleinflugzeuge (AIK). Final Report. Hamburg Aircraft Research Program (Lufo Hamburg 2), HH 107 B "AIK," Helmut-Schmidt-University/University of the Federal Armed Forces Hamburg, 2008b). The project was conducted at the Helmut-Schmidt-University/University of the Federal Armed Forces Hamburg in cooperation with Innovint—Aircraft Interior GmbH. Financial support by the city of Hamburg is gratefully acknowledged.

12.1 Aircraft Interior Noise and ANC with Audio Entertainment

12.1.1 Comments on Aircraft Interior Noise

The reduction of interior noise levels in passenger aircraft has been a focus of research in active control for many years (Elliott et al. 1990), (Johansson et al. 1999) and (Borchers et al. 1997). This is caused by the fact that propeller-driven aircraft are prone to high noise levels caused by propeller rotation.

As mentioned in (Wilby 1996): "All the active noise implementations have one factor in common. They all involve the control of periodic signals and utilize a reference signal taken from the propeller or engine using a tachometer or accelerometer."

T. Kletschkowski, *Adaptive Feed-Forward Control of Low Frequency Interior Noise*, Intelligent Systems, Control and Automation: Science and Engineering 56,
DOI 10.1007/978-94-007-2537-9_12, © Springer Science+Business Media B.V. 2012

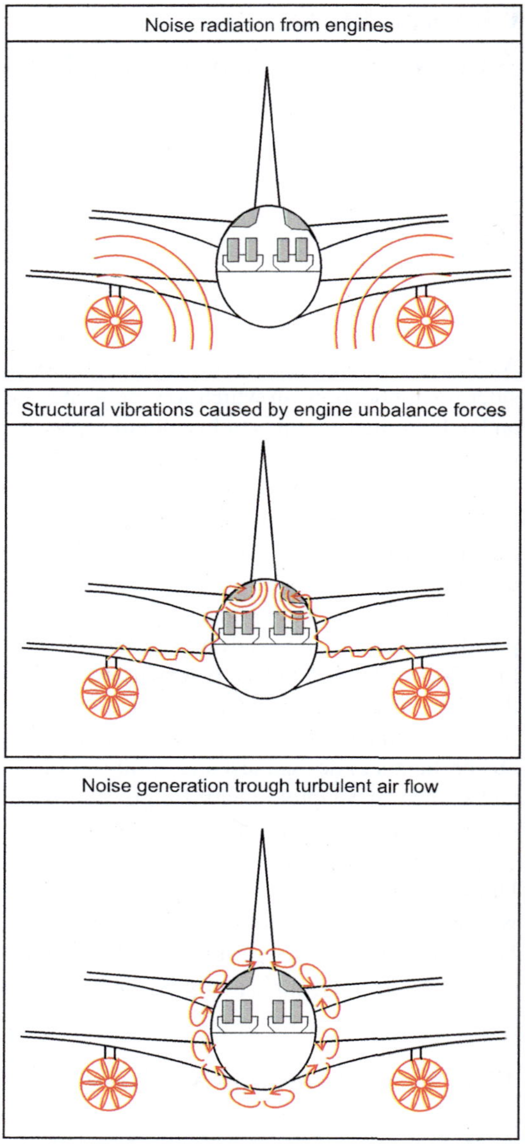

Fig. 12.1 Typical sources for jet aircraft interior noise (adapted from Gardinio 2007). *Top*: Sound radiation from engine. *Middle*: Structural vibrations caused by unbalance forces of engines that are in this case directly attached to the fuselage. *Bottom*: Boundary layer noise exciting the fuselage of a jet powered aircraft

This statement express the fact that active noise reduction in propeller aircraft has typically been addressed with feed-forward control, because a well correlated reference signal is accessible by monitoring the engine.

The situation is different for jet powered aircraft, because the noise field in a jet aircraft cabin is generated by the engines as well as by the airflow over the aircraft surfaces. As illustrated by Fig. 12.1 it is possible to distinguish between cabin noise that is (i) radiated from the engines and transmitted through the fuselage,

Fig. 12.1 (top), (ii) caused by structural vibrations resulting from engine unbalance forces, Fig. 12.1 (middle), and (iii) excited by the turbulent boundary layer, Fig. 12.1 (below).

Backward radiated broadband jet noise, generated by the engines, results in an acoustic field on the airframe that is time and spatially random. Forward radiated noise from a jet engine fan inlet consists of broadband as well as of dominant tonal components at various frequencies. The latter is also known as compressor buzz-saw noise and occurs particularly at take off and landing. However, tonal components in the cabin noise spectra may also result from engine unbalance forces.

As reported in (Wilby and Mixson 1991), aerodynamic cabin noise is generated by the airflow over the aircraft surfaces. For smaller aircraft, airflow noise is important at higher frequencies. For larger, jet-powered, well-streamlined aircraft, high speed flow generates significant levels of turbulent boundary layer noise that is usually the most important source of cabin noise for these types of aircraft. Results of in-flight measurements of the fluctuating pressure acting on the fuselage surface beneath the boundary layer taken from a large jet aircraft operating at speeds from 138 m/s to 242 m/s at an altitude of 7620 m clarified that the pressure was broadband and contributed significantly to the cabin noise between 100 Hz up to frequencies above 2 kHz. It has been found that an increasing airspeed (from Mach 0.45 to Mach 0.78) resulted in an increasing pressure spectral density of about 7 dB.

Active control in jet aircraft has also been focused on the rejection of tonal disturbances. One particular application in which the signals of ten error microphones and four active mounts were applied to reduce the cabin noise in a Citation II caused by engine unbalance forces has been described in (Ivers et al. year unknown). It was found that the interior noise was globally reduced by 6.1 dB; the compressor tone (also known as N_1 tone) was reduced by 16.5 dB.

12.1.2 Comments on Hybrid ANC-Audio Systems

The concept of a hybrid ANC-audio system was presented in (Kuo and Finn 1993). The authors report on the integration of three-dimensional ANC into an existing audio system in order to reduce annoying low frequency noise.

For this purpose it is necessary to remove the musical interference in the sound pressure $p_1(t)$ and in the sound pressure $p_2(t)$ measured at the error microphones as illustrated by Fig. 12.2.

The discrete-time error signals $e_1(n)$ and $e_2(n)$, obtained after performing the compensation, are free of the audio signals $q_L(t)$ and $q_R(t)$. Hence, the control signals $q_1(t)$ and $q_2(t)$ that are superimposed on the audio signals have only an effect on the noise, but not on the audio entertainment. Computer simulations were conducted to verify the capabilities of interference suppression and on-line modeling. It turned out that the musical interference suppression process reduces the interference of musical signals in the updates of the ANC filter. Furthermore, it was found that this technique also provides an optimal method for on-line modeling of secondary paths.

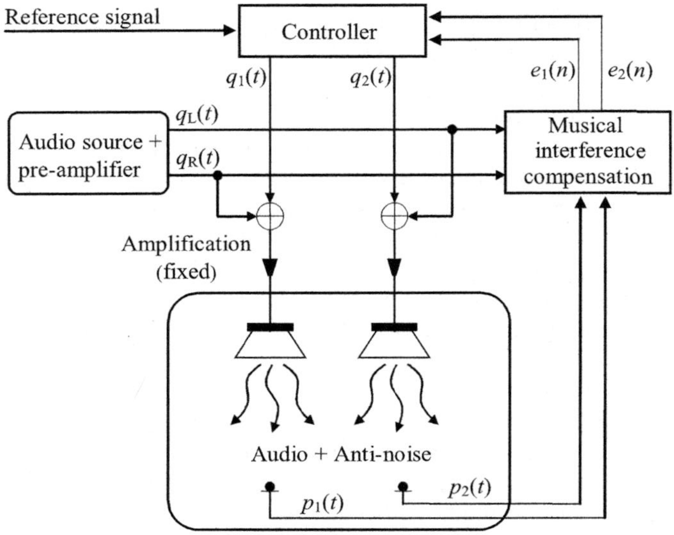

Fig. 12.2 Musical interference compensation in a multi-channel ANC-system (adapted from Kuo and Finn 1993)

Engineering applications are to be found in (Sano et al. 2001), (Gan and Kuo 2001) and (Gan and Kuo 2002). The first reference proposes commercial application for active suppression of low-frequency road noise in automobiles combined with an audio system. The combined ANC-audio system was tested in a station wagon. This system reduced (very) low-frequency road noise in the front seat by about 10 dB (the reduction was measured in the frequency range 35 Hz $< f <$ 45 Hz) and improved the audio listening experience while driving.

The second and third reference report on an integrated approach in designing a noise reduction headset for audio and communication applications. The proposed system uses a single microphone per ear cup, thus it is a compact solution that can easily be integrated into existing audio and communication devices to form an integrated feedback ANC-headset.

Simulation were conducted successfully showing that such an ANC-headset, can remove the disturbing noise and, at the same time, allow the desired speech or audio signal to pass through without cancellation. Nowadays, noise canceling head sets are components-of-the-shelf products, see (Harris 2008).

12.2 Feasibility Study on a Very Light Jet ANC-Audio System

Active control of broadband disturbances in the jet aircraft cabin is still a challenge, since many complex and uncorrelated sources such as boundary layer noise and turbine-induced jet noise are involved in creating a broadband disturbance as mentioned in Sect. 12.1.1. However, because of cabin sizes in the range of large automobiles, ANC-systems for small jet aircraft, especially very light jets (VLJ), may retain a feasible effort in terms of algorithm complexity and system topology. It should be kept in mind that an active system, especially in such small aircraft, must neither create excessive payload nor take up much space to create a local zone of quiet around the passengers head.

For these reasons, a *feasibility study* on a combined ANC-audio system for VLJs was conducted within the Hamburg aeronautics research framework (Lufo Hamburg). As mentioned in the motivation to this chapter, the project work was carried out at the Helmut-Schmidt-University/University of the Federal Armed Forces Hamburg in cooperation with Innovint—Aircraft Interior GmbH. This two year research project included

- *Noise field analysis* based on in-flight measurements in a typical VLJ cabin,
- *Anti-noise field analysis* in a VLJ mock-up,
- *Domain-specific design* for the combination of audio entertainment and ANC,
- *System integration and functional testing* of the ANC-audio system performance.

These design tasks were processed to analyze the feasibility of designing an ANC-audio system that is capable of meeting the following **requirement**: *Remove the low frequency disturbing noise around the head of a VLJ passenger by a local ANC-system. At the same time allow the desired speech or audio signal to pass through without cancellation.*

12.2.1 In-flight Measurements and Noise Field Analysis

Description of In-flight Measurements in a VLJ

To perform the *noise field analysis task* which requires the evaluation of typical cabin noise in a VLJ, in-flight measurement data were recorded in a Cessna Citation Bravo as reported in (Pabst et al. 2008a). The aircraft engines are of type Pratt and Whitney PW 530A. They are directly mounted on the aft fuselage.

The cabin of this particular aircraft, shown in Fig. 12.3 (left), has a lounge configuration and is equipped with 7 seats. The cabin dimensions are approximately $1.4 \times 5 \times 1.5$ meters (height, length and width). A view into the cabin in is presented in Fig. 12.3 (middle). The measurements were divided into three parts:

- Constant readout of signals from five microphones at fixed positions throughout the entire flight from takeoff to landing,
- Step-wise sound-field mapping with an array of five microphones at ear level of a seated passenger throughout the cabin area during the cruise state,
- Constant readout of sound pressure at the ears of an artificial head.

The five microphones used in the fixed positions as well as in the array were of type B&K 4188. Two of the fixed microphones were placed in the gangway,

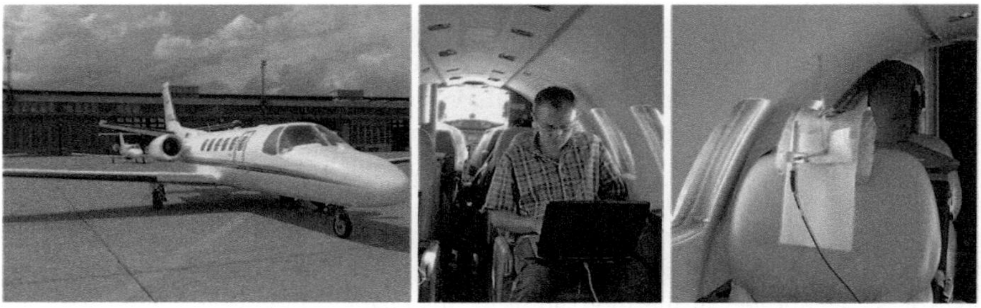

Fig. 12.3 In-flight measurement in a very light jet. Aircraft is shown *left*. View into cabin shows Dipl.-Ing. Kai Simanowski performing measurements (*middle*). Test equipment is shown *right*

whereas three of them were attached to the seat headrest as Fig. 12.3 (right). The distance from microphone to headrest was 15 cm and 65 cm from microphone to the seat area, as required by DIN ISO 5129 (2003). Two frontends (type: B&K 3560B) were used for data acquisition. The artificial head (type: HEAD Acoustics, HMS III.0) was used to record aurally compensated data. As all components were battery-powered, no further power sources were required.

The sound-field mapping was done from back to front for each side of the cabin. Data acquisition was done in equal intervals of 0.1 m at 1.18 m height (measured from the gangway floor). The microphones were aligned in a small line array, starting 0.05 m from the cabin wall, firmly attached to a rod which was moved along the gangway. Data from the complete seat area was collected, excluding the lavatory/luggage compartment. The mapping was done at 24000 ft with 268 KIAS (knots indicated air speed). Please notice that all air conditioning was turned off during data acquisition, to exclude noise generated by internal sources.

Test Results and Conclusions

Off-line data analysis was performed to analyze the recorded data. The results of this post-processing were presented in (Pabst et al. 2008b) and (Pabst et al. 2009). It turned out that the total SPL varied between 84 dB and 95 dB. The lowest values were found at the center of the front cabin and the highest at rear positions close to the fuselage as shown in Fig. 12.4.

It can be seen that the total SPL decreases in flight direction as well as in cross direction (from the cabin side wall to the gangway). Further considering Fig. 12.5, the spectral shape of the ANC-relevant part of the disturbance is colored, broadband noise with few narrowband components, concentrated below 300 Hz. The low-frequency components are mainly responsible for the difference in SPL of the front and rear cabin areas.

The spectral shape over time is (for the whole flight) shown in Fig. 12.6. It was calculated from data that were recorded at a rear seat using maximum window overlap and 1 Hz frequency resolution. It indicates that the disturbance is nearly station-

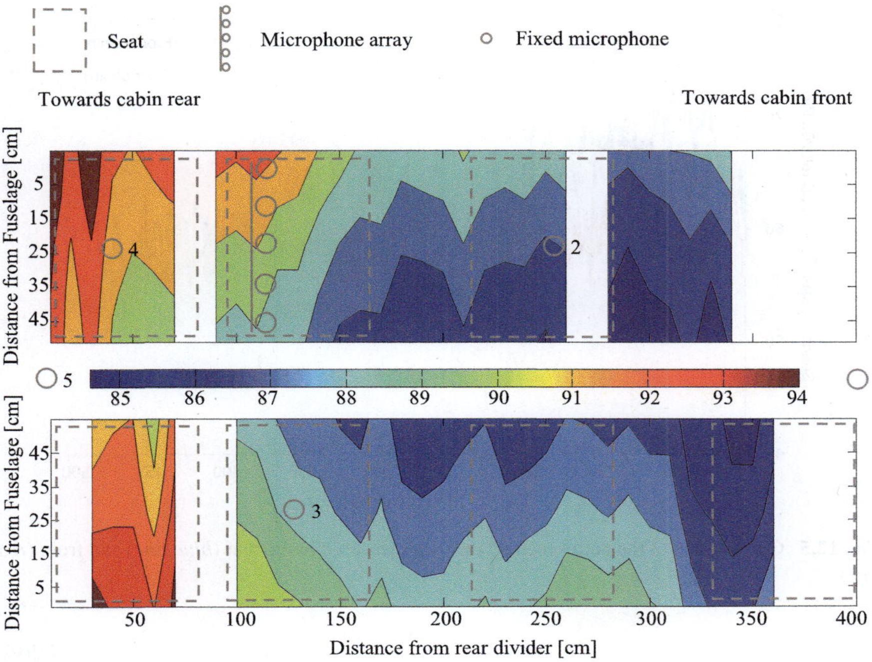

Fig. 12.4 Mapping of the total cabin SPL (scale is given in dB rel. 20^{-5} Pa without any weighting) during cruise for the left (*top*) and right (*below*) cabin halves. Microphone *1* and *5* were placed in the gangway (from Pabst et al. 2008b)

ary during cruise. This indication was confirmed by analyzing ten minutes of the "cruise-sequence."

This time span was used to determine both mean and variance of the recorded signal. It was found that the mean of the signal strives against zero, whereas the variance strives against a constant value, compare Fig. 12.7.

Considering that the main noise contribution during cruise was (i) nearly stationary, and (ii) found in the low frequency range below 300 Hz (especially in the rear cabin), it was concluded that the application of local ANC is possible. Because the primary sound field was analyzed in a VLJ cabin, it was possible to observe, and report, the acoustic properties of the operational environment in which the future ANC-audio system will operate. Hence, SML 1 was achieved at the end of this design task.

12.2.2 Anti-noise Field Analysis Using a VLJ Mock-up

In the desired low frequency range, a standard loudspeaker is the first choice as a secondary source for an active noise reduction system. The positioning of such

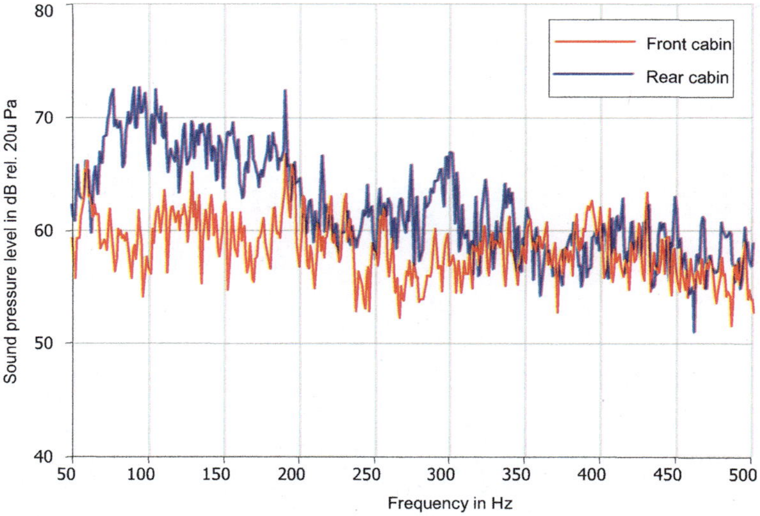

Fig. 12.5 Comparison of the auto spectral densities measured in the rear (*blue line*) and front (*red line*)

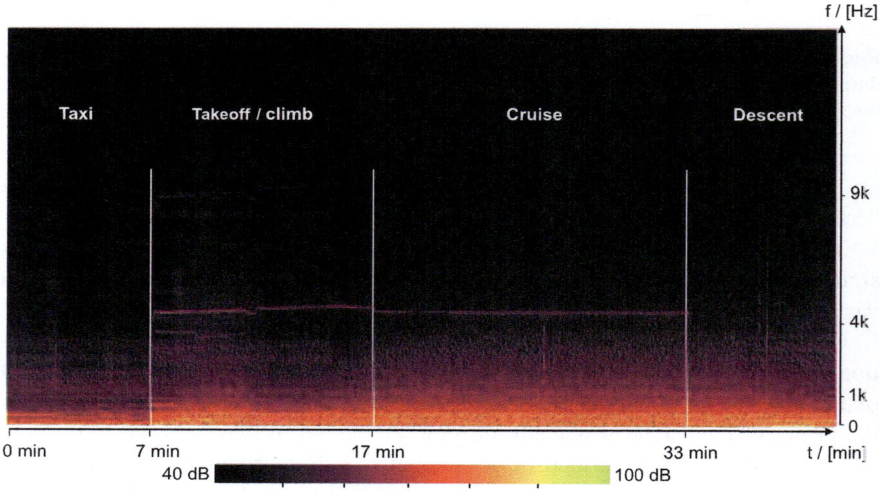

Fig. 12.6 Periodogram showing the spectral shape over time for the in-flight measurements

a speaker in an aircraft cabin is restricted, as is the number of speakers. In order to determine actuator positions that can be used to reject broadband noise at the ears of a sitting passenger, TPA (compare Sect. 7.3.2) was performed. This *anti-noise field analysis* was based on experimental data that were measured in a VLJ mock-up. The latter is shown in Fig. 12.8.

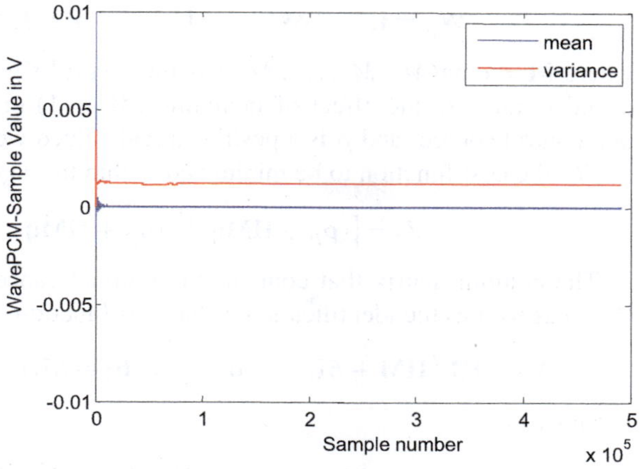

Fig. 12.7 Mean and variance of sound pressure signal measured in a VLJ during cruise

The mock-up was designed, manufactured and assembled by the industrial partner. It represents the dimensions of a typical VLJ and is made of wood. It has a length of approximately 5.7 m and is 1.6 m high. It incorporates a simplified cabin interior, compare Fig. 12.8 (right), made of foam and wood. Original cabin materials were used for the lining, carpets and seats.

Simplifications were made for the avionics in the cockpit as well as for the seats as reported in (Pabst et al. 2008a). An average value of 0.23 s was measured for the reverberation time inside the mock-up.

A pragmatic and efficient method to choose an optimal subset from a given finite number of possible actuator locations is the sequential search, see e.g. (Elliott 2001), using experimental estimates of the frequency responses measured between all possible actuator- and sensor locations as well as the measured disturbance at the sensors. Please notice that fixed sensor positions were specified at the headrest (two sensors per seat), because local ANC around the passengers head is desired. For this reason and considering the identifications

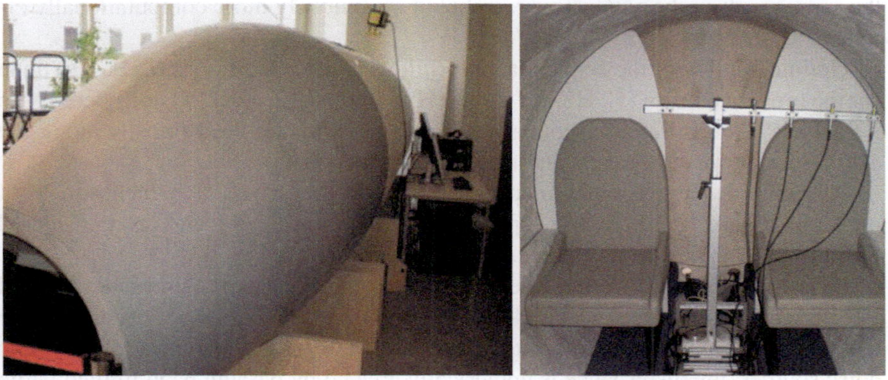

Fig. 12.8 Acoustic mock-up of VLJ. *Left*: View against flight direction. *Right*: View into cavity

$$\mathbf{W}_p \equiv \mathbf{I}, \qquad \mathbf{W}_q \equiv \beta \mathbf{M}^{-2}, \qquad \mathbf{P} \equiv \mathbf{I}, \qquad \mathbf{Q} = \mathbf{M}, \tag{12.1}$$

where $\mathbf{M} = \text{diag}(M_1, M_2, \ldots, M_M)$ is the control signal weighting matrix (introduced to analyze the effect of including ($M_i = 1$) or excluding ($M_i = 0$) a certain control source) and β is a positive a real valued weighting parameter, compare (7.57), the cost function to be minimized within the TPA was given by

$$J_0 = \left[(\mathbf{p}_p + \mathbf{HMq})\right]^H (\mathbf{p}_p + \mathbf{HMq}) + \beta \mathbf{q}^H \mathbf{q}. \tag{12.2}$$

The column matrix that contains the optimal canceling signals is, considering (6.13) as well as the identification of the cost function parameters

$$\mathbf{A} \equiv \mathbf{MH}^H \mathbf{HM} + \beta \mathbf{I}, \qquad \mathbf{b} = \mathbf{b}_1 = \mathbf{b}_2 \equiv \mathbf{MH}^H \mathbf{p}_p, \qquad c \equiv \mathbf{p}_p^H \mathbf{p}_p, \tag{12.3}$$

given by

$$\mathbf{q}_{opt}(\mathbf{M}, \beta) = -\left[\mathbf{MH}^H \mathbf{HM} + \beta \mathbf{I}\right]^{-1} \mathbf{MH}^H \mathbf{p}_p. \tag{12.4}$$

Equation (12.4) clarifies that $\mathbf{q}_{opt}$ depends on the actuator locations that are specified by $\mathbf{M}$. The estimate of the optimal nominal controller performance is for each frequency line according to (6.147) defined as

$$NR_{J_0} = 10 \log_{10} \left(\frac{J_0[\mathbf{q}_{opt}(\mathbf{M})]}{\mathbf{p}_p^H \mathbf{p}_p} \right). \tag{12.5}$$

Using (12.5) it is possible to calculate an IPE for a multi-channel ANC-system that uses a certain set of M actuators to reduce the noise measured at the (in these analysis fixed) position of L error sensors assuming a perfect coherence between disturbance and reference signals.

When analyzing a subset of speakers for a fixed number of sensors the transfer function matrix $\mathbf{H}$ and the column matrix $\mathbf{q}$ are reduced accordingly. Instead of determining the best locations for a specific fixed number of speakers by analyzing all possible combinations (exhaustive search), a set starting with the best single speaker is sequentially filled with speakers that respectively create the best reduction with the ones previously chosen. As noted in (Elliott 2001) this will typically not deliver the same result as the exhaustive search, but is obviously more computationally efficient.

Optimization with the sequential algorithm, summarized in Fig. 12.9, was done for the broadband case (50 Hz $< f <$ 500 Hz). In the VLJ mock-up considered here a total set of 46 different secondary source locations was used (mostly along the cabin floor, lower side panels and front- and rear-dividers).

As mentioned above, eight microphones (M1–M8) were fixed in the headrests of the seats. Figure 12.10 presents a projection of all transducer locations (considered for optimization) to the x–y-plane.

Please notice that different orientations of the loudspeaker membranes as well as different z-coordinates were taken into account to obtain the set of 46 possible installations as reported in (Sachau et al. 2008b). In order to compensate for multiple noise sources, the disturbance is numerically created by passing band limited white noise through one of the secondary sources, excluding that one from the search set.

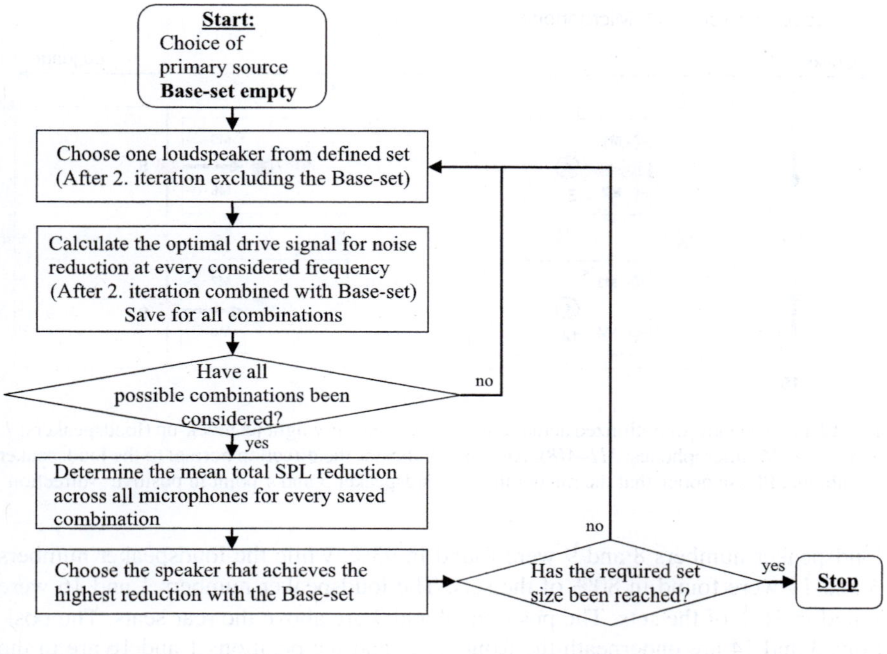

Fig. 12.9 Flow diagram of the actuator location selection procedure applying a sequential search algorithm (from Pabst et al. 2009)

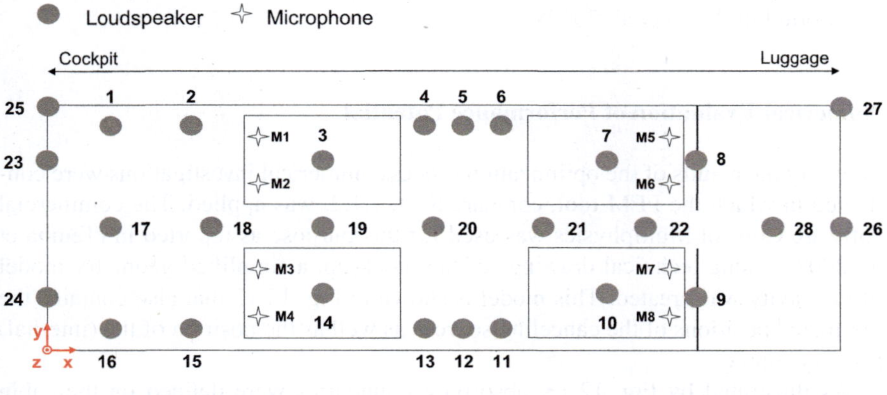

Fig. 12.10 Top view of actuator locations projected to the *x–y*-plane that were considered for optimization in the very light jet mock-up (location numbers *1–28*, microphones *M1–M8*)

An optimization result was determined using each position as a disturbance source, leading to 46 sets of optimal speaker locations. The occurrence of each speaker location over all sets was then examined.

The final set was chosen according to the highest number of occurrences consisting of six speakers (1, 3, 8, 9, 14, 16) arranged as shown in Fig. 12.11. The

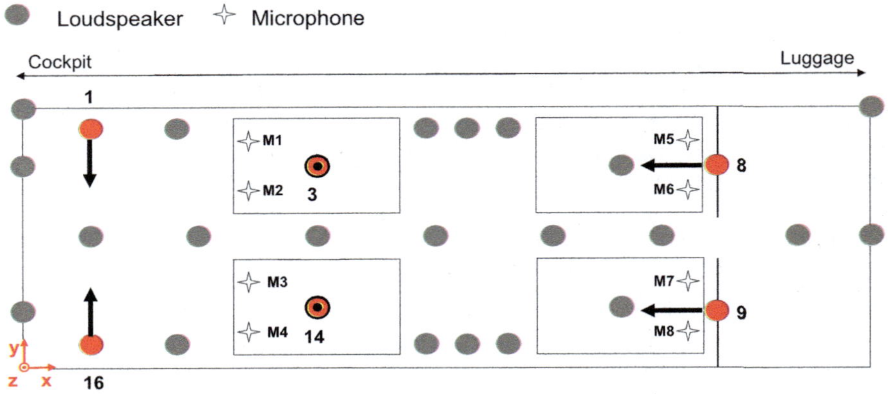

Fig. 12.11 Top view of optimized actuator locations in the very light jet mock-up (loudspeakers: *1*, *3*, *8*, *9*, *14*, *16*; microphones: *M1*–*M8*). An *arrow* indicates the direction normal to the loudspeaker membrane. Please notice that the membranes of loudspeaker *3* and *4* point in positive z-direction

loudspeaker numbers 8 and 9 were found in 93%, while the loudspeaker numbers 3 and 14 were found in 80% of the sets. The loudspeaker numbers 1 and 16 were found in 30% of the sets. The positions 8 and 9 are above the rear seats. The positions 3 and 14 are underneath the front seats, and the positions 1 and 16 are in the lower side-panels.

The IPE calculated from (12.5) for this set predicts a noise reduction of the mean total SPL by 16 dB over all disturbance sources in the considered frequency range as reported in (Pabst et al. 2009).

Numerical Evaluation of Performance Potential

To verify the results of the optimization process, numerical investigations were conducted in which the FEM tool, compare Sect. 7.1.2, was applied. The commercial software Comsol Multiphysics was used for this purpose as reported in (Teuma et al. 2007). Using technical drawings of the mock-up, a simplified geometry model of the cavity was created. This model is shown in Fig. 12.12 that also contains the optimized positions of the canceling sources as well as the position of the (internal) primary source.

As illustrated by Fig. 12.13, absorbing boundaries were defined on the cabin back sides and the cabin side-wall, to consider acoustic absorption caused by the lining material. The complex reflection coefficients R_1, R_2 and R_3 specified on these surfaces were used to prescribe a normal impedance using the inverse form of (3.51). Furthermore, a complex wave number, compare (3.75), was taken into account to model viscous damping effects.

The geometry model of the VLJ mock-up was discretized using tetrahedron elements with quadratic shape functions as reported in (COMSOL 2008). A number of at least 3 elements per wavelength for time-harmonic analysis up to 500 Hz was

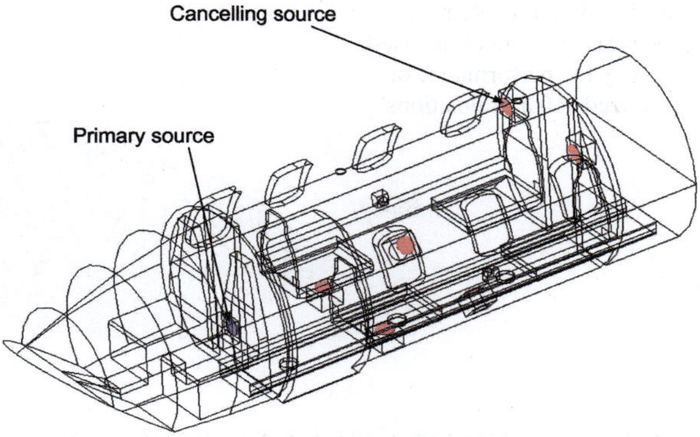

Fig. 12.12 Geometry model of a very light jet mock-up and actuator position in the cavity (adapted from Teuma and Kletschkowski 2009)

realized. Therefore, the largest element edge length was limited to 0.22 m. The finite element model used for verification is shown in Fig. 12.14. It consists of 60989 elements and 92162 degrees of freedom. By choosing the mesh parameters, attention was given to generate a regular mesh.

Experimental data measured in the VLJ mock-up were used to adjust the finite element model parameter (speed of sound c, damping ratio ξ, and reflection coefficients R_i with $i = 1, 2, 3, \ldots$). For this purpose, the sound field in the mock-up was mapped (as in the in-flight test) using broadband excitation with band limited white noise ($f < 800$ Hz) generated by the primary noise source position shown in Fig. 12.12. The resulting sound field was measured at 232 points located in one plane. The height of the specified measurement plane (approximately 1.15 m) was adjusted to the ear level of a sitting passenger. Because the normal acceleration at the loudspeaker membrane was measured too, it was possible to determine the FRFs between all 232 measurement points and the normal acceleration of the primary source.

Finite element model parameters that were adjusted for two excitation frequencies are shown in Table 12.1. Details about the adjusting procedure were presented in (Teuma et al. 2008) and in (Sachau et al. 2008b) and will not be discussed here. However, it is necessary to motivate the values for the speed of sound listed in Table 12.1, because these values represent an overestimation of 3% for $f = 100$ Hz

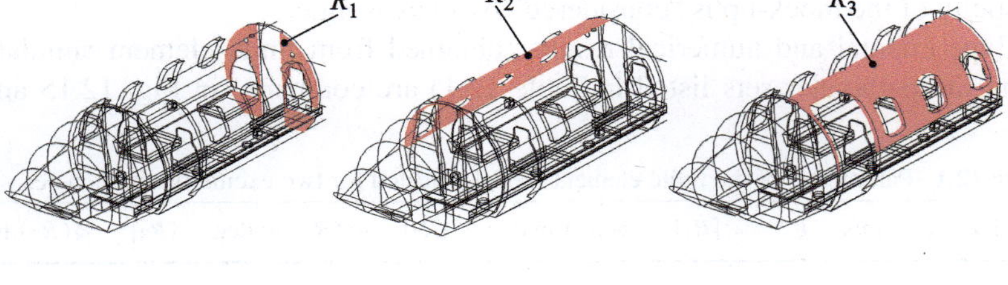

Fig. 12.13 Surfaces modeled as absorbing boundaries at the back side of the cabin (*left*) and the cabin sidewall (*middle, right*), adapted from (Teuma et al. 2008)

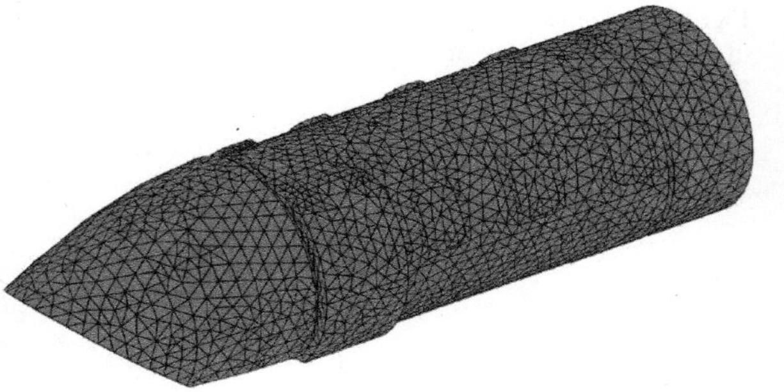

Fig. 12.14 Finite element model of a VLJ mock-up used to verify the performance of optimized actuator positions

and an overestimation of 9% for $f = 300$ Hz, if a "correct" value of $c = 340$ m/s is accepted.

It is obvious that such an overestimation also leads to an overestimation of the wave length that may be calculated as $\lambda_\alpha = \alpha \cdot cf^{-1}$, where $\alpha > 1$ is a proportionality factor that represents the deviation from the "correct" value for the speed of sound. Please notice that the same overestimation in wave length is obtained, if the "correct" value is used to express the speed of sound, but the excitation frequency is lowered such as $f_{red} = \alpha^{-1} \cdot f$.

This relation can be explained as follows. The system response of the (purely) acoustic finite element model will certainly differ from the vibro-acoustic system response of the VLJ mock-up. Hence, it will be impossible to match numerical results with experimental data considering the same excitation frequency. However, it will be possible to identify similar acoustic mode shapes that appear at adjacent frequencies, if the coupling is weak as reported in (Sandberg and Göransson 1987).

A motivation to this approach that is based on physical parameters can be given for the first acoustic mode. If the air filled cavity of the VLJ mock-up is represented by a lumped system with stiffness $k_c = \rho_\infty c^2 S/L$ (where S is the cross section of the cavity and L is its length) and mass $m = \rho_\infty SL$ and, if the boundaries are assumed to be sound hard, the first natural frequency of the cavity can be approximated as $f_0 \approx 0.5 \cdot cL^{-1}\pi^{-1}$. This approximation is reduced to $f_{0,red} = \alpha_0^{-1} \cdot f_0$, if the elasticity of interior wall is taken into account by the structural stiffness k_s used to model a series connection of elastic springs that is represented by the proportionality factor $\alpha_0^2 = (1 + k_c \cdot k_s^{-1})$. Because α_0 is greater than one, the frequency of the first acoustic mode is approximated for a lower frequency, if $k_s \neq 0$. Thus, the elasticity of the mock-up is "considered" by increasing c.

Experimental and numerical results (obtained from finite element simulations using the parameter sets listed in Table 12.1) are compared in Fig. 12.15 and in

Table 12.1 Parameters of the finite element model adjusted for two excitation frequencies

f in Hz	c in m/s	ξ	$\lvert R_1\rvert$	$\sphericalangle(R_1)$ in deg	$\lvert R_2\rvert$	$\sphericalangle(R_2)$ in deg	$\lvert R_3\rvert$	$\sphericalangle(R_3)$ in deg
100	350	0.10	0.80	0.0	0.80	0.0	0.80	0.0
300	370	0.09	0.99	0.0	0.99	35.0	0.99	35.0

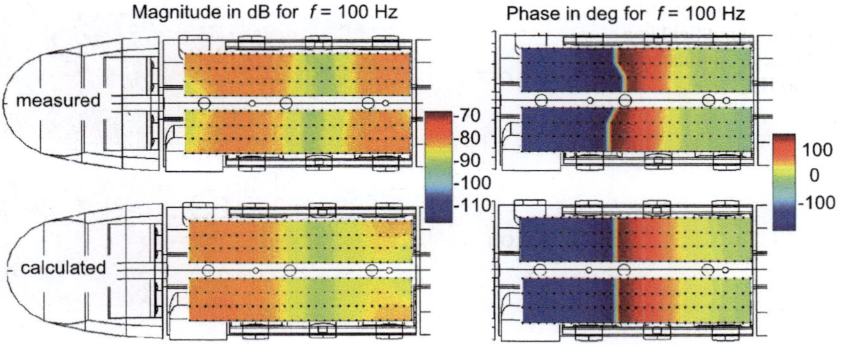

Fig. 12.15 Comparison of measured and calculated FRF for $f = 100$ Hz. Magnitude is shown in dB rel. 20 Pa $\cdot$ s^2 m^{-1} (*left*); phase is shown in degree (*right*)

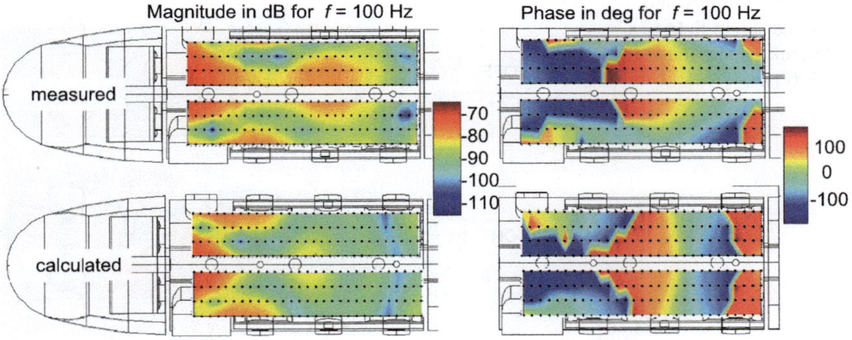

Fig. 12.16 Comparison of measured and calculated FRF for $f = 300$ Hz. Magnitude is shown in dB rel. 20 Pa $\cdot$ s^2 m^{-1} (*left*); phase is shown in degree (*right*)

Fig. 12.16. These illustrations present linear interpolations for both magnitude and phase of the FRFs that were determined by relating the pressure response at all 232 microphone positions to the acceleration of the primary source. The evaluation points are indicated by black dots.

It can bee seen that the acoustic finite element model is capable of describing the interior noise field with minor deviations from the measured data for $f = 100$ Hz. For the second excitation frequency ($f = 300$ Hz), the deviations in magnitude and phase become more significant.

However, also for this frequency the numerical values of simulated and measured data are still on the same scale and the qualitative shape of the FRFs is nearly identical. Hence, it was possible to use the finite element model of the VLJ to verify the optimization results for these particular frequencies.

The validated numerical model of the VLJ mock-up was applied to verify the performance of the ANC-system considering 6 loudspeakers (located at optimal positions) and 8 microphones (two microphones per seat) as reported in (Teuma and

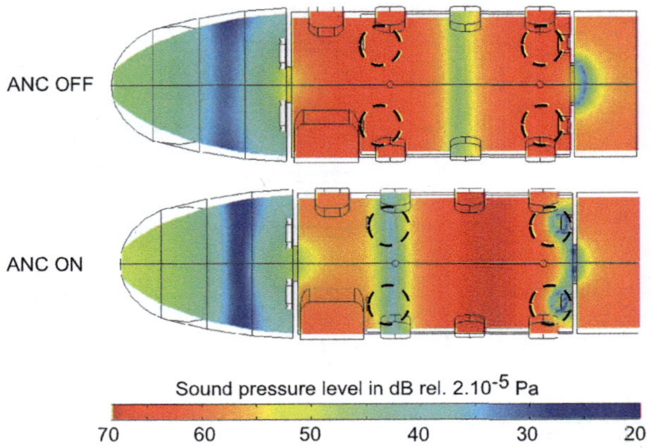

Fig. 12.17 Primary and controlled sound field in a VLJ mock-up simulated for $f = 100$ Hz

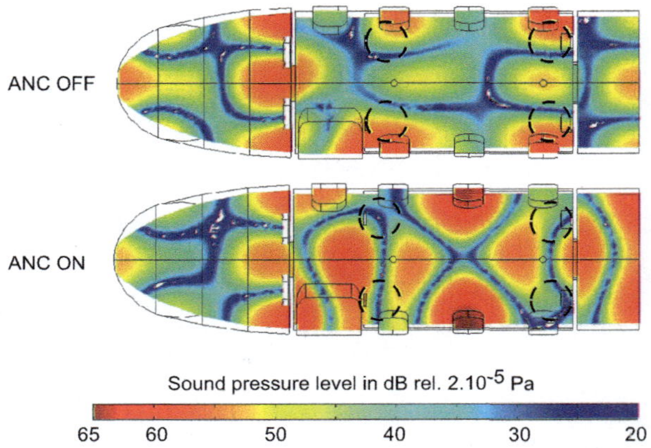

Fig. 12.18 Primary and controlled sound field in a VLJ mock-up simulated for $f = 300$ Hz

Kletschkowski 2009). To determine an IPE, the steady state of the control system was analyzed for $f = 100$ Hz and $f = 300$ Hz.

Equation (6.93) was applied to determine the optimal normal acceleration for each secondary source (their positions are shown in Fig. 12.12) that were used to specify *Neumann* BCs according to (3.24). A weighting parameter of $\beta = 0.072$ was used for this purpose.

Simulation results for both the uncontrolled and the controlled sound fields are shown in Fig. 12.17 and in Fig. 12.18. They were determined in a plane 1 m above floor level. In both figures the primary noise field is shown top, whereas the controlled sound field is shown below. The domains marked with the circles indicate the areas around the error microphones (control points).

The results indicate that ANC can be applied successfully at low frequencies (100 Hz) as well as close to the mid frequency range (300 Hz), if the canceling sources are located at the positions determined by the selective optimization procedure.

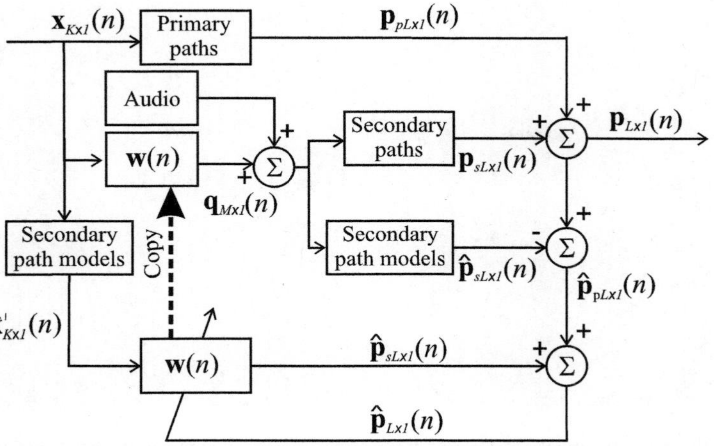

Fig. 12.19 Block-diagram of the modified FxLMS algorithm for ANC with audio entertainment

12.2.3 Design of the ANC-Audio Controller

It is obvious that a future ANC-audio system for VLJs can require a feedback control algorithm. However, to evaluate the feasibility of this combined approach, the less complicated feed-forward approach was sufficient. For this reason, the *controller design task* was focused on the MFxLMS (summarized in Table 6.3) to assess the real-time ANC performance in combination with audio entertainment in the VLJ mock-up. This algorithm was implemented and adapted for the multichannel usage considering one reference signal ($K = 1$), eight error signals ($L = 8$) and six control signals ($M = 6$).

To merge the in-flight audio-system and the ANC-system, the low frequency components of the audio signal could simply be added to the controller output, as shown in Fig. 12.19. However, two main issues arise in this case. The audio may interfere with the ANC-system inhibiting a correct adaptation of the filter and furthermore, the ANC-system may also cancel parts of the audio signal as well as the disturbing noise.

Nonetheless, high quality audio in the passenger area can be achieved if the error signal is freed of audio signal components before being used in the filter adaptation process. This can be done in a straightforward manner. As mentioned above, the pre-amplified audio signal is added to the control signal and, because of the modified structure, it is subtracted from the error signal along with the cancellation signal.

The extra secondary path model which is required for the audio compensating FxLMS algorithm proposed in (Kuo and Morgan 1996) is already contained in the modified structure. Considering that driving signals for the canceling sources are the sum of the audio- and controller-output, assuming perfect plant models (leading to $\hat{\mathbf{p}}_s(n) = \mathbf{p}_s(n)$), and implying that the audio signal is not correlated with reference signal, the initial disturbance $\hat{\mathbf{p}}_p(n)$ as without audio is obtained.

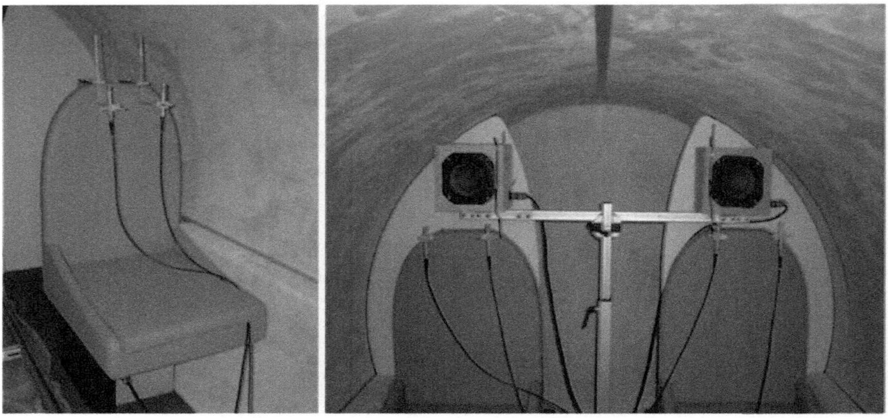

Fig. 12.20 Acoustic mock-up interior with seats, electro-dynamical loudspeakers and microphones mounted close to the backrests (from Pabst et al. 2009)

Because a system concept based on 8 headrest microphones, 6 canceling loudspeakers located at optimal positions and the MFxLMS-algorithm was worked out, SML 2 was achieved after completion of the *controller design task*

12.2.4 Integration and Test of the ANC-Audio Controller

System Integration

Using the numerically determined actuator positions an experimental ANC-system was integrated in the VLJ mock-up as shown in Fig. 12.20. It can be seen that two microphones were located at each headrest. The canceling speakers were located on the positions found by selective optimization.

The control algorithm was implemented on a dSPACE hardware platform (type 1006, 2.6 GHz, 16 bit I/O) for a multichannel system size of $K = 1$ reference signal, $L = 8$ error microphones and $M = 6$ secondary sources using a sampling frequency of 4 kHz. The algorithm was implemented in the power normalized, leaky form with off-line plant modeling for the secondary paths.

The layout of the integrated ANC-audio system is shown in Fig. 12.21. The primary noise field was created with a separate speaker, also located in the cabin, compare Fig. 12.12. For these experiments, the noise signal was passed directly through to the controller (internal referencing). The audio signal input to the controller was pre-filtered with a generic low-pass filter (cut-off frequency 1 kHz). A high-pass filtered version of the audio signal was sent to a separate audio system.

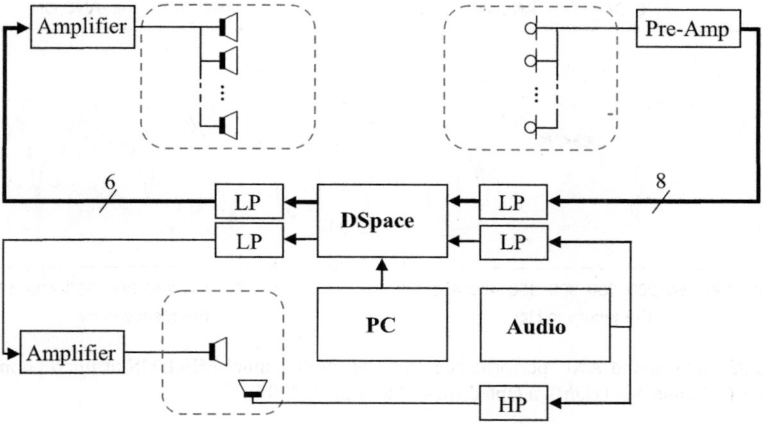

Fig. 12.21 Overview of the experimental system layout used for real-time testing of a combined ANC-audio system in a very light jet mock-up (from Pabst et al. 2009)

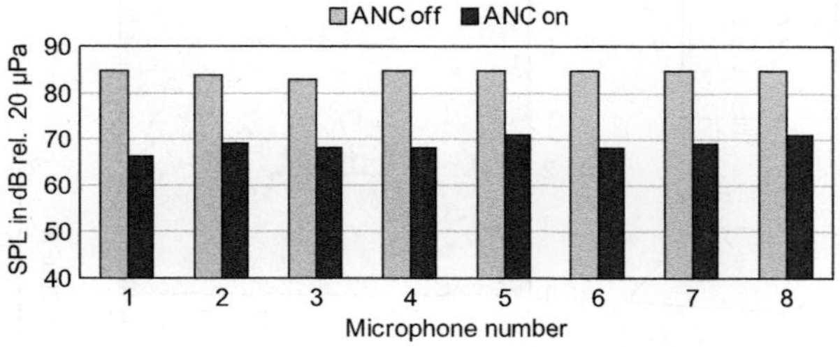

Fig. 12.22 Total sound pressure level (50 Hz $< f <$ 500 Hz) with and without active control shown for all microphone positions (adapted from Pabst et al. 2009)

Experimental Evaluation of ANC Performance

The experimental results shown here were created with a broadband primary noise field (50 Hz $< f <$ 800 Hz) 45 dB above the cabin background noise. The latter was of given by approximately 40 dB.

Best results were obtained using a filter length of 300 coefficients for the adaptive filters and 200 coefficients for the secondary path models. The SPL measured at microphones M1–M8 in the uncontrolled and in the controlled sound field are shown in Fig. 12.22. It can be seen that the initial disturbance was lowered by about 15 dB at all error sensors.

An example of the spectral shape of the controlled- and uncontrolled sound pressure is given in Fig. 12.23. The test results presented in this figure clarify that the control success was concentrated in the frequency range below 250 Hz.

The area of effect of the active noise system around the error sensors was mapped with a rectangular array of 16 microphones. Two planes, one vertically, one horizontally aligned, were analyzed for each seat. As an example, the zone of quiet around

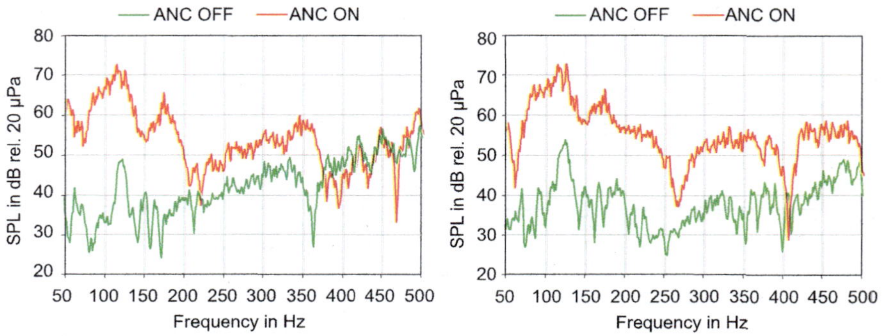

Fig. 12.23 Broadband ANC performance with the multichannel MFxLMS at microphone positions M1 (*left*) and M8 (*right*), adapted from (Pabst et al. 2009)

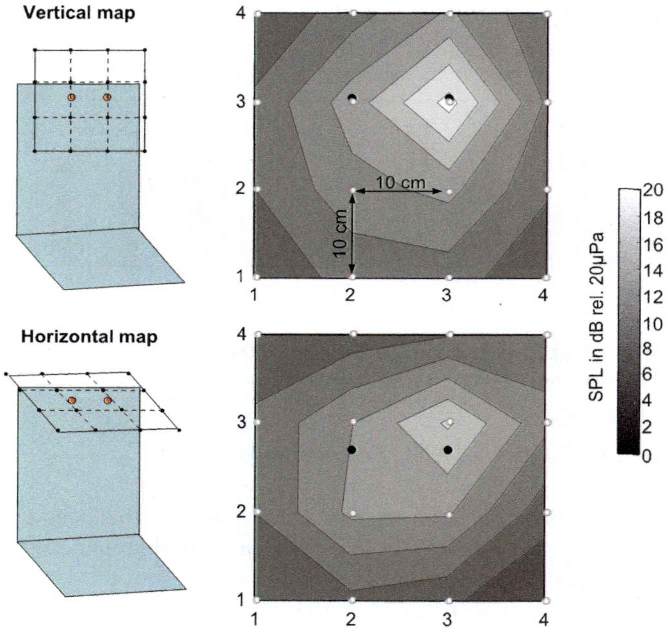

Fig. 12.24 Attenuation mapped at microphones M7 and M8, vertically and horizontally. Measurement points shown as *white dots*. *Black dots* represent the actual position of the error sensors that were not included by mapping

the microphone pair M7, M8 is shown in Fig. 12.24. Because of the linear interpolation between the measurement points the highest values of attenuation are not exactly at the positions of error sensors (their signals were excluded from the interpolation) in the vertical plot.

As it can be seen, the attenuation is higher for the microphone that is closer to the gangway. The zone with a minimum of 10 dB attenuation (total SPL reduction) is extended to approximately 0.15–0.2 m around the error microphones. Please notice that the mapping was done without placing an artificial head and torso in the seat. For the rear seats a slightly more confined quiet zone (approximately 0.1 m for 10 dB drop in attenuation) was detected which is due to the close positioning of the actuators and sensors in the rear cabin.

Evaluation of Combined Audio Active Noise Control

The controller was now tested with combined ANC and audio playback. As described in the experimental setup, the low-pass filtered audio signal was input to the ANC controller whereas the high-pass filtered part was passed on to a separate audio system.

The effect of turning on the controller can be seen in the top chart of Fig. 12.25. After approximately 2 s the ANC-system is turned on and noise reduction is observable in the desired frequency range. Furthermore, the audio signal remains clearly visible as can be seen when comparing with the unfiltered reference spectrogram of Fig. 12.25 (below).

When comparing the audio signal elements in the spectrogram of the cabin microphone and the reference, the effect of the 1 kHz low- and high-pass filters as well as the acoustic cabin properties must be kept in mind. During audio playback no disturbance of the ANC-system is detectable. The noise reduction can therefore be interpreted as a direct improvement of the signal to noise ratio thereby enabling the use of lower playback SPLs in the low frequency range for the audio system.

However, an objective evaluation of the change in audio quality is complex in this case as most analysis tools and algorithms are based on evaluating audio coding schemes and therefore, are not suitable for use with the noise levels encountered here (ITU-R 1999). This leaves large scale listening tests as a viable option.

12.2.5 ANC-Audio System Design Process Summary

The development and considerations in the system design of an experimental ANC controller combined with an audio system for a VLJ cabin have been presented. Having determined the noise characteristics in the target environment, the feasibility of a broadband multichannel ANC-system was analyzed with success. Actuator placement was discussed with a variant of the sequential search algorithm.

Using a control system with 8 microphones and 6 speakers, broadband active noise control in the frequency range 50 Hz $< f <$ 500 Hz was achieved with an average attenuation of 15 dB at the microphones and a zone of quiet in 0.1 m–0.2 m distance to the microphones.

A system concept was formulated and the feasibility of ANC in combination with audio entertainment (done with the MFxLMS-algorithm) was proven in real-time

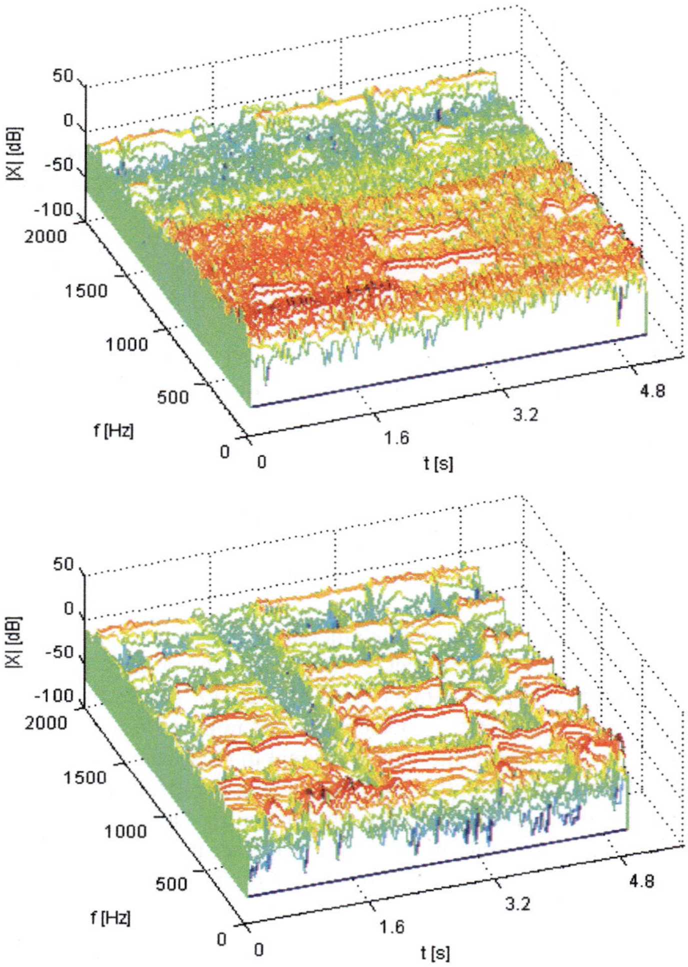

Fig. 12.25 *Top*: Cabin microphone signal with ANC (switched on after approximately 2 s). *Below*: Original unfiltered audio signal only (from Pabst et al. 2009)

experiments. For this reason, SML 3 was achieved. However, to increase the SML, further improvements such as successful testing of a feedback control approach that is independent of reference signals is required.

Chapter 13
Noise Source Localization in an Aircraft Cabin

Abstract Noise field analysis is usually the first design task of an ANC-system design process whose procedure has been discussed in Chap. 8. Such an analysis may also include noise source identification in order to develop an ANC concept that is capable of shielding the interior effectively against exterior noise. For this purpose it is essential to determine both positions and source strengths of the primary noise sources. These information are required for an appropriate placement and control of actuators and sensors. A powerful ANC-system design tool that can be used for noise source identification is the IFEM. This tool was described in Sect. 7.2. The upcoming chapter reports on investigations in which the IFEM has been applied to identify acoustic hot spots in a cross-section of a long-range airliner. The sound field in this cavity that represents the basic system was excited by internal as well as by external sources and mapped with a custom-built microphone array. A matching finite element model was developed. As it will be shown in the following chapter, the acoustic variables were calculated from the measured data with success.

13.1 Acoustic Ground Tests in an Aircraft Cross-Section

The application of the IFEM requires measurements in the cavity first. This section presents the results of acoustic ground tests (Simanowski et al. 2009) that were performed to prepare a validation of this noise source localization technique.

13.1.1 Description of the Aircraft Mock-up

The cross-section of a long-range airliner used for the ground tests is shown in Fig. 13.1 (left). It is a part of an aircraft mock-up that is located at the Technologiezentrum Hamburg Finkenwerder (THF).

This mock-up consists of an A340 cross-section (length: 4.2 m, diameter: 5.64 m, total weight including passengers: 5 t). It is designed to study the vibro-acoustic behavior of cross-section number 16, a cabin part in which the sound field is mainly excited by fan-noise generated by the engines as illustrated by Fig. 13.2.

T. Kletschkowski, *Adaptive Feed-Forward Control of Low Frequency Interior Noise*, Intelligent Systems, Control and Automation: Science and Engineering 56, DOI 10.1007/978-94-007-2537-9_13,

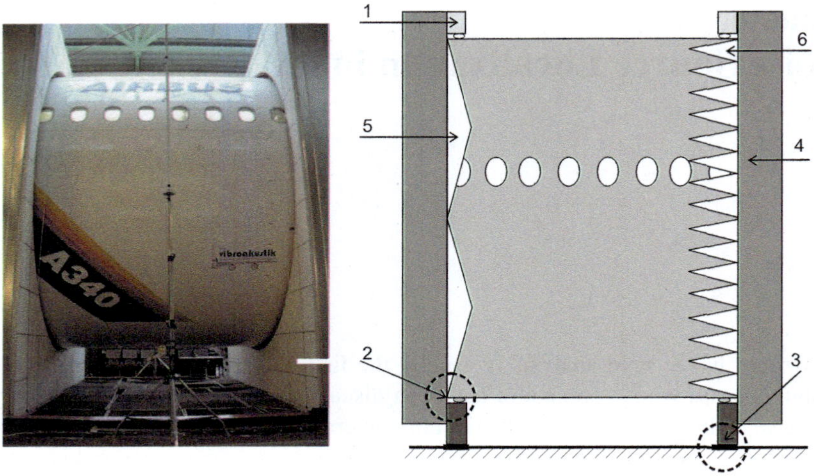

Fig. 13.1 Vibro-acoustic cabin mock-up based on an A340 cross-section (*left*) and plan of configuration (*right*) with *1*—mounting frame, *2*—elastic isolation between frame and mounts, *3*—elastic isolation between mounts and ground, *4*—outer acoustic termination of cross-section, *5*—inner termination based on reflectors, *6*—inner termination based on sound absorbers

As illustrated by Fig. 13.1 (right), the A340 cross-section is supported by a sophisticated mounting system that provides a good-approximation of the vibro-acoustic cabin-behavior under in-flight conditions. Depending on the requirements of a specific measurement campaign, the inner termination can be based on reflectors or on absorbing elements. If the latter are used, it is possible to suppress longitudinal acoustic modes. For these reasons, the THF-mock-up enables experimental investigations under improved laboratory conditions that are, according to Chap. 8, required to reach higher SMLs up to SML 6. For more information on this acoustic test rig see (Sachau et al. 2005b).

13.1.2 Description of Ground Test Experiments and Results

During the measuring campaign the cavity was fully equipped with lining, luggage bins (so called hat racks or overhead storage compartments) and five rows of seats. The mock-up was terminated at the cut faces with sound absorbing foam wedges to attenuate sound propagation in flight direction. This cabin configuration including the microphone array that was used for sound field mapping is shown in Fig. 13.3.

Correct mapping of the sound field requires a sufficient density of measurement points. The microphone array shown in Fig. 13.3 was based on a microphone spacing of 0.17 m. Using this spatial resolution it was possible to obtain six measurement points per wavelength (referring to the cabin cross-dimension) up to 330 Hz. The spacial distribution of the resulting 152 measurement points in the cross-section is shown in Fig. 13.4. Because measurements were performed at 22 array positions,

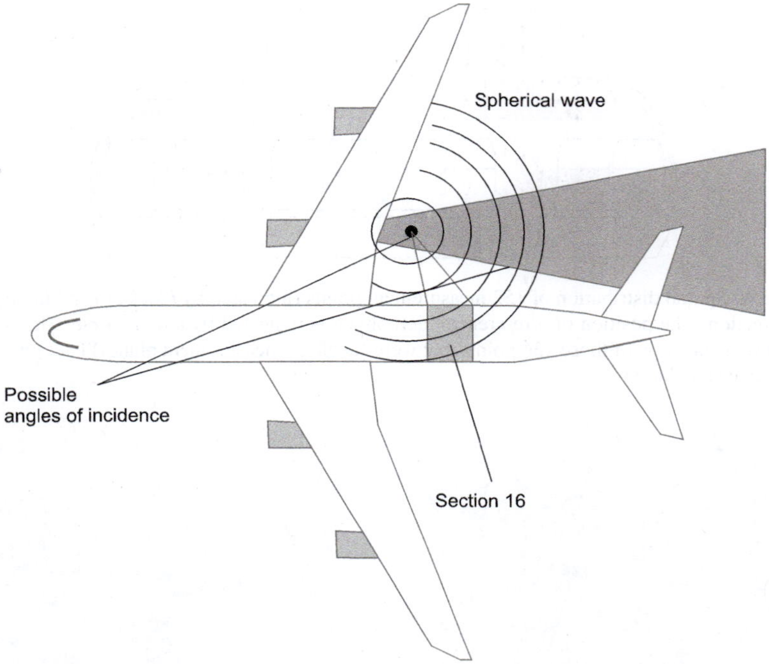

Fig. 13.2 Motivation of spherical wave propagation along the A340 cabin for the standing aircraft

Fig. 13.3 Cabin configuration including microphone array for sound field mapping during ground tests in the A340 cross-section

the total number of measurement points was 3344. To ensure high accuracy, B&K front ends (type: 3560 B and D) and B&K 1/4″ array microphones (type: 4935 and 4958) were used.

The cavity was excited by interior as well as by exterior noise sources using broadband disturbances with a total SPL of at least 80 dB at the reference micro-

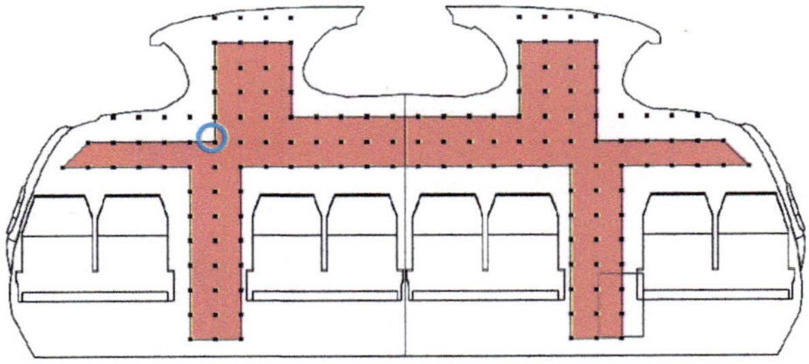

Fig. 13.4 Spatial distribution of 152 measurement points (indicated by *black circles*) in one cabin cross-section. The position of a reference microphone is indicated by a *blue circle*. The *red area* specifies the position of those 136 points that were (in all 22 measurement planes) later on used for the inverse calculation

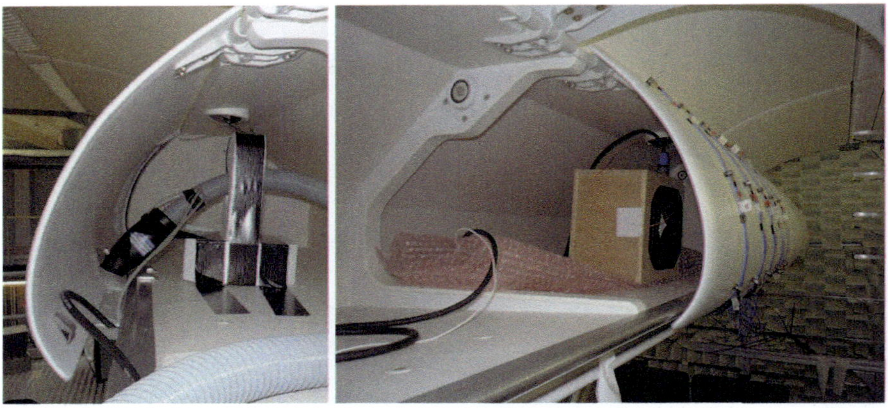

Fig. 13.5 View into hat rack with mounted velocity measurement adapter and extension hose of a volume velocity source (*left*) and view into hat rack with mounted electro-dynamical loudspeaker (*right*), adapted from (Simanowski et al. 2009)

phone, compare Fig. 13.4. This microphone was located close to the position of the internal sound sources.

In total three different excitation methods were used. To validate the IFEM approach, internal excitation was in the first step realized by a volume velocity source (VVS, B&K OmniSource type: 4295 with velocity measurement adapter and extension hose) that provides a direct acoustic excitation. As shown in Fig. 13.5 (left), the VVS was mounted in the hat rack with its orifice at the position of a hat rack opening grip. In a second step, an electro-dynamic loudspeaker placed in a had rack was used. For referencing the FRFs, a miniature accelerometer (B&K type: 4517) was mounted on the center of the loudspeaker membrane and used as reference. This arrangement is shown in Fig. 13.5 (right). To simulate sound transmission through the cabin side wall, external excitation was realized by public-address (PA) loudspeakers in a third step. These external sources were placed symmetrically on both

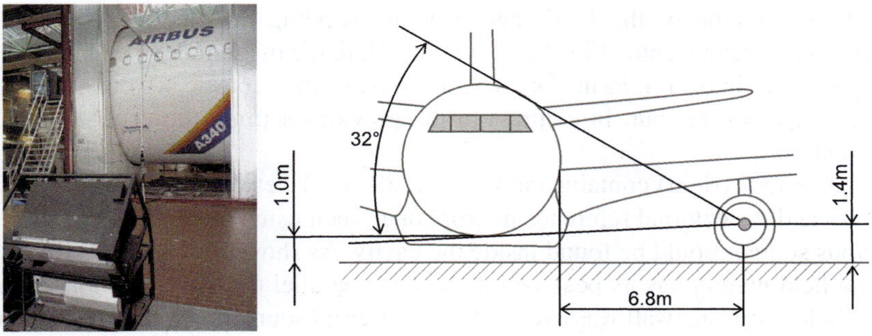

Fig. 13.6 Simulation of external jet-noise using PA-loudspeakers, see (Simanowski et al. 2009). External microphone shown in this figure was used as reference for calculation of frequency response functions

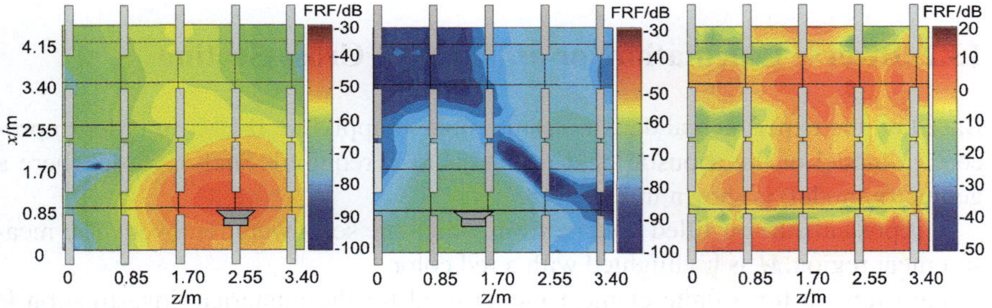

Fig. 13.7 FRFs measured for $f = 120$ Hz using the VVS (*left*), the electro-dynamical loudspeaker (*middle*), and the PA-system (*right*). Seats are represented by *gray rectangles*

sides of the mock up in a distance of 6 m. Figure 13.6 (left) shows this set-up in which an external microphone close to a window was used as reference.

The overall background noise level in the cabin was measured with less then 50 dB. Hence, the signal-to-noise ratio was at least 30 dB. The sound pressure level and FRFs were measured up to 800 Hz using a frequency resolution of 1 Hz. Averaging was performed over 100 time records with overlap, leading to duration of about 10 seconds for each measurement. For post processing and visualization, the data was exported as text file and reorganized with Matlab to get a 4-D matrix (X-position, Y-position, Z-position, value at frequency) for each sound excitation set up. Then 2-D slices were mapped with filled isoplanes (Matlab function "contourf") and saved as emf-images or assembled to create an avi movie file.

The experimental results for an excitation frequency of $f = 120$ Hz measured in a horizontal plane just above the back rests are shown in Fig. 13.7. The magnitudes of the FRFs (given in dB rel. 1) between the mapping microphones and one of the VVS-microphones are presented Fig. 13.7 (left). The vertical position of the VVS orifice, about 0.5 meters above this plane, is marked by the loudspeaker symbol. It can be seen that the cavity was efficiently excited. The system response peaks close to the VVS and is lowered, especially at the left termination of the mock-up.

The magnitude of the FRFs between the mapping microphones and the accelerometer are presented Fig. 13.7 (middle). Here the magnitude of each presented FRF is shown in dB rel. $\mathrm{Pa\,m^{-1}\,s^{2}}$. The response of the acoustic system peaks again close to the source, but, in contrast to the previous setting, the excitation is less efficient.

Figure 13.7 (right) contains the FRFs in dB rel. 1 between the mapping microphones and the external reference microphone. Then using the PA loudspeakers, no obvious sources could be found inside the cavity. As shown in Fig. 13.7 (right), the sound field mainly has its peak and nodal lines parallel to flight direction, because the whole cabin sidewall was excited by the external sound pressure. The excitation frequency was close to the third acoustic mode[1] in cross direction. The shape of this mode can be found in Fig. 13.7 (right), especially the three nodal lines that are parallel to the cabin sidewall.

13.2 Inverse Calculation for an Aircraft Cross-Section

Based on the experimental data it was possible to apply the IFEM. In order to proceed, a model of the acoustic field in the cabin is required. Such a model (here a geometry model) is shown in Fig. 13.8.

It represents the air filled volume from which the seats were removed. The measurement region M is highlighted with a red color.

An example for a finite element model used for the numerical investigation is also shown in Fig. 13.8. This model consists of an unstructured finite element mesh. Tetrahedron elements with linear shape functions were used for spatial discretization. Details on both model verification and model validation of are reported in (Weber 2009).

The results presented in this section were derived using two different discretization levels, compare Table 13.1, leading to an under-determined least square problem. As mentioned in Sect. 13.1.2 a total number of 2992 measurement points were taken into account. Figure 13.4 presents the distribution of these measurement points. The CGLS algorithm summarized in (7.40) was applied to solve the modified version of the inverse problem that is specified by (7.36). The number of iterations k was determined by the NCP approach, compare (7.42).

The results are presented in Fig. 13.9, 13.10, and 13.11 in which the distribution of the normal component of the mean active intensity on the interior boundary is shown. The post-processor of the commercial finite element software Comsol in which the normal vector is defined to point out of the fluid was used to generate this plots. For this reason a source is indicated by a negative sign, whereas a sink is indicated by a positive sign.

---

[1]If the cavity is approximated by a rectangular enclosure with sound hard walls, having a width of $L = 4.2$ m, the frequency of the third acoustic mode can be approximated by $f_{n3} \approx 3 \cdot 343\ \mathrm{m\,s^{-1}}/(2 \cdot 4.2\ \mathrm{m}) = 122.5$ Hz.

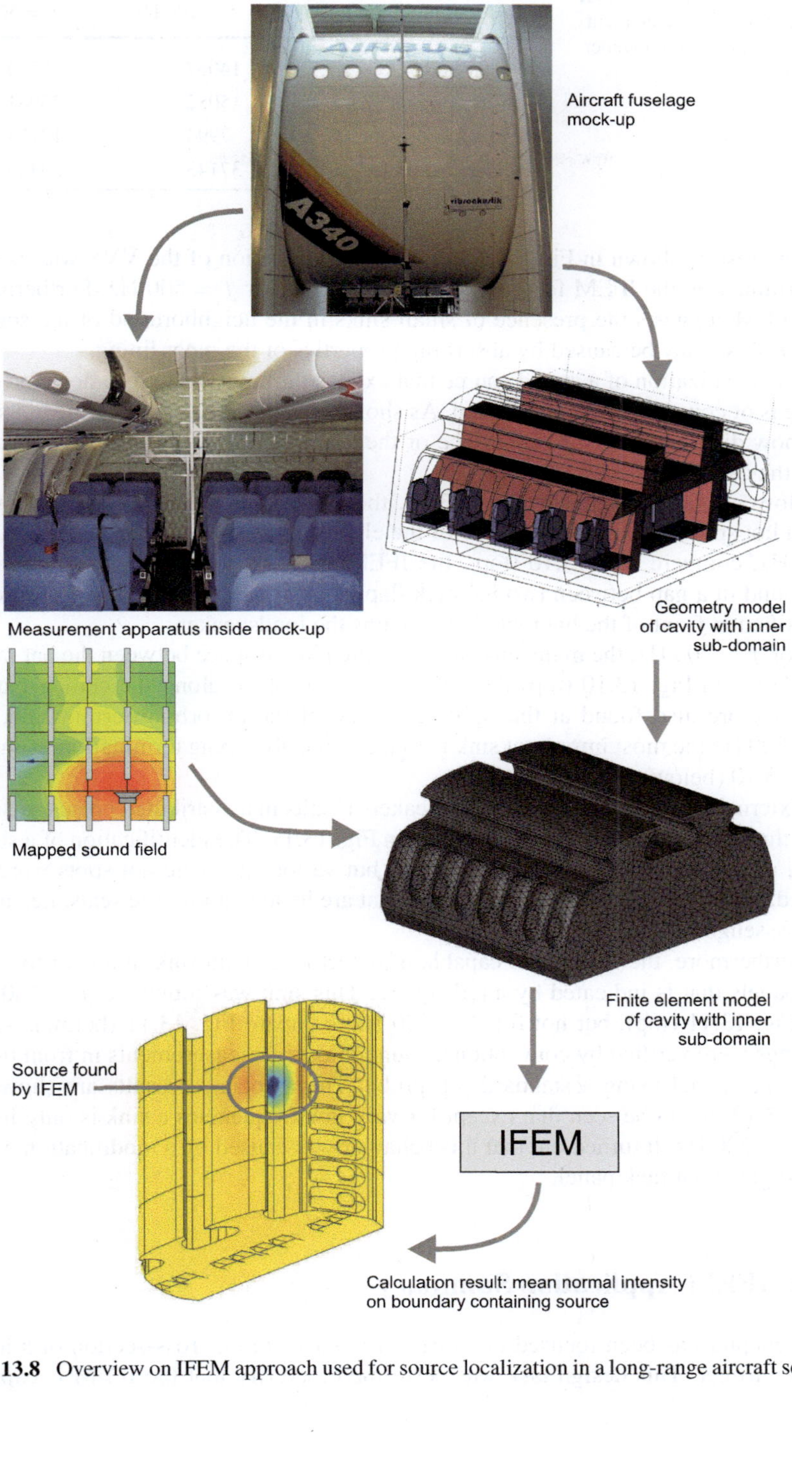

Fig. 13.8 Overview on IFEM approach used for source localization in a long-range aircraft section

Table 13.1 Number of nodes in the applied finite element models according to (Weber 2009)

Number of nodes on/in	$f = 200$ Hz	$f = 500$ Hz
Boundary (B)	14067	17284
Transition region (T)	15082	23981
Measurement region (M)	7994	12174
Total number of nodes	37143	53439

The results shown in Fig. 13.9 clarify that the position of the VVS was exactly determined by the IFEM for $f = 200$ Hz as well as for $f = 500$ Hz. Furthermore, the IFEM suggests the presence of small sinks in the neighborhood of the source. These sinks may be caused by absorbing properties of the inner lining.

The localization of a sound source that excites the cavity via vibro-acoustic coupling is of course more sophisticated. As shown in Fig. 13.10, the acoustic hot spots are now distributed along the surface of the vibrating hat racks that radiate sound into the cabin.

However, as indicated by Fig. 13.10, the IFEM was capable of detecting the main hot spots in the neighborhood of the electro-dynamical loudspeaker. For $f =$ 200 Hz, compare Fig. 13.10 (top), the IFEM suggests that the main source is to be found in a gap between two hat rack flaps. For $f = 500$ Hz, the main source is identified in front of the hat rack that contains the loudspeaker.

For $f = 200$ Hz, the main sinks are identified in the space between the hat racks. As shown in Fig. 13.10 (top) these sinks are distributed along the ceiling panels, but they are also found at the right hand side of the absorbing termination. For $f = 500$ Hz the most important sink is found at the absorbing termination, compare Fig. 13.10 (below).

External excitation using PA-loudspeakers results in a distributed noise transmission through the cabin sidewall as shown in Fig. 13.11. The identification of a single main source was (of course) not possible, but various acoustic hot spots were detected, especially in front of those panels that are located above the seats, i.e. above the passengers head.

Furthermore, the IFEM was capable of detecting a main sink in one of the ceiling panels that is indicated by a red square. This sink was found for $f = 200$ Hz, see Fig. 13.11 (top), but not for $f = 500$ Hz, compare Fig. 13.11 (below). These findings were verified by conventional sound intensity measurements in front of the particular panel using a standard p-p probe. The associated results are shown in Fig. 13.12. It can be seen that except for very low frequencies a sink is only found for $f = 200$ Hz. It turned out that this behavior was caused by a modification of the investigated hat rack panel.

13.3 IFEM Application Summary

This chapter has been focused on *noise field analysis* in a cross-section of a long-range airliner. This design task was performed to verify that the IFEM is capable

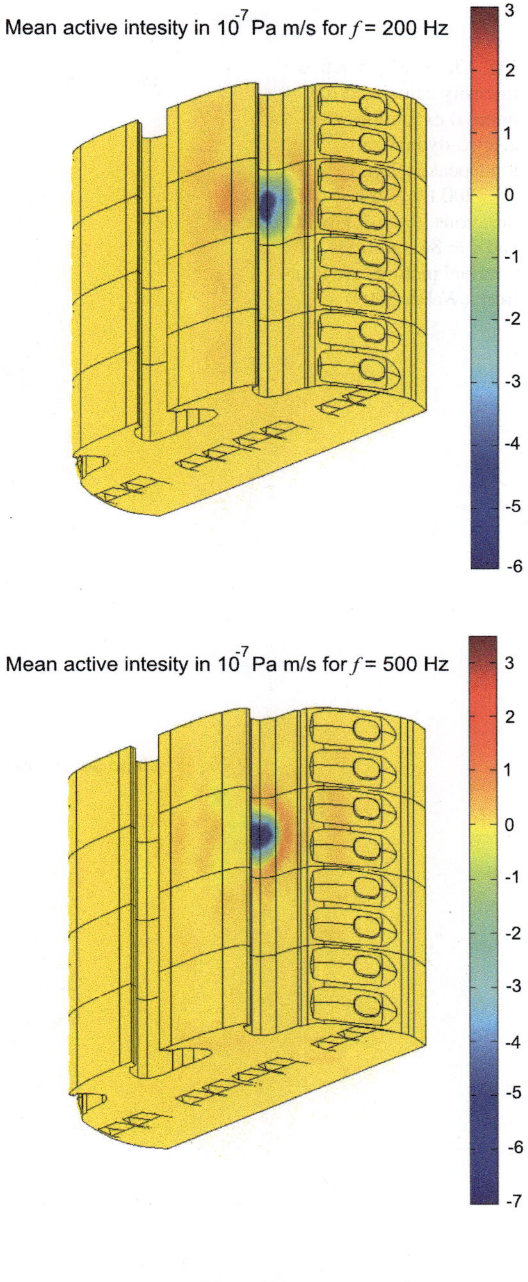

Fig. 13.9 Mean active intensity calculated for internal excitation with VVS. *Top*: $f = 200$ Hz and $k = 469$ iterations. *Below*: $f = 500$ Hz and $k = 111$ iterations. Original pictures are taken from (Weber 2009)

of detecting acoustic hot spots in an operational environment. For this purpose the cavity was excited by internal as well as external sources.

The resulting interior noise field was mapped with a custom made microphone array. A total number of 2992 points was taken into account for sound pressure

Fig. 13.10 Mean active intensity calculated for internal excitation with electro-dynamical loudspeaker. *Top*: $f = 200$ Hz and $k = 645$ iterations. *Below*: $f = 500$ Hz and $k = 84$ iterations. Original pictures are taken from (Weber 2009)

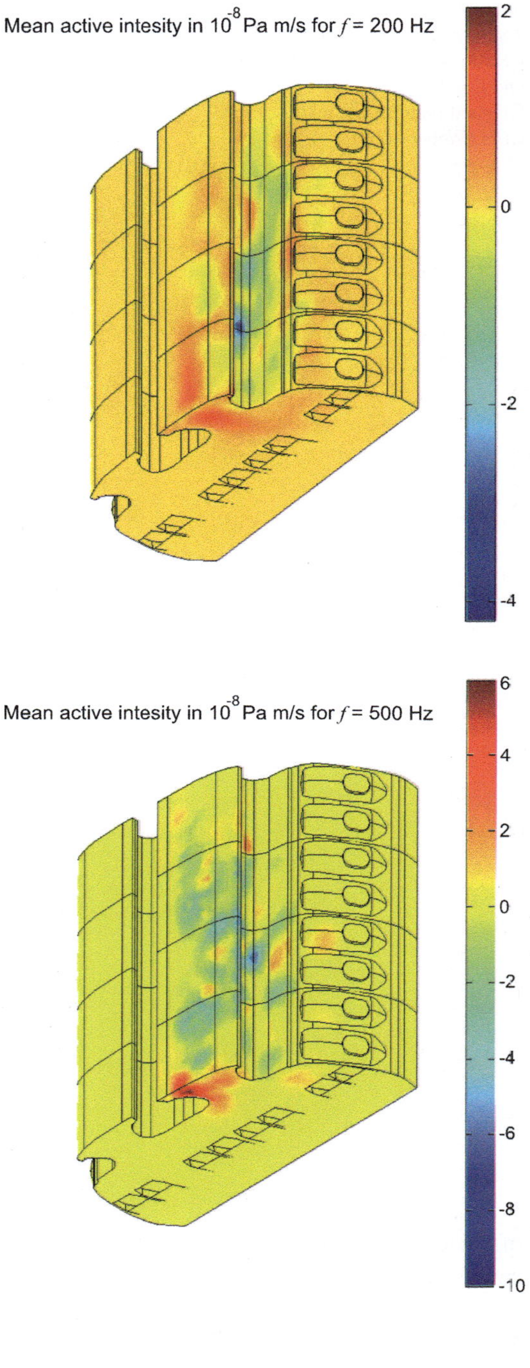

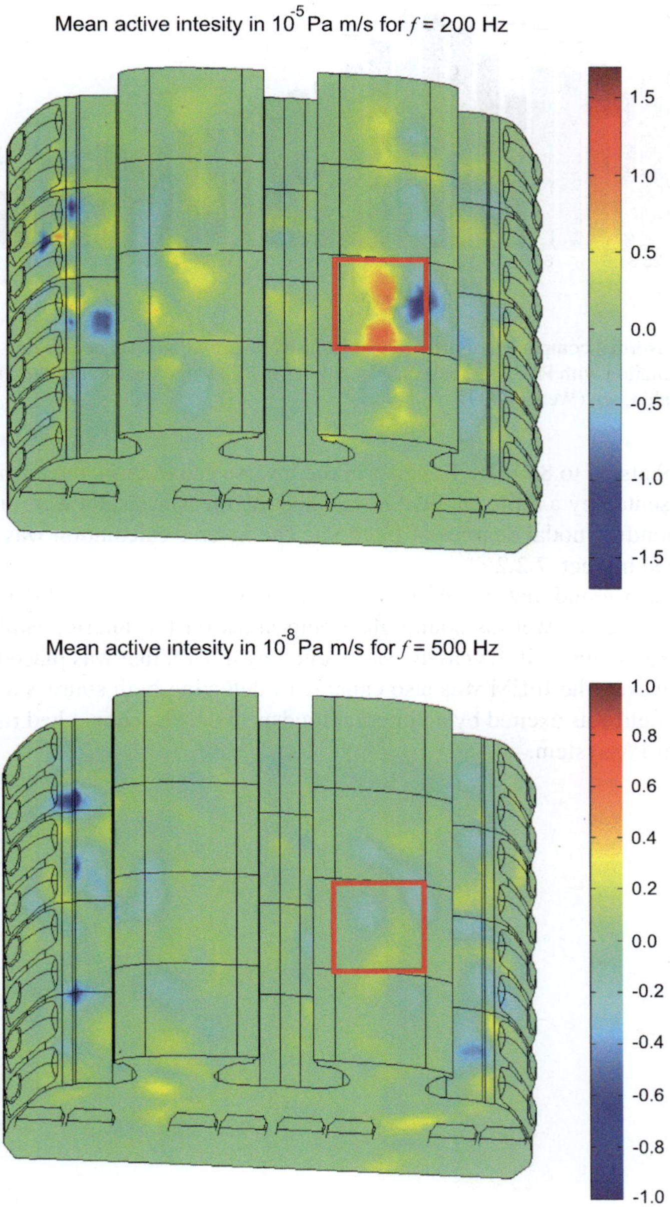

Fig. 13.11 Mean active intensity calculated for external excitation. *Top*: $f = 200$ Hz and $k = 644$ iterations. *Below*: $f = 500$ Hz and $k = 178$ iterations. Pictures are taken from (Weber 2009)

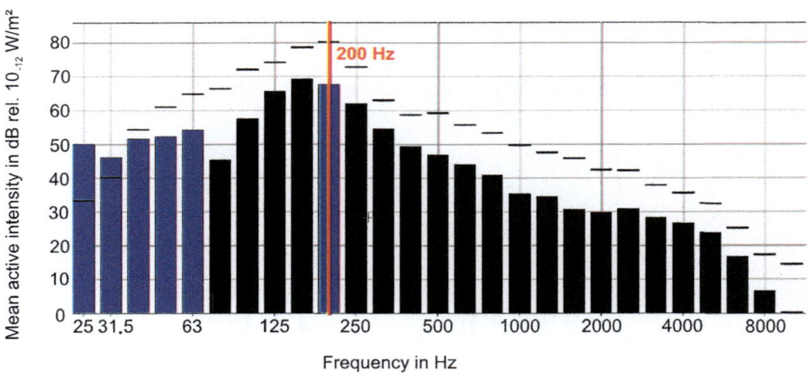

Fig. 13.12 Normal component of mean active intensity measured in front of a ceiling panel during external excitation with PA-loudspeakers. *Black*: Intensity flow into the cabin. *Blue*: Intensity flow out of the cabin, see (Weber 2009)

measurements up to 800 Hz. To perform the inverse calculation, the air filled cavity was represented by a finite element model and the measured data were interpolated to corresponding nodal degrees of freedom. The inverse calculation was performed as described in Sect. 7.2.2.

It has been found that this ANC-system design tool is capable of detecting both acoustic sources as well as sound absorbing areas on the interior wall. The best results were obtained, if the cavity was excited by a VVS that was placed inside the cabin. However, the IFEM was also capable of detecting both sources and sinks, if the sound field was excited by an internal loudspeaker (placed in a had rack) and by an external PA-system.

Part V
Summary

Chapter 14
Summary

Noise is a problem to which we cannot turn a deaf ear. German law (BImSchG 2002) has accepted it as a form of environmental pollution, and it is a fact that the noise exposure in enclosed sound fields is a limiting factor for both work security and human comfort. As a consequence, the analysis of interior noise treatments and the design of noise control systems have become one of the major tasks in the engineering design process.

In order to reduce interior noise to an acceptable level, passive treatments (e.g. adding high damping material) or passive systems (such as vibration absorbers) are commonly applied. These solutions pose problems, especially when they are used to reduce low frequency interior noise; they add unwanted mass and the space they require can be out of scale. In the past decades research has been done to avoid these problems by changing from *passive* treatments to *active* noise reduction systems. Nowadays, active noise control (ANC) is an evolving technology that is applied in car interiors, aircraft cabins, and personal hearing protection systems. However, even if the potential of ANC is widely accepted, it is a fact that only noise canceling headphones have made the technological breakthrough and became a commercial "off-the-shelf product". This raises following questions: What problems prevent ANC from making an overall breakthrough? And how can these problems be solved?

In this book it is emphasized that (i) a general design methodology for ANC-systems has not yet been established and (ii) the existing literature mainly focuses on one particular sub-domain of ANC (e.g. engineering acoustics, electro-acoustics or signal processing for ANC-systems).

A solution to these problems can be provided by accepting an ANC-system as an example of a mechatronic system. As such, it consists of a basic system (the air filled volume of the enclosed sound field), sensors (e.g. microphones), actuators (e.g. electro-dynamical loudspeakers), and an information processing unit (such as a digital signal processor). This implies that ANC can be understood as a special sub-domain of *mechatronics*. The intention of this book therefore is to:

1. Present a mechatronic approach to adaptive feed-forward control of low frequency interior noise that combines the disciplines of engineering acoustics,

T. Kletschkowski, *Adaptive Feed-Forward Control of Low Frequency Interior Noise*, Intelligent Systems, Control and Automation: Science and Engineering 56,
DOI 10.1007/978-94-007-2537-9_14,

electro-acoustics and adaptive signal processing and presents them in a unified manner,
2. Propose a design methodology for ANC-systems that fits into the frame of an existing general design methodology for mechatronic systems as well as to demonstrate its practical use,
3. Discuss established and novel tools for (a) numerical simulation, (b) sound source localization, and (c) performance estimation in the ANC-system design process.

In the first part of this book, an illustrative introduction has been provided, in which the acoustic approach to interior active noise control has been motivated by a simple aircraft model. This model has been used to compare both the noise reduction potential and the required control effort of ANC to those of active vibration control (AVC) and to those of active structural acoustic control (ASAC).

Although it has been shown that all three active approaches can suppress acoustic resonances, ANC requires the significantly smaller control effort. This result has been used to motivate this purely acoustic approach to interior active noise control.

In order to present a mechatronic theory of ANC, the second part of this book focuses on the description of the basic system as well as on the characterization of sensors, actuators and signal processing used for ANC. Because the air filled volume of an enclosed sound field represents the basic system, it has been sufficient to restrict the discussion to acoustic field variables, acoustic field equations and the associated boundary conditions. Solutions of the Helmholtz equation are presented for one-dimensional enclosed sound fields. The review of sensors is restricted to condenser microphones, sound intensity probes and accelerometers. The functional principles of electro-dynamical loudspeakers and flat panel speakers are addressed to illustrate the operational characteristic of typical actuators used for ANC. Both optimal and adaptive control of tonal and broadband disturbances is discussed in great detail. After a mathematical preparation, six different acoustic control strategies known as

- Global control of acoustic potential energy,
- Local control of acoustic potential energy,
- Minimization of total power input,
- Local control of sound pressure and particle velocity,
- Maximization of power absorption,
- Maximization of acoustic absorption,

have been reviewed and evaluated in order to discuss the physical basis for interior active noise control. The control strategies have been benchmarked using a traveling wave model that has been applied to analyze controlled sound fields in a finite duct. The control strategies have been compared according to their noise reduction potential. Furthermore, the control effort of every control strategy has been analyzed. In contrast to previous publications, the influence of different damping ratios on the results has been taken into account.

In agreement with existing textbooks (none of which contains a comparison of all six control strategies), it has been shown that each strategy is capable of reducing the sound pressure magnitude in acoustic resonances. However, two strategies (local control of acoustic potential energy and maximization of acoustic absorption power) can cause significant amplification. This is especially true, if these two strategies are applied to anti-resonance frequencies of the uncontrolled system.

In addition to the findings presented in existing textbooks, it has been shown how the amount of viscous damping influences the noise reduction potential and the control effort of a particular control strategy. Using diagrams, it has been explained that an increase in viscous damping causes a decrease of the noise reduction that can be realized with ANC. It has also been shown that the control effort is also reduced, if the amount of viscous damping is increased. This is true for all strategies, except for minimizing the total power input.

Signal processing for ANC has been discussed for tonal noise as well as for broadband disturbances. Control of tonal noise has been described in frequency domain. Special attention has been paid to optimal control based on modified cost functions and parametric controller design. Different approaches have been derived by identifying weighting matrices used to modify the objective function. It has been shown that it is possible to include remote sensor control approaches (such as virtual sensing and sound design at control microphones) as well as robust controller design into this framework.

Optimal and adaptive control of broadband disturbances in time-domain has been described in a unified manner considering both multiple inputs and multiple outputs. To make the adaptive control algorithms "ready-to-use", algorithmic formulations have been presented for four different time-domain approaches. Such a consistent summary is (as mentioned in Part I of this book) either missing in existing textbooks on ANC or restricted to a limited number of signals (one reference signal, two error signals and two canceling signals).

The third part of the book focuses on ANC-system design tools and proposes a new mechatronic design methodology for ANC-systems. Because the present work is restricted to interior noise, the finite element method has been considered as a tool for numerical simulation. The inverse finite element method has been introduced as a novel tool for sound source localization in standing wave fields. This procedure requires sound pressure measurements in the interior. Then, these data are associated with the nodes of an acoustic finite element model that represents the investigated cavity. If all sources are located on the interior boundary, the equation system resulting from this finite element model can be resorted in such a way that the boundary values of sound pressure and normal component of particle velocity can be reconstructed.

Because it is also necessary to evaluate the noise reduction potential of ANC-systems directly at the beginning as well as during the system design process, four tools for initial performance estimation have been discussed. These performance estimation tools are the

- Coherence analysis,
- Transducer placement analysis,
- Correlation analysis,
- Impulse response analysis.

Because single-channel coherence analysis (that is nowadays implemented on standard FFT-analyzers) only requires two microphones and the presence of the primary noise field, this tool will be the most important one in the early stage of the design process. However, to study the influence of a certain transducer setting on the control profit, transducer placement analysis becomes more important, when the design process proceeds. The correlation analysis tool can be used to determine the theoretical optimum of active noise reduction. Impulse response analysis can be applied to study the influence of the control filter length on the control profit. The underlying equations of these performance estimators are not new. What is new is their survey, presentation and exemplary use as ANC-system design tools.

The methodology for ANC-system design presented in this book describes ANC-system design by a matrix model. This model considers a hierarchy of five different design steps that have been introduced as the

1. Feasibility study,
2. First specification,
3. Weight, cost, and efficiency study,
4. Sensitivity analysis,
5. Robustness study.

It has been outlined that different design tasks must be handled in each design step to increase the maturity level of the ANC-system. These tasks are the

- Noise field analysis,
- Analysis of anti-noise field,
- Domain specific design, including:
  - Sensor design,
  - Actuator design,
  - Controller design,
- Functional testing.

To assess the ANC-system maturity level, a set of nine system maturity levels (SML) ranging from SML 1 (physical behavior of the operational environment in which the future ANC-system will operate are observed and reported) to SML 9 (ANC-system established through successful service in operational environment) has been introduced. The definitions of these SMLs are inspired by the technology readiness levels defined by the National Aeronautics and Space Administration (NASA). It has also been shown that the proposed design methodology can be embedded into the general design methodology for mechatronic systems [VDI 2206]. For this reason, a mechatronic methodology for ANC-system design has been proposed—for the first time!

Examples of ANC-system design processes that (i) fit into the frame of the proposed design methodology and (ii) include examples for the use of the proposed ANC-system design tools, have been presented in Part IV. With these examples the book reports on applied research projects that have been conducted at the Helmut-Schmidt-University/University of the Federal Armed forces Hamburg to study

1. **Tonal active noise control in a semi-closed interior:** The design process of a distributed multi-channel ANC-system that enables robust adaptive control in a special working area of a propeller driven aircraft has been discussed. This system will be applied to reduce the noise exposure associated with the blade pass frequency and its two first harmonics using remote error microphones. It has been shown that it is possible to successfully develop such a system according to the proposed design methodology, when (i) both error microphones and canceling sources are located at optimal positions and (ii) parametric controller design is applied to consider uncertainties in the primary disturbance and in the secondary transfer paths.
2. **A sound intensity probe with active free field:** It has been shown that a sound tube containing two microphones and terminated at one end by a loudspeaker can be used in a standing wave field to determine the normal component of the mean intensity radiated under free field conditions. To achieve this, the acoustic absorption inside the tube must be maximized by active control. A first realization of this novel sound source localization tool has been the outcome of this design process. It has been successfully tested and calibrated.
3. **Broadband active noise control around human head:** To illustrate how an ANC-system should be developed to reach the physical limits of active control, a project on adaptive control of broadband noise in bedrooms has been reviewed. Because of systematic system design according to the proposed design methodology, a noise reduction of up to 18 dB (as predicted by coherence analysis) has been realized at the error microphones for broadband excitation in the frequency range of 80 Hz–480 Hz using an external reference sensor.
4. **The combination of active noise control with audio entertainment:** The development and considerations in the system design of an experimental ANC controller combined with an audio system for a light jet cabin have been presented. A multi-channel ANC-audio system (8 error microphones and 6 canceling speakers) has been designed and integrated into a very light jet mock-up. Actuator placement has been discussed using a sequential search algorithm. Applying broadband active noise control in a frequency range of 50 Hz–500 Hz, an average attenuation of 15 dB has been achieved at the error microphones, and a zone of quiet in 0.1 m–0.2 m distance to the microphones has been generated. It has been shown that the combination of audio and ANC can be successfully realized, if the MFxLMS is used to perform the active control task.
5. **Inverse noise source localization in the cabin of a long range airliner:** Because sound source localization can be essential for the noise reduction realized by ANC, the performance of the inverse finite element method has been analyzed. For this purpose acoustic ground tests have been performed in the cabin of a long range airliner. The cabin noise field has been excited by internal as well as by external sources. In all experiments it has been confirmed that this novel sound source localization tool is capable of identifying both acoustic hot spots and sinks located on the interior boundary of enclosed sound fields.

All these examples prove that the novel design methodology is a useful tool. It enables systematic design of ANC-systems on the basis of a unified mechatronic theory of interior active noise control.

References

Akpan UO, Brennan DP, Koko TS, Beslin O, Masson P, Renault S (1999) Active noise and vibration control literature survey: controller technologies. DREA-CR-1999-177, Contractor Report, Defence Research Establishment Atlantic, Dartmouth NS (CAN); MARTEC Ltd, Halifax NS (CAN); Sherbrooke Univ, Sherbrooke QUE (CAN)

Anderssohn R, Marburg S, Großmann C (2006) FEM-based reconstruction of sound pressure field damped by partially absorbing boundary conditions. Mech Res Commun 33:851–859

ANSYS (2010) Product description ANSYS. http://www.ansys.com/. Cited 09 June 2010

Antoniou A (1979) Digital filters: analysis and design. McGraw-Hill, New York

Baker S (1955) An acoustic intensity meter. J Acoust Soc Am 27(2):269–273

Ballou G (2009) Electroacoustic devices: microphones and loudspeakers. Focal Press, Elsevier, Oxford

Baranek LL (1954a) Acoustics. McGraw-Hill, New York

Baranek LL (1954b) Loudspeakers and microphones. J Acoust Soc Am 26(5):618–629

Bathe KJ (1996) Finite element procedures. Prentice Hall, London

Benassi L, Elliott SJ (2003) Active vibration isolation using an inertial actuator with local displacement feedback control. J Sound Vib 278:705–724

Berbbia CA (1978) The boundary element method for engineers. Pentech Press, London

Berglund B, Hassmén P (1996) Sources and effects of low-frequency noise. J Acoust Soc Am 99(5):2985–3002

Billoud G (2001) Active control at Lord corporation—a reality. Technical paper, Lord Corporation, LL-6508

BImSchG (2002) Gesetz zum Schutz vor schädlichen Umwelteinwirkungen durch Luftverunreinigungen, Geräusche, Erschütterungen und ähnliche Vorgänge (Bundes-Immissionsschutzgesetz—BImSchG)

Bjarnason E (1992) Active noise cancellation using a modified form of the filtered-X LMS algorithm. In: Vanderwalle J, Boite R, Moonen M, Oosterlink A (eds) Signal processing VI: theory and applications. Elsevier, Amsterdam, pp 1053–1056

Björck A (1996) Numerical methods for least squares problems. SIAM, Philadelphia

Bose KN (1985) Digital filters—theory and applications. Elsevier, New York

Böhme S, Sachau D, Breitbach H (2006) Optimization of actuator and sensor positions for an active noise reduction system. In: 13th annual symposium on smart structures and materials, San Diego, CA, USA. Proc of SPIE, vol 6171, pp 61710N-1–61710N-11

Böhme S, Sachau D, Kletschkowski T, Breitbach H (2007) Mock-up of a loadmaster area for acoustic ground tests. In: Proc of CEAS 2007, 1st CEAS European air and space conference. Berlin, Germany

Böhme S, Sachau D, Breitbach H (2009) Concepts for global noise control in cabins. In: Proc of IMAC XXVI a conference and exposition on structural dynamics, Orlando, FL, USA, February 4–7

T. Kletschkowski, *Adaptive Feed-Forward Control of Low Frequency Interior Noise*, Intelligent Systems, Control and Automation: Science and Engineering 56, DOI 10.1007/978-94-007-2537-9,

Borchers IU, Emborg U, Sollo A, Waterman EH, Paillard J, Larsen PN, Venet G, Goeransson P, Martin V (1997) Advanced study for active noise control in aircraft (ASANCA), fourth aircraft interior noise workshop. NASA Langley Research Center, USA, pp 129–141

Bouchard M, Quednau S (2000) Multichannel recursive-least-squares algorithms and fast-transversal-filter algorithms for active noise control and sound reproduction systems. IEEE Trans Speech Audio Process 8(5):606–618

Breitbach H, Sachau D (2009) Active acoustic solutions for a high speed turboprop transport aircraft. In: Proc of ACTIVE 2009, Ottawa, Canada, August 20–22

Breitbach H, Sachau D, Böhme S (2006) Acoustic challenges of the A400M for active systems. In: 13th annual symposium on smart structures and materials, San Diego, CA, USA. Proc of SPIE, vol 6171, pp 617104-1–617104-8

Breitbach H, Storm U, Stothers I (2009) Development of active noise reduction systems for a high speed turboprop transport aircraft. In: Proc of AC—adaptronic congress, Berlin, Germany, May 19–20

B&K (1996) Microphone handbook, vol 1: theory. Brüel & Kjaer, Nærum http://www.bksv.com/. Cited 10 May 2010

Cadzov JA, van Landingham HF (1985) Signals, systems, and transforms. Prentice Hall, New Jersey

Champeney DC (1987) A handbook of Fourier theorems. Cambridge University Press, New York

Chargin M, Gartmeier O (1990) A finite element procedure for calculating fluid-structure interaction using MSC/NASTRAN. NASA Technical Memorandum 102857

Chassaing R (2005) Digital signal processing and applications with the C6713 and C6416 DSK. Wiley, Canada

Christensen JJ, Hald J (2004) Beamforming. Technical Review 1, Brüel & Kjaer, 1–35

COMSOL (2008) COMSOL Multiphysics 3.5. Manual. Pressure acoustics, application mode description

COMSOL (2010) COMSOL Multiphysics. Product description COMSOL http://www.comsol.de/. Cited 09 June 2010

Cremer L, Heckl M, Petersson BAT (1995) Structure-borne sound. Springer, Berlin

Crocker MJ, Arenas JP (2003) Fundamentals of direct measurement of sound intensity an practical applications. Acoust Phys 49(2):199–214

Curtis ARD, Nelson PA, Elliott SJ, Bullmore AJ (1987) Active suppression of acoustic resonance. J Acoust Soc Am 81(3):624–631

Danzer WF, Huber F (1994) System Engineering—Methoden und Praxis. Verlag Industrieelle Organisation, Zürich

DEGA (2006) Akustische Wellen und Felder; DEGA Deutsche Gesellschaft für Akustik e.V. http://www.dega-akustik.de/publikationen/online-publikationen. Cited 05 May 2010

Diniz PSR (2008) Adaptive filtering—algorithms and practical implementations. Springer, New York

DIN 1314 (1990) Referenzzustand, Normzustand, Normvolumen; Begriffe und Werte. DIN Deutsches Institut für Normung e.V., Germany

DIN 1311-1 (2000) Schwingungen und Schwingungsfähige Systeme. Teil 1: Grundbegriffe, Einteilung (ICS 17.160;01.040.17). DIN Deutsches Institut für Normung e.V., Germany

DIN 18041 (2004) Hörsamkeit in kleinen bis mittelgroßen Räumen. Deutsche Fassung DIN 18041. DIN Deutsches Institut für Normung e.V., Germany

DIN EN 61672-1 (2003) Elektroakustik—Schallpegelmesser. Teil 1: Anforderungen (IEC 61672-1:2002); Deutsche Fassung EN 61672-1:2003. DIN Deutsches Institut für Normung e.V., Germany

DIN EN 60268-4 (2004) Elektroakustische Geräte. Teil 4: Mikrofone (IEC 602684:2004); Deutsche Fassung EN 60268-4:2004. DIN Deutsches Institut für Normung e.V., Germany

DIN EN 60268-5 (2004) Elektroakustische Geräte. Teil 5: Lautsprecher (IEC 60268-5:2003); Deutsche Fassung EN 60268-5:2003. DIN Deutsches Institut für Normung e.V., Germany

DIN EN 60268-16 (2004) Elektroakustische Geräte. Teil 16: Objektive Bewertung der Sprachverständlichkeit durch den Sprachübertragungsindex (IEC 60268-16:2003); Deutsche Fassung EN 60268-16:2003. DIN Deutsches Institut für Normung e.V., Germany

DIN ISO 5129 (2003) Akustik—Messung des Schalldruckpegels in Luftfahrzeugen während des Fluges; DIN ISO 5129:2003-10 (ISO 5129:2001). DIN Deutsches Institut für Normung e.V., Germany

Douglas SC (1997) Fast exact filtered-X LMS and LMS algorithms for multichannel active noise control. IEEE, Munich, 0-8186-7919-0/9, pp 399–402

Douglas SC (1999) Fast implementations of the filtered-X LMS and LMS algorithms for multichannel active noise control. IEEE Trans Speech Audio Process 7(4):454–465

Drenckhan J, Sachau D (2004) Identification of sound sources using inverse FEM. In: Proc of ICSV11, 11th int congress on sound and vibration, St. Petersburg, Russia, July 5–8

Drenckhan J, Sachau D (2005) Influence of smoothing to the inverse finite element method for acoustic hot-sport identification. In: Proc of InterNoise 2005, 34th international congress and exposition on noise control engineering, Rio de Janeiro, Brazil, August 7–10

Elliott Aviation (2010) Moline, IL 61266-0100, USA. http://www.elliotaviation.com. Cited 31 July 2010

Elliott SJ (2001) Signal processing for active noise control. Academic Press, London

Elliott SJ (2002) Adaptive methods in active control. In: Tokhi O, Veres S (eds) Active sound and vibration control. Institution of Electrical Engineers, London, pp 57–72

Elliott SJ (2008) A review of active noise and vibration control in road vehicles. ISVR Technical Memorandum No 981, University of Southampton, December 2008

Elliott SJ, Garcia-Bonito J (1995) Active cancellation of pressure and pressure gradient in a diffuse sound field. J Sound Vib 186(4):696–704

Elliott SJ, Joseph P, Bullmore AJ, Nelson PA (1988) Active cancellation at a point in a pure tone diffuse sound field. J Sound Vib 120(1):183–189

Elliott SJ, Nelson PA, Stothus LM, Boucher CC (1990) In-flight experiments on the active control of propeller-induced cabin noise. J Sound Vib 140(2):219–238

Elliott SJ, Joseph P, Nelson PA, Johnson ME (1991) Power output minimization and power absorption in the active control of sound. J Acoust Soc Am 90(5):2501–2512

Emborg U (1998) Cabin noise control in the Saab 2000 high-speed turboprop aircraft. In: Proc of ISMA 1998, int conference on noise and vibration engineering, Sept 16–18. Katholieke Universiteit Leuven, Belgium

EN ISO 10534-1 (2001) Akustik—Bestimmung des Schallabsorptionsgrades und der Impedanz in Impedanzrohren Teil 1: Verfahren mit Stehwellenverhältnis (ISO 10534-1:1996) Deutsche Fassung EN ISO 10534-1:2001. DIN Deutsches Institut für Normung e.V., Germany

EN ISO 10534-2 (2001) Akustik—Bestimmung des Schallabsorptionsgrades und der Impedanz in Impedanzrohren Teil 2: Verfahren mit Übertragungsfunktion (ISO 10534-2:1998). Deutsche Fassung EN ISO 10534-2:2001. DIN Deutsches Institut für Normung e.V., Germany

Farhang-Boroujeny B (1998) Adaptive filters—theory and applications. Wiley, New York

Fahy F (1989) Sound intensity. Elsevier, London

Fahy F (2003) Foundations of engineering acoustics. Academic Press, Amsterdam

Fahy F, Gardonio P (2007) Sound and structural vibration. Elsevier, Amsterdam

Fliege N (1991) Systemtheorie. Teubner, Stuttgart

Freienstein H, Guicking D (1996) Experimentelle Untersuchung von linearen Lautsprecheranordnungen als aktive Absorber in einem Kanal. In: Proc of DAGA 96, Bonn, Germany, pp 112–113

Frequency Devices Inc. (2009) Analog and digital products design/selection guide. Frequency Devices Inc., Ottawa, Canada. http://www.freqdev.com Cited 30 Sep 2009

Fuller WA (1976) Introduction to statistical time series. Wiley, New York

Fuller CC, Elliott SJ, Nelson PA (1996) Active control of vibration. Academic Press, London

Fuller CR, Maillarda JP, Mercadalb M, von Flotow AH (1997) Control of aircraft interior noise using globally detuned vibration absorbers. J Sound Vib 203(5):745–761

Gade S (1982) Sound intensity (theory). Technical Review 3, Brüel & Kjaer, 3–39

Gan WS, Kuo SM (2001) Integrated headsets using the adaptive feedback active noise control system. In: Proc of ICSV11, 8th int congress on sound and vibration, Hong Kong, July 2–6

Gan WS, Kuo SM (2002) An integrated audio and active noise control headset. IEEE Trans Consum Electron 48(2):242–247

Gaul L, Kögl M, Wagner M (1997) Boundary element methods for engineers and scientist—an introductory course with advanced topics. Springer, Berlin

Garcia-Bonito J, Elliott SJ, Boucher CC (1998) Generation of zones of quiet using a virtual microphone arrangement. J Acoust Soc Am 101(6):3498–3616

Gardinio P (2007) Active control of cabin noise. Workshop on revolutionary aircraft for quiet communities, Hamtion, VA, USA, July 24–26

Genelec (2010) 1029A data sheet. http://www.genelec.com/. Cited 05 August 2010

Gerner C (2005) Optimale aktive Geräuschreduzierung in Flugzeugkabinen für hohe tonale Lärmpegel. Dissertation, Helmut-Schmidt-University/University of the Federal Armed Forces Hamburg

Gerner C, Sachau D, Breitbach H (2004a) Active noise control in an aircraft cabin. In: Proc of IMAC-XXII, conference on structural dynamics 20040126-29, Dearborn, MI, USA, January 26–29

Gerner C, Sachau D, Breitbach H (2004b) Aircraft interior ANC with flat panel speakers. In: Industrial and commercial applications of smart structures technologies, Bellingham, WA, USA. Proc of SPIE, vol 5388, pp 266–275

Gerner C, Sachau D, Breitbacu H (2005) Optimization of actuator and sensor positions for active noise reduction (ANR). In: Proc of ICSV12, twelfth international congress on sound and vibration, Lisbon, Portugal, July 11–14

Girod B, Rabenstein R, Stenger A (2005) Einführung in die Systemtheorie. Teubner, Stuttgart

Golub GH, Kahan W (1965) Calculating the singular values and pseudo-inverse of a matrix. J Soc Ind Appl Math Ser B Numer Anal 2(2):205–224

Golub GH, Loan CFv (1996) Matrix computation. Johns Hopkins University Press, London

Guicking D (2002) An overview of ASVC: from laboratory curiosity to commercial products. In: Tokhi O, Veres S (eds) Active sound and vibration control. Institution of Electrical Engineers, London, pp 3–23

Guicking D, Karcher K (1984) Active impedance control for one-dimensional sound. J Vib Acoust Stress Reliab Des 106:393–396

Hald J (1989) STSF a unique technique for scan-based near-field acoustic holography without restrictions on coherence. Technical Review 1, Brüel & Kjaer, 1–50

Halvorsen WG, Emborg U (1989) Interior noise control of the Saab 340. SAE Technical Paper Series, 891080

Hamann M (1980) Ein Beitrag zur Realisierung eines Schallintensitätsmessverfahrens. Dissertation, University of Hannover

Hanke M (1996) Limitations of the L-curve method for ill-posed problems. BIT Numer Math 36:287–301

Hansen CH (2003) Does active noise control have a future? In: Proc of Wespac 8, the 8th Western Pacific acoustics conference, acoustics on the move, Melbourne, Australia

Hansen PC, O'Leary DP (1993) The use of the L-curve in the regularization of discrete ill-posed problems. SIAM J Sci Comput 14(6):1487–1503

Hansen PC, Kilmer M, Kjeldsen RH (2006) Exploiting residual information in the parameter choice for discrete ill-posed problems. BIT Numer Math 46:41–59

Harashima F, Tomizuka M, Fukuda T (1996) Mechatronics—what is it, why and how? An editorial. IEEE/ASME Trans Mechatron 1(1):1–4

Harris CM (1975) Handbook of noise control. McGraw-Hill, New York

Harris M (2008) Test bench: noise-cancelling headphones. The Sundy Times, London

Hasegawa S, Tabata T, Kinoshita A, Hideki H (1992) The development of an active noise control system for automobiles. Society Automotive Eng., Tech. Paper, 922086

Havelock D, Kuwano S, Vorländer M (2008a) Handbook of signal processing in acoustics, vol 1. Springer, New York

Havelock D, Kuwano S, Vorländer M (2008b) Handbook of signal processing in acoustics, vol 2. Springer, New York

Haykin S (1996) Adaptive filter theory. Prentice Hall, London

Heckl M, Müller H (1995) Taschenbuch der Technischen Akustik. Springer, Berlin

Henn H, Sinambary GR, Fallen M (1984) Ingenieurakustik. Vieweg, Wiesbaden

Hess W (1989) Digitale Filter—Eine Einführung. Teubner, Stuttgart

Holland JH (1975) Adaption in natural and artificial systems: an introductory analysis with applications to biology, control and artificial intelligence. University of Michigan Press, Ann Arbor

Holland KR, Nelson PA (2004) Sound source characterization: the focused beamformer vs the inverse method. In: Proc of ICSV11, 11th int congress on sound and vibration, St. Petersburg, Russia, July 5–8

Holters M, Zölzer U (2008) DSP-Implementierung eines $1 \times 2 \times 2$ ANC-Systems und Konzeptstudie für ANC bei instationären Signalen. Unterauftrag zum Forschungsvorhaben Aktive Lärmminderung in Wohn- und Schlafräumen. Final Report for a Research Project (Az: 24483)—financed by the Deutsche Bundesstiftung Umwelt, Helmut-Schmidt-University/University of the Federal Armed Forces Hamburg

Honig ML, Messerschmidt DG (1984) Adaptive filters—structure, algorithms and applications. Kluwer Academic, Boston

Hundeck C (2008) Moderne Verfahren zur Schallquellenortung mit Arraysystemen. Lärmbekämpfung—Zeitschrift für Akustik Schallschutz und Schwingungstechnik 3(2):55–70

ITU-R (1999) Method for the objective measurement of perceived audio quality. ITU-R recommendations, BS. 1387

Ivers D, Southward S, Heath M, Buchner S (year unknown) Active mounting system flight test results. Technical paper, Lord Corporation, LL-6009

Jacobsen F, de Bree HE (2005) A comparison of two different sound intensity measurement principles. J Acoust Soc Am 188(3):1510–1517

Johansson S (2000) Active control of propeller—induced noise in aircraft. Algorithms & methods. Dissertation, Blekinge Institute of Technology, BTH Ronneby, Sweden, ISBN: 91-631-0172-6

Johansson S, Lago T, Nordebo S, Claesson I (1999) Control approaches for active noise control of propeller-induced cabin noise evaluated from data from a dornier 328 aircraft. In: Proc of 6th int congress on sound and vibration, Copenhagen, Denmark, pp 1611–1618

Johnson RJ (1991) Digitale Signalverarbeitung. Hanser, München in Cooperation with Prentice Hall International, London

Jones JD, Fuller CR (1986) An experimental investigation of the interior noise control effects of propeller synchrophasing. NASA Contractor Report 178185

Kang SW, Kim YH (1997) Active intensity control for the reduction of radiated duct noise. J Sound Vib 201(5):595–611

Kletschkowski T, Sachau D (2007a) Lokale aktive Lärmminderung in Schlafräumen. Lärmbekämpfung—Zeitschrift für Akustik Schallschutz und Schwingungstechnik 2(5):174–177

Kletschkowski T, Sachau D (2007b) Noise source identification using a sound intensity probe with active free-field. In: Proc of 8th conference on active noise and vibration control methods, Krakow-Krasiczyn, Poland, June 11–14

Kletschkowski T, Sachau D (2008) Design and calibration test of an active sound intensity probe. Adv Acoust Vib 2008, Article ID:574806

Kletschkowski T, Sachau D (2009) Broadband active noise control around human's head—determination and achievement of physical limit. In: Proc of NAG/DAGA, int conf on acoustics, including the 35th German annual conference on acoustics (DAGA), Rotterdam, Mart 23–26

Kletschkowski T, Sachau D, Böhme S, Breitbach H (2005) Optimized active noise control of semi-closed aircraft interiors. Int J Aeroacoust 6(1):61–71

Kletschkowski T, Sachau D, Böhme S (2007) Erholsamer Schlaf dank aktiven Gegenschalls. In: Mechatronik 2007—Innovative Produktentwicklung, Wiesloch bei Heidelberg, May 23–24. VDI-Berichte Nr. 1971

Kletschkowski T, Simanowski K, Sachau D (2008) Aktive Lärmminderung in Wohn- und Schlafräumen. Final Report for a Research Project (Az: 24483)—financed by the Deutsche Bundesstiftung Umwelt, Helmut-Schmidt-University/University of the Federal Armed Forces Hamburg

Klünder C (2004) Verbesserung eines adaptiven Systems zur aktiven Reduzierung von Propellerlärm. Studienarbeit, Helmut-Schmidt-University/University of the Federal Armed Forces Hamburg

Kochan K (2009) Robuste aktive Schallreduktion in Flugzeugen. Dissertation, Helmut-Schmidt-University/University of the Federal Armed Forces Hamburg

Kochan K, Sachau DK (2009) Robust parameter design of an adaptive multi-channel active noise controller. In: Proc of NAG/DAGA, int conf on acoustics, including the 35th German annual conference on acoustics (DAGA), Rotterdam, Mart 23–26

Kochan K, Sachau D, Kletschkowski T (2008) Active noise control in a semi-closed aircraft cabin. In: Proc of ISMA 2008, int conference on noise and vibration engineering, Sept 15–17. Katholieke Universiteit Leuven, Belgium

Kochan K, Weizel E, Sachau D (2009a) Evaluation of the stability and performance of an uncertain active noise control system. In: Proc of ACTIVE 2009, Ottawa, Canada, August 23–26

Kochan K, Sachau D, Breitbach H (2009b) Robust controller design of an active noise control system for a propeller aircraft. In: Proc of INTER-NOISE 2009, Ottawa, Canada, August 23–26

Kuo SM, Finn BM (1993) An integrated audio and active noise control system. In: Proc of ISCAS, Chicago, May 3–6, pp 2529–2532

Kuo SM, Morgan DR (1996) Active noise control systems—algorithms and DSP implementations. Wiley, Canada

Kuo SM, Morgan DR (1999) Active noise control: a tutorial review. Proc IEEE 87(6):943–973

Liu L, Kuo SM (2009) Virtual sensing techniques and their applications. In: Proc of the 2009 IEEE int conference on networking, sensing and control, Okayama, March 26–29, pp 31–36

Kuttruff H (2004) Akustik: Eine Einführung. Hirzel, Stuttgart

Lajos F (1973) Zufallsschwingungen und ihre Behandlung. Springer, Berlin

Lücker R (1980) Grundlagen digitaler Filter—Eine Einführung in die Theorie linearer zeitdiskreter Systeme und Netzwerke

Lueg P (1936) Process of silencing sound oscillations. US Patent: No 2,043,416. Application: 8 March 1934. Patented: 9 June 1936. Priority (Germany): 27 January 1933

Lueg P (1937) Verfahren zur Dämpfung von Schallschwingungen. German Patent: No 655,508 Filed: 27 January 1933. Patented: 30 December 1937

Mackay A, Kenchington S (2004) Active control of noise and vibration—a review of automotive applications. In: Proc of ACTIVE 04, Williamsburg, VA, USA, September 20–22

Mankins JC (1995) Technology readiness levels: a white paper. Control engineering practice. NASA, Office of Space Access and Technology, Advanced Concepts Office

Matlab (2010) Active noise control using a filtered-X LMS FIR adaptive filter. In: Product overview of filter design toolbox 4.7. The Mathworks. http://www.mathworks.com/products/filterdesign/. Cited 04 June 2010

Maynard JD, Williams EG, Lee Y (1985) Nearfield acoustic holographie: I. Theory of generalized holography and the development of NAH. J Acoust Soc Am 78(4):395–1413

Meirovitch L (1990) Dynamics and control of structures. Wiley, New York

Microflown (2009) Datasheet ultimate sound probe. Microflown technologies. http://www.microflown.com. Cited 07 Sept 2009

Moreau D, Cazzolato B, Zander A, Petersen C (2008) A review of virtual sensing algorithms for active noise control. Algorithms 1:69–99

Morese PM, Ingard KU (1961) Linear acoustic theory. In: Flügge S (ed) Encyclopedia of physics XI/1. Acoustics 1, 1st edn. Springer, Berlin

Morese PM, Ingard KU (1968) Theoretical acoustics. McGraw-Hill, New York

Moschytz G, Hofbauer M (2000) Adaptive Filter—Eine Einführung in die Theorie mit Aufgaben und Matlab-Simulationen auf CD-ROM. Springer, Berlin

Möser M (2005) Technische Akustik. Springer, Berlin

Möser M (2010) Messtechnik der Akustik. Springer, Heidelberg

Munn JM (2003) Virtual sensors for active noise control. PhD Thesis, The University of Adelaide, Australia

Müller I (1994) Grundzüge der Thermodynamik. Springer, Berlin
NASA (2007) System engineering handbook. NASA/SP-2007-6105 Rev 1. http://ntrs.nasa.gov. Cited 22 June 2010
NASTRAN (2010) MSC NATSRAN. Product description MSC software. http://www.mscsoftware.com. Cited 09 June 2010
Nelson PA, Elliott SJ (1992) Active control of sound. Academic Press, London
Nelson PA, Curtis ARD, Elliott SJ, Bullmore J (1987a) The active minimization of harmonic enclosed sound fields. Part I: Theory. J Acoust Soc Am 117(1):1–13
Nelson PA, Curtis ARD, Elliott SJ, Bullmore J (1987b) The active minimization of harmonic enclosed sound fields. Part II: A computer simulation. J Acoust Soc Am 117(1):15–33
Nelson PA, Curtis ARD, Elliott SJ, Bullmore J (1987c) The active minimization of harmonic enclosed sound fields. Part III: Experimental verification. J Acoust Soc Am 117(1):35–58
Nelson PA, Hammond JK, Joseph P, Elliott SJ (1990) Active control of stationary random sound fields. J Acoust Soc Am 87(3):963–975
Oppenheim AV, Willsky AS (1989) Signale und Systeme—Lehrbuch. VCH, Weinheim
Ordufia-Bustamant F, Nelson PA (1992) An adaptive controller for the active absorption of sound. J Acoust Soc Am 91(5):2740–2747
Pabst O, Kletschkowski T, Sachau D (2008a) Audio interior for light aircraft. In: Proc of CEAS 2007, 1st CEAS European air and space conference, Berlin, Germany
Pabst O, Kletschkowski T, Sachau D (2008b) Active noise control in light jet aircraft. In: Proc acoustics '08—organized by the Acoustical Society of America, the European Acoustics Association, and the Societe Francaised Acoustique, Paris, June 29–July 04
Pabst O, Kletschkowski T, Sachau D (2009) Combined active noise control and audio in a light jet. In: Proc of ICSV16, the sixteenth int conf on sound and vibration, Krakow, Poland, July 5–9
Paige CC, Saunders MA (1982) LSQR: an algorithm for sparse linear equations and sparse least squares. ACM Trans Math Softw 8(1):43–71
Pareto V (1971) Manual of political economy. Kelley, New York
Parkins JW, Tichy J, Nelson PA, Elliott SJ, Bullmore AJ (2000) Narrowband and broadband active control in an enclosure using the acoustic energy density. J Acoust Soc Am 118(1):192–203
Pawelczyk M (2004) Adaptive noise control algorithms for active headrest system. Control Eng Pract 12:1101–1112
Pawelczyk M, Elliott JS, Rafaely B (1997) Active noise control using feedback. Fixed and adaptive controllers. ISVR Technical Memorandum 822, University of Southampton
Petersen D, Zander AC, Cazzolato BS, Hansen CH (2005) Optimal virtual sensing for active noise control in a rigid-walled acoustic duct. J Acoust Soc Am 118(5):3086–3093
Press WH, Teukolsky SA, Vetterling WT, Flannery BP (1994) Numerical recipes in FORTRAN. The art of scientific computing, 2nd edn. Cambridge University Press, New York
QinetiQ (2007) Trim panel speaker technology. QinetiQ. http://www.qinetiq.com. Cited 07 Sept 2009
Rafaely B, Elliott SJ, Garcia-Bonito J (1999) Broadband performance of an active headrest. J Acoust Soc Am 116(2):787–793
Sachau D, Drenckhan J (2006) Sound source localization in cabins by inverse finite element analysis. In: Proc of DAGA '06—32. Deutsche Jahrestagung für Akustik, Braunschweig, Mart 20–23
Sachau D, Gerner C (2004) Hardware optimisation for tonal ANR. In: Proc of ACTIVE 04, Williamsburg, VA, USA, September 20–22
Sachau D, Drenckhan J, Schäfer I (2005a) Schallintensitätsdetektor sowie Verfahren zum Messen der Schallintensität. Patent. DE102004009644A1 22.09.2005
Sachau D, Drenckhan J, Kletschkowski T, Petersen S (2005b) Entwicklung von Messtechniken zur Lärmquellenidentifizierung in Kabinen. Final Report, Hamburg Aircraft Research Program (Lufo Hamburg 1), TUT-34, Helmut-Schmidt-University/University of the Federal Armed Forces Hamburg
Sachau D, Böhme S, Kletschkowski T, Kochan K (2007a) Vorrichtung und Verfahren zur aktiven Schalldämmung in einem geschlossenen Raum. Patent Application, DE 10 2007 012 611 A1

Sachau D, Kletschkowski T, Gerner C (2007b) Noise source identification and optimized active noise control in aircraft cabins. In: Proc of AST 2007, Workshop on aircraft system technologies, Hamburg, Mart 31–April 1

Sachau D, Kletschkowski T, Kochan K (2008a) Active noise reduction in bedrooms. In: Proc of IMAC XXVI a conference and exposition on structural dynamics, Orlando, FL, USA, February 4–7

Sachau D, Pabst O, Teuma FT, Kletschkowski T (2008b) Audio Interior für Kleinflugzeuge (AIK). Final Report, Hamburg Aircraft Research Program (Lufo Hamburg 2), HH 107 B "AIK", Helmut-Schmidt-University/University of the Federal Armed Forces Hamburg

Sachau D, Böhme S, Kletschkowski T, Breitbach H, Simanowski K (2008c) Aktiver Schallblocker Patent Application, DE 10 2006 023630.0 und EP 1 858 004 A2

Sachau D, Weber M, Kletschkowski T (2009) Schlussbericht zu Lufo IV "MONSOS". Final Report, German Aircraft Research Program (Lufo IV), Helmut-Schmidt-University/University of the Federal Armed Forces Hamburg

Sage AP (1992) System engineering. Wiley, New York

Sandberg G, Göransson P (1987) A symmetric finite element formulation for acoustic fluid-structure interaction analysis. J Sound Vib 123(3):507–515

Sano H, Inone T, Terai K, Nakamura Y (2001) Active control system for low-frequency road noise combined with an audio system. IEEE Trans Speech Audio Process 9(7):755–763

Santag S, Ziese T (2003) Validierung von FE-Berechnungen des LMWS-Testaufbaus durch experimentelle Untersuchungen. Project work, Helmut-Schmidt-University/University of the Federal Armed Forces Hamburg

Sayed AH (2003) Fundamentals of adaptive filtering. Wiley, Hoboken

Scheuren J, Shirmacher R, Hobelsberger J (2002) Active design of automotive engine sound. In: Proc of InterNoise 2002, 34th international congress and exposition on noise control engineering, Dearborn, MI, USA, August 19–21

Sennheiser (2010) Aviation headsets with NoiseGard active noise compensation—introduction. Published 2005. http://www.sennheiser.com. Cited 31 July 2010

Shea R (1966) Amplifier handbook. McGraw-Hill, New York

Simanowski K, Kletschkowski T, Sachau D, Samtleben B (2009) Acoustic ground tests in a cross-section of a long-range airliner for validation of the inverse finite element method. In: Proc of NAG/DAGA 2009, int conf on acoustics, including the 35th German annual conference on acoustics (DAGA), Rotterdam, Mart 23–26

Skudrzyk E (1971) The foundations of acoustics, basic mathematics and basic acoustics. Springer, Wien

Sommer J (2005) Projektierung eines experimentellen ANR-Systems zur Reduzierung von tonalem Propellerlärm. Diploma Thesis, Helmut-Schmidt-University/University of the Federal Armed Forces Hamburg

Sommer J, Sachau D, Gerner C (2005) Control algorithm for an experimental active noise reduction (ANR) system to reduce tonal propeller noise. In: Proc of ICSV12, twelfth international congress on sound and vibration, Lisbon, Portugal, July 11–14

Strang G (1993) Introduction to linear algebra. Cambridge University Press, Wellesley

Sundararajan D (2008) A practical approach to signals and systems. Wiley Eastern, Singapore

SYSNOISE (2010) LMS SYSNOISE. Product description LMS engineering innovation. http://www.lmsintl.com/SYSNOISE. Cited 09 June 2010

Teuma FT, Kletschkowski T (2009) Numerical investigations and validation of an acoustic finite element model for aircraft interiors. In: Proc of ICEDyn 2009, int conf on structural engineering dynamics, Ericeira, Portugal, June 22–24

Teuma FT, Kletschkowski T, Sachau D (2007) Numerical model of a very light jet cavity. In: Proc of COMSOL conference 2007, Grenoble, France, October 23–24

Teuma FT, Kochan K, Kletschkowski T, Sachau D (2008) Validation of the acoustic finite element model of a very light jet cavity mock-up. In: Proc of COMSOL conference 2008, Hannover, Germany

Tikhonov AN (1963) Solution of incorrectly formulated problems and the regularization method. Dokl Akad Nauk SSSR 151:501–504. Translated in Sov Math 4:1035–1038

Ultra Electronics (2009) Greenford Middlesex, UK. http://www.ultraquiet.com/demo/index.htm. Cited 10 Jan 2009

Ultra Electronics (2010) Greenford Middlesex, UK. http://www.ultra-controls.com. Cited 31 July 2010

Utsumi M (1985) Reduction of noise transmission in a duct by termination impedance control of a sidebranch resonator. J Vib Acoust 123:289–296

Vapnyarskii BI (2001) Lagrange multipliers. In: Hazewinkel M Encyclopaedia of mathematics. Springer, Berlin. http://eom.springer.de/L/l057190.htm. Cited 6 December 2009

VDI 2206 (2004) Entwicklungsmethodik für mechatronische Systeme (Design methodology for mechatronic systems). Verein Deutscher Ingenieure, Düsseldorf

VDI 2064 (2008) Aktive Schwingungsisolierung—Entwurf. Verein Deutscher Ingenieure, Düsseldorf

Vogel CR (1996) Non-convergence of the L-curve regularization parameter selection method. Inverse Probl 12:535–547

von Estorff O, Coyette JP, Migeot JL (2000) Governing formulations of the BEM in acoustics. In: von Estorff O (ed) Boundary elements in acoustics, advances and applications. WIT Press, Southampton

Wan EA (1996) Adjoint LMS: an efficient alternative to the filtered-X LMS and multiple error LMS algorithms. In: Proc of IEEE int conf acoust, speech, signal processing, Atlanta, GA

Weber M (2009) Inverse Schallquellenortung in Flugzeugkabinen. Dissertation, Helmut-Schmidt-University/University of the Federal Armed Forces Hamburg

Weber M, Sachau D, Kletschkowski T (2008a) Identifikation von Schallquellen mittels inverser FEM mit realen Messdaten. In: Proc of DAGA '08-34. Deutsche Jahrestagung für Akustik, Dresden, Mart 10–13

Weber M, Kletschkowski T, Samtleben B (2008b) Identification of noise sources by means of inverse finite element method. In: Proc of COMSOL conference '08, Hannover, Nov 4–6

Weber M, Kletschkowski T, Samtleben B (2008c) Identification of noise sources by means of inverse finite element method using measured data. In: Proc of acoustics '08—organized by the Acoustical Society of America, the European Acoustics Association, and the Societe Francaised Acoustique. Paris, June 29–July 04

Weber M, Kletschkowski T, Sachau D, Samtleben B (2009a) Tracking down the noise—identification of noise sources in aircraft cabins using simulated acoustics. Physics' Best. Wiley-VCH, Weinheim, pp 32–34

Weber M, Sachau D, Kletschkowski T (2009b) Noise source identification in a cross-section of a long-range airliner by means of the inverse finite element method. In: Proc of NAG/DAGA 2009, int conf on acoustics, including the 35th German annual conference on acoustics (DAGA), Rotterdam, Mart 23–26

Weber M, Sachau D, Kletschkowski T (2009c) Inverse identification of noise sources in aircraft interiors. In: Proc of AST 2009 workshop on aircraft system technologies, Hamburg, Mart 26–27

Weizel E (2009) Robuste Auslegung eines ANC-Reglers mit Gewichtungsmatrix. Diploma Thesis. Helmut-Schmidt-University/University of the Federal Armed Forces Hamburg

Wendt O (1995) Tourenplanung durch Einsatz naturanaloger Verfahren. Dt. Univ.-Vlg. Gabler, Wiesbaden

Widrow B, Stearns SD (1985) Adaptive signal processing. Prentice Hall International, London

Wilby JF (1996) Aircraft interior noise. J Sound Vib 190(3):545–564

Wilby JF, Mixson JS (1991) Interior noise. In: Hubbard HH (ed) Aeroacoustics of flight vehicles: theory and practice, vol 2: noise control. NASA Reference Publication vol 1258. WRDC, Technical Report 90-3052

Wu SF (2008) Methods for reconstructing acoustic quantities based on acoustic pressure measurements. J Acoust Soc Am 124(5):2680–2697

Ziemer RE, Tranter WH, Fannin DR (1983) Signals and systems, continuous and discrete. Macmillan, New York

Zienkiewicz OC, Taylor RL (2000) Finite element method, vol 1—the basis. Elsevier, Amsterdam

Zölzer U (2008) Digital audio signal processing. Wiley, Chichester
Zurmühl R (1964) Matrizen und ihre technischen Anwendungen. Springer, Berlin
Zwicker E, Zollner M (1984) Elektroakustik. Springer, Berlin
Zwillinger D (1989) Handbook of differential equations. Academic Press, San Diego

Index

T. Kletschkowski, *Adaptive Feed-Forward Control of Low Frequency Interior Noise*,
Intelligent Systems, Control and Automation: Science and Engineering 56,
DOI 10.1007/978-94-007-2537-9, © Springer Science+Business Media B.V. 2012

9789400725362